Lecture Notes in Physics

Volume 839

For further volumes:
http://www.springer.com/series/5304

The Lecture Notes in Physics

The series Lecture Notes in Physics (LNP), founded in 1969, reports new developments in physics research and teaching—quickly and informally, but with a high quality and the explicit aim to summarize and communicate current knowledge in an accessible way. Books published in this series are conceived as bridging material between advanced graduate textbooks and the forefront of research and to serve three purposes:

- to be a compact and modern up-to-date source of reference on a well-defined topic
- to serve as an accessible introduction to the field to postgraduate students and nonspecialist researchers from related areas
- to be a source of advanced teaching material for specialized seminars, courses and schools

Both monographs and multi-author volumes will be considered for publication. Edited volumes should, however, consist of a very limited number of contributions only. Proceedings will not be considered for LNP.

Volumes published in LNP are disseminated both in print and in electronic formats, the electronic archive being available at springerlink.com. The series content is indexed, abstracted and referenced by many abstracting and information services, bibliographic networks, subscription agencies, library networks, and consortia.

Proposals should be sent to a member of the Editorial Board, or directly to the managing editor at Springer:

Christian Caron
Springer Heidelberg
Physics Editorial Department I
Tiergartenstrasse 17
69121 Heidelberg/Germany
christian.caron@springer.com

Luis Álvarez-Gaumé · Miguel Á. Vázquez-Mozo

An Invitation to Quantum Field Theory

Luis Álvarez-Gaumé
Theory Unit
Physics Department
CERN
Geneva 23
Switzerland
e-mail: Luis.Alvarez-Gaume@cern.ch

Miguel Á. Vázquez-Mozo
Departamento de Física Fundamental
Universidad de Salamanca
Plaza de la Merced s/n
37007 Salamanca
Spain
e-mail: vazquez@usal.es

ISSN 0075-8450 e-ISSN 1616-6361
ISBN 978-3-642-23727-0 e-ISBN 978-3-642-23728-7
DOI 10.1007/978-3-642-23728-7
Springer Heidelberg Dordrecht London New York

Library of Congress Control Number: 2011937783

Printed on acid-free paper

Springer is part of Springer Science+Business Media (www.springer.com)

To my mother and to the memory of my father
(L.A.-G.)

A mis padres
Für Kerstin
(M.A.V.-M.)

Preface

This book is intended to provide an introduction to quantum field theory at an elementary level. The reader is supposed to know special relativity, electromagnetism and quantum mechanics. Quantum field theory is a vast subject that provides us with the basic tools to understand the physics of the elementary constituents of matter. There are excellent textbook expositions of the subject in the literature (see the references to Chap. 1), and it is not our intention to write one more. We have selected a representative sample of topics containing some of the more innovative and challenging concepts and presented them without too many technical details. Few proofs are included, the concepts are exhibited by working out examples and analogies. We have been careful to include all numerical factors in the equations, although the reader is often not required to understand more than their general features. Adequate references are provided where one can find all the necessary technical details. We prime the discussion of the main ideas over the mathematical details necessary to obtain the final results, which often require a more in-depth presentation of the subject. As its title indicates, this book tries to motivate the reader to study quantum field theory, not to provide a thorough presentation.

The guiding principle for the topics chosen was to present some basic aspects of the theory that contain some conceptual subtleties, or at least we found them subtle when learning the subject ourselves. We have paid special attention to the realization of symmetries in particle physics. The notion of symmetry is central in modern physics, and we present its many different aspects: global and local symmetries, explicit, spontaneously broken, anomalous continuous symmetries, discrete symmetries. We give a detailed account of the standard model of the strong, weak and electromagnetic interactions, our current understanding of the origin of mass, the general features of renormalization theory, as well as a cursory description of effective field theories and the problem of naturalness in physics. Sometimes the presentation gets a bit more abstract, as in the chapters on discrete symmetries (Chap. 11) and effective field theories (Chap. 12). We have delayed on purpose the study of discrete space-time symmetries in order to develop all the necessary background needed to explore some of their fascinating consequences.

In particular we present an outline of the first principles derivation of the CPT theorem and the spin-statistics connection. Among the few Feynman diagrams evaluated in full detail we have chosen Compton scattering in the Thomson limit to understand polarisation in the cosmic microwave background radiation and its sensitivity to primordial gravitational waves.

By lack of space and purpose, few proofs have been included. Instead, very often we illustrate a concept or property by describing a physical situation where it arises. Full details and proofs can be found in the many textbooks in the subject, and in particular in the ones provided in the bibliography. We should nevertheless warn the reader that we have been a bit cavalier about references. Our aim has been to provide mostly a (not exhaustive) list of reference for further reading. We apologize to those authors who feel misrepresented.

The book grew out of lectures at the CERN—Latin-American schools of High Energy Physics held in Malargüe (2005), Medellín (2009) and Natal (2011), and of undergraduate and graduate courses at the University of Salamanca (2005–2010). We would like to thank in particular Gilvan Alves, Teresa Dova, Miriam Gandelman, Christophe Grojean, Nick Ellis, Egil Lillestøl, Marta Losada and Enrico Nardi, for the opportunity to present this material and for the wonderful atmosphere they created during the schools. We are also grateful to Bob Jaffe who motivated us (through our Springer editor Christian Caron) to turn the original notes into book form. We also want to thank Chris for his patience and understanding. We would like to thank José L. F. Barbón and Agustín Sabio Vera for reading a preliminary version of this book and their many suggestions.

We have learned so much about quantum field theory from so many colleagues that it is difficult to list them all. However we would like to thank in particular: Sidney Coleman, Daniel Z. Freedman, David Gross, Roman Jackiw, Julius Wess and Edward Witten from whom we have learned a great deal about the subject. Any remaining misconceptions in the text are entirely our fault.

Geneva and Salamanca
June 2011

Luis Álvarez-Gaumé
Miguel Á. Vázquez-Mozo

Contents

Chapter 1
Why Do We Need Quantum Field Theory After All?

Quantum field theory is the basic tool to understand the physics of the elementary constituents of matter (see [1–15] for an incomplete list of textbooks in the subject). It is both a very powerful and a very precise framework: using it we can describe physical processes in a range of energies going from the few millions electrovolts typical of nuclear physics to the thousands of billions of the Large Hadron Collider (LHC). And all this with astonishing precision.

In this first chapter our aim is to explain why quantum mechanics is not enough and how quantum field theory is forced upon us by special relativity. We will review a number of riddles that appear in the attempt to extend the results of quantum mechanics to systems where relativistic effects cannot be ignored. Their resolution requires giving up the quantum mechanical description of a single particle to allow for the creation and annihilation of particles. As we will see, quantum fields provide the right tool to handle this.

1.1 Relativistic Quantum Mechanics

In spite of the impressive success of quantum mechanics in describing atomic physics, it was immediately clear after its formulation that its relativistic extension was not free of difficulties. These problems were clear already to Schrödinger, whose first guess for a wave equation of a free relativistic particle was the Klein–Gordon equation[1]

$$\left(\frac{\partial^2}{\partial t^2} - \nabla^2 + m^2\right)\psi(t, \mathbf{x}) = 0. \tag{1.1}$$

This equation follows directly from the relativistic "mass-shell" identity $E^2 = \mathbf{p}^2 + m^2$ using the correspondence principle

[1] We use natural units $\hbar = c = 1$. A summary of the units and conventions used in the book can be found in Appendix A.

L. Álvarez-Gaumé and M. Á. Vázquez-Mozo, *An Invitation to Quantum Field Theory*, Lecture Notes in Physics 839, DOI: 10.1007/978-3-642-23728-7_1,

$$\begin{aligned} E &\to i\frac{\partial}{\partial t}, \\ \mathbf{p} &\to -i\nabla. \end{aligned} \tag{1.2}$$

Plane wave solutions to the wave equation (1.1) are readily obtained

$$\psi(t,\mathbf{x}) = e^{-ip_\mu x^\mu} = e^{\mp iE_{\mathbf{p}}t + i\mathbf{p}\cdot\mathbf{x}} \tag{1.3}$$

with

$$E_{\mathbf{p}} = \sqrt{\mathbf{p}^2 + m^2}. \tag{1.4}$$

In order to have a complete basis of functions, we must include both signs in the exponent. The probability density is read from the time component of the conserved current

$$j_\mu = \frac{i}{2}\Big(\psi^*\partial_\mu\psi - \partial_\mu\psi^*\psi\Big), \tag{1.5}$$

Since $j^0 = E$, we find that it is not positive definite.

A complete, properly normalized, continuous basis of solutions of the Klein–Gordon equation (1.1) labelled by the momentum $\mathbf{p}$ is given by

$$\begin{aligned} f_p(t,\mathbf{x}) &= \frac{1}{(2\pi)^{\frac{3}{2}}\sqrt{2E_{\mathbf{p}}}} e^{-iE_{\mathbf{p}}t + i\mathbf{p}\cdot\mathbf{x}}, \\ f_{-p}(t,\mathbf{x}) &= \frac{1}{(2\pi)^{\frac{3}{2}}\sqrt{2E_{\mathbf{p}}}} e^{iE_{\mathbf{p}}t - i\mathbf{p}\cdot\mathbf{x}}. \end{aligned} \tag{1.6}$$

Defining the inner product

$$\langle\psi_1|\psi_2\rangle = i\int d^3x\Big(\psi_1^*\partial_0\psi_2 - \partial_0\psi_1^*\psi_2\Big),$$

the states (1.6) form an orthonormal basis

$$\begin{aligned} \langle f_p|f_{p'}\rangle &= \delta(\mathbf{p}-\mathbf{p}'), \\ \langle f_{-p}|f_{-p'}\rangle &= -\delta(\mathbf{p}-\mathbf{p}'), \end{aligned} \tag{1.7}$$

$$\langle f_p|f_{-p'}\rangle = 0. \tag{1.8}$$

The wave functions $f_p(t,\mathbf{x})$ describe states with momentum $\mathbf{p}$ and energy $E_{\mathbf{p}} = \sqrt{\mathbf{p}^2 + m^2}$. On the other hand, the wave functions $f_{-p}(t,\mathbf{x})$ not only have negative scalar product but they correspond to negative energy states

$$i\frac{\partial}{\partial t} f_{-p}(t,\mathbf{x}) = -\sqrt{\mathbf{p}^2 + m^2}\, f_{-p}(t,\mathbf{x}). \tag{1.9}$$

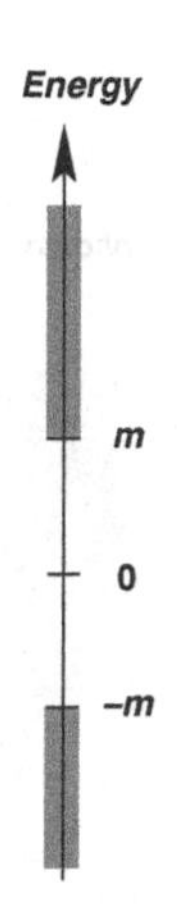

Fig. 1.1 Spectrum of the Klein–Gordon wave equation

Therefore the energy spectrum of the theory satisfies $|E| > m$ and is unbounded from below (see Fig. 1.1). Although in the case of a free theory the absence of a ground state is not necessarily a fatal problem, once the theory is coupled to the electromagnetic field this is the source of all kinds of disasters, since nothing can prevent the decay of any state by the emission of electromagnetic radiation.

The problem of the instability of the "first-quantized" relativistic wave equation can be heuristically tackled in the case of spin-$\frac{1}{2}$ particles, described by the Dirac equation

$$\left(-i\beta\frac{\partial}{\partial t} + \alpha \cdot \nabla - m\right)\psi(t, \mathbf{x}) = 0, \tag{1.10}$$

where α and β are 4×4 matrices

$$\alpha^i = \begin{pmatrix} 0 & i\sigma_i \\ -i\sigma_i & 0 \end{pmatrix}, \quad \beta = \begin{pmatrix} 0 & \mathbf{1} \\ \mathbf{1} & 0 \end{pmatrix}, \tag{1.11}$$

with σ_i the Pauli matrices (see Appendix A) and the wave function $\psi(t, \mathbf{x})$ has four components: it is a Dirac spinor, an object that will be studied in more detail in Chap. 3. The wave equation (1.10) can be thought of as a kind of "square root" of the Klein–Gordon equation (1.1), since the latter can be obtained as

$$\begin{aligned}&\left(-i\beta\frac{\partial}{\partial t} + \alpha \cdot \nabla - m\right)^\dagger \left(-i\beta\frac{\partial}{\partial t} + \alpha \cdot \nabla - m\right)\psi(t, \mathbf{x}) \\ &= \left(\frac{\partial^2}{\partial t^2} - \nabla^2 + m^2\right)\psi(t, \mathbf{x}).\end{aligned} \tag{1.12}$$

An analysis of Eq. (1.10) along the lines of the one presented for the Klein–Gordon equation leads again to the existence of negative energy states and

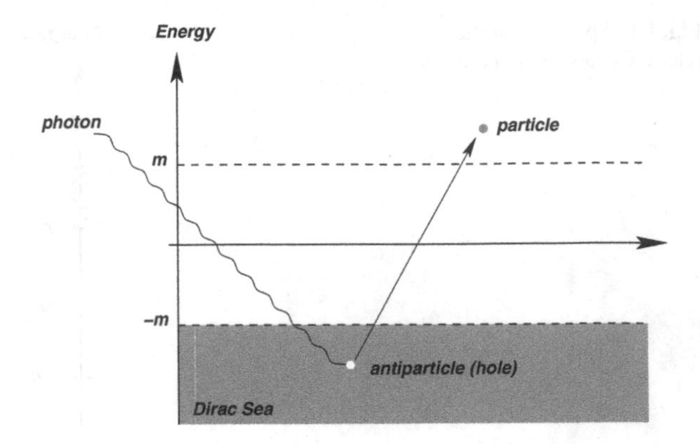

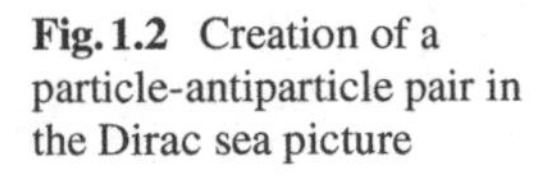
Fig. 1.2 Creation of a particle-antiparticle pair in the Dirac sea picture

a spectrum unbounded from below as in Fig. 1.1. Dirac, however, solved the instability problem by pointing out that now the particles are fermions and therefore they are subject to Pauli's exclusion principle. Hence, each state in the spectrum can be occupied by at most one particle, so the states with $E = m$ can be made stable if we assume that *all* the negative energy states are filled.

If Dirac's idea restores the stability of the spectrum by introducing a stable vacuum where all negative energy states are occupied, the so-called Dirac sea, it also leads directly to the conclusion that a single-particle interpretation of the Dirac equation is not possible. Indeed, a photon with enough energy ($E > 2m$) can excite one of the electrons filling the negative energy states, leaving behind a "hole" in the Dirac sea (see Fig. 1.2). This hole behaves as a particle with equal mass and opposite charge that is interpreted as a positron, so there is no escape to the conclusion that interactions will produce particle-antiparticle pairs out of the vacuum.

1.2 The Klein Paradox

In spite of the success of the heuristic interpretation of negative energy states in the Dirac equation this is not the end of the story. In 1929 Oskar Klein stumbled into an apparent paradox when trying to describe the scattering of a relativistic electron by a square potential using Dirac's wave equation [16] (for pedagogical reviews see [17–19]). In order to capture the essence of the problem without entering into unnecessary complication we will study Klein's paradox in the context of the Klein–Gordon equation.

Let us consider a square potential with height $V_0 > 0$ of the type showed in Fig. 1.3. A solution to the wave equation in regions I and II is given by

$$\begin{aligned} \psi_I(t,x) &= e^{-iEt+ip_1x} + Re^{-iEt-ip_1x}, \\ \psi_{II}(t,x) &= Te^{-iEt+ip_2x}, \end{aligned} \tag{1.13}$$

where the mass-shell condition implies

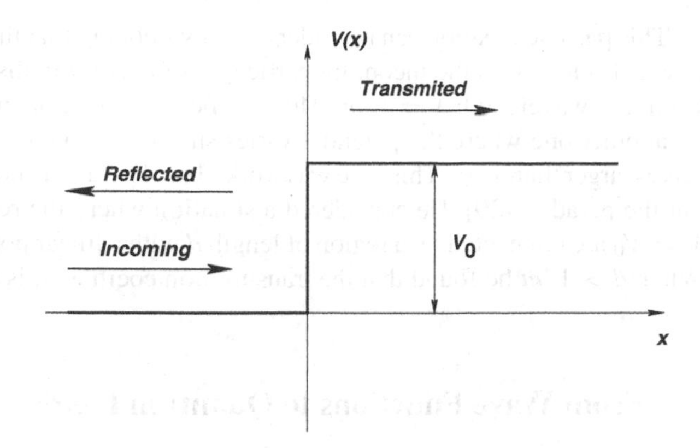

Fig. 1.3 Illustration of the Klein paradox

$$p_1 = \sqrt{E^2 - m^2}, \quad p_2 = \sqrt{(E - V_0)^2 - m^2}. \tag{1.14}$$

The constants R and T are computed by matching the two solutions across the boundary $x = 0$. The conditions $\psi_I(t, 0) = \psi_{II}(t, 0)$ and $\partial_x \psi_I(t, 0) = \partial_x \psi_{II}(t, 0)$ imply that

$$T = \frac{2p_1}{p_1 + p_2}, \quad R = \frac{p_1 - p_2}{p_1 + p_2}. \tag{1.15}$$

At first sight one would expect a behavior similar to the one encountered in the nonrelativistic case. If the kinetic energy is bigger than V_0 both a transmitted and reflected wave are expected, whereas when the kinetic energy is smaller than V_0 one only expects to find a reflected wave, the transmitted wave being exponentially damped within a distance of a Compton wavelength inside the barrier.

This is indeed what happens if $E - m > V_0$. In this case both p_1 and p_2 are real and we have a partly reflected, and a partly transmitted wave. In the same way, if $V_0 - 2m < E - m < V_0$ then p_2 is imaginary and there is total reflection.

However, in the case when $V_0 > 2m$ and the energy is in the range $0 < E - m < V_0 - 2m$ a completely different situation arises. In this case one finds that both p_1 and p_2 are real and therefore the incoming wave function is partially reflected and partially transmitted across the barrier. This is a shocking result, since it implies that there is a nonvanishing probability of finding the particle at any point across the barrier with negative kinetic energy ($E - m - V_0 < 0$)! This weird result is known as Klein's paradox.

As with the negative energy states, the Klein paradox results from our insistence in giving a single-particle interpretation to the relativistic wave function. In fact, a multiparticle analysis of the paradox [17] shows that what happens when $0 < E - m < V_0 - 2m$ is that the reflection of the incoming particle by the barrier is accompanied by the creation of particle-antiparticle pairs out of the energy of the barrier (notice that the condition implies that $V_0 > 2m$, the threshold for the creation of a particle-antiparticle pair).

This particle creation can be understood by noticing that the sudden potential step in Fig. 1.3 localizes the incoming particle with mass m in distances smaller than its Compton wavelength $\lambda = 1/m$. This can be seen by replacing the square potential by another one where the potential varies smoothly from 0 to $V_0 > 2m$ in distance scales larger than $1/m$. This case was worked out by Sauter shortly after Klein pointed out the paradox [20]. He considered a situation where the regions with $V = 0$ and $V = V_0$ are connected by a region of length d with a linear potential $V(x) = V_0 x/d$. When $d > 1/m$ he found that the transmission coefficient is exponentially small.[2]

1.3 From Wave Functions to Quantum Fields

The creation of particles is impossible to avoid whenever one tries to localize a particle of mass m within its Compton wavelength. Indeed, from the Heisenberg uncertainty relation we find that if $\Delta x \sim 1/m$, the fluctuations in the momentum will be of order $\Delta p \sim m$ and fluctuations in the energy of order

$$\Delta E \sim m \tag{1.16}$$

can be expected. Therefore, in a relativistic theory, the fluctuations of the energy are enough to allow for the creation of particles out of the vacuum. In the case of a spin-$\frac{1}{2}$ particle, the Dirac sea picture shows clearly how, when the energy fluctuations are of order m, electrons from the Dirac sea can be excited to positive energy states, thus creating electron–positron pairs.

It is possible to see how the multiparticle interpretation is forced upon us by relativistic invariance. In non-relativistic quantum mechanics observables are represented by self-adjoint operator that in the Heisenberg picture depend on time. Therefore measurements are localized in time but are global in space. The situation is radically different in the relativistic case. Since no signal can propagate faster than the speed of light, measurements have to be localized both in time and space. Causality demands then that two measurements carried out in causally-disconnected regions of space–time cannot interfere with each other. In mathematical terms this means that if $\mathscr{O}_{R_1}$ and $\mathscr{O}_{R_2}$ are the observables associated with two measurements localized in two causally-disconnected regions R_1, R_2 (see Fig. 1.4), they satisfy

$$[\mathscr{O}_{R_1}, \mathscr{O}_{R_2}] = 0, \quad \text{if}(x_1 - x_2)^2 < 0, \quad \text{for all} \quad x_1 \in R_1, \quad x_2 \in R_2. \tag{1.17}$$

Hence, in a relativistic theory, the basic operators in the Heisenberg picture must depend on the space–time position x^μ. Unlike the case in non-relativistic quantum mechanics, here the position $\mathbf{x}$ *is not* an observable, but just a label, similarly to the case of time in ordinary quantum mechanics. Causality is then imposed microscopically by requiring

---

[2] In Sect. 13.1 we will see how, in the case of the Dirac field, this exponential behavior can be associated with the creation of electron–positron pairs due to a constant electric field (Schwinger effect).

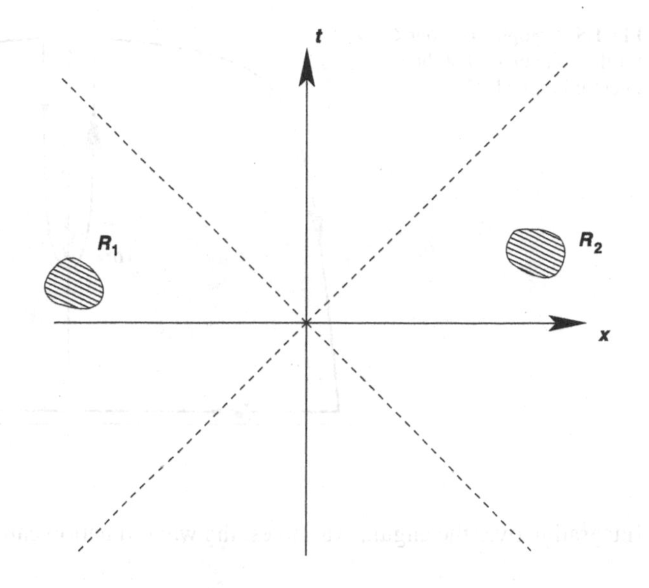

Fig. 1.4 Two regions R_1, R_2 that are causally disconnected

$$[\mathscr{O}(x), \mathscr{O}(y)] = 0, \quad \text{if } (x-y)^2 < 0. \tag{1.18}$$

A smeared operator $\mathscr{O}_R$ over a space–time region R can then be defined as

$$\mathscr{O}_R = \int d^4x \, \mathscr{O}(x) f_R(x) \tag{1.19}$$

where $f_R(x)$ is the characteristic function associated with R,

$$f_R(x) = \begin{cases} 1 & x \in R \\ 0 & x \notin R \end{cases}. \tag{1.20}$$

Equation (1.17) follows now from the microcausality condition (1.18).

Therefore, relativistic invariance forces the introduction of quantum fields. It is only when we insist in keeping a single-particle interpretation that we crash against causality violations. To illustrate the point, let us consider a single particle wave function $\psi(t, \mathbf{x})$ that initially is localized in the position $\mathbf{x} = 0$

$$\psi(0, \mathbf{x}) = \delta(\mathbf{x}). \tag{1.21}$$

Evolving this wave function using the Hamiltonian $H = \sqrt{-\nabla^2 + m^2}$ we find that the wave function can be written as

$$\psi(t, \mathbf{x}) = e^{-it\sqrt{-\nabla^2+m^2}} \delta(\mathbf{x}) = \int \frac{d^3k}{(2\pi)^3} e^{i\mathbf{k}\cdot\mathbf{x} - it\sqrt{k^2+m^2}}. \tag{1.22}$$

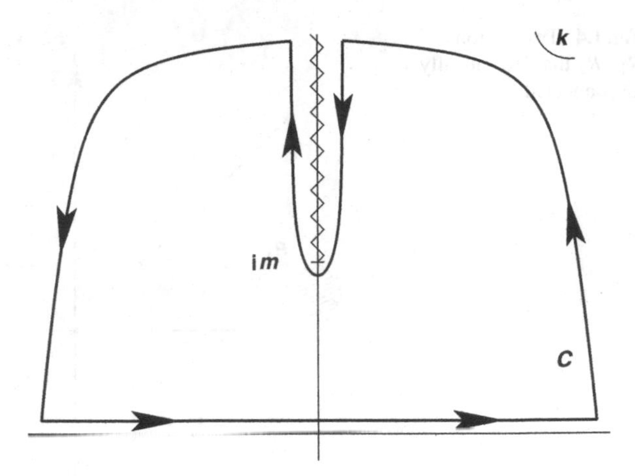

Fig. 1.5 Complex contour C for the computation of the integral in Eq. (1.23)

Integrating over the angular variables, the wave function can be recast in the form

$$\psi(t,\mathbf{x}) = \frac{-i}{4\pi^2|\mathbf{x}|}\int_{-\infty}^{\infty} kdke^{ik|\mathbf{x}|}e^{-it\sqrt{k^2+m^2}}. \tag{1.23}$$

The resulting integral can be evaluated using the complex integration contour C shown in Fig. 1.5. The result is that, for any $t > 0$, $\psi(t,\mathbf{x}) \neq 0$ for any $\mathbf{x}$. If we insist in interpreting the wave function $\psi(t,\mathbf{x})$ as the probability density of finding the particle at the location $\mathbf{x}$ at the time t, the probability leaks out of the light cone, thus violating causality.

The bottom line of the analysis of this chapter is clear: a fully relativistic quantum theory must give up the idea of describing the system in terms of the wave function of a *single* particle. As a matter of fact, relativistic quantum mechanics is, at best, a narrow boundary area. It might be a useful tool to compute the first relativistic corrections in certain quantum systems. However it runs into serious trouble as soon as one tries to use it for a full-fledged relativistic description of the quantum phenomena. Next we will see how quantum field theory provides the right framework to handle these problems.

References

1. Bjorken, J.D, Drell, S.D: Relativistic Quantum Fields. McGraw-Hill, New York (1965)
2. Itzykson, C., Zuber, J.-B.: Quantum Field Theory. McGraw-Hill, New York (1980)
3. Aitchinson, I.J.R.: An Informal Introduction to Gauge Field Theories. Cambridge University Press, Cambridge (1982)
4. Mandl, F., Shaw, G.: Quantum Field Theory. Wiley, New York (1984)
5. Ramond, P.: Field Theory: A Modern Primer. Addison-Wesley, Reading (1990)

6. Sterman, G.: An Introduction to Quantum Field Theory. Cambridge University Press, Cambridge (1995)
7. Peskin, M.E, Schroeder, D.V: An Introduction to Quantum Field Theory. Addison Wesley, Reading (1995)
8. Weinberg, S.: The Quantum Theory of Fields. vol. 1–3, Cambridge University Press, Cambridge (1995)
9. Deligne, P. et al. (eds) Quantum Fields and Strings: A Course for Mathematicians. American Mathematical Society, Princeton (1999)
10. Gribov, V.N., Nyiri, J.: Quantum Electrodynamics. Cambridge University Press, Cambridge (2001)
11. Zee, A.: Quantum Field Theory in a Nutshell. 2nd edn. 2010 Princeton (2003)
12. DeWitt, B.S.: The Global Approach to Quantum Field Theory. vols. 1 and 2, Oxford University Press, Oxford (2003)
13. Nair, V.P.: Quantum Field Theory. A Modern Perspective. Springer, New York (2005)
14. Maggiore, M.: A Modern Introduction to Quantum Field Theory. Oxford University Press, Oxford (2005)
15. Banks, T.: Modern Quantum Field Theory. Cambridge University Press, Cambridge (2008)
16. Klein, O.: Die Reflexion von Elektronen an einem Potentialsprung nach der relativischen Dynamik von Dirac. Z. Phys. **53**, 157 (1929)
17. Holstein, B.R.: Klein's paradox. Am. J. Phys. **66**, 507 (1998)
18. Dombey, N., Calogeracos, A.: Seventy years of the Klein paradox. Phys. Rep. **315**, 41 (1999)
19. Dombey, N., Calogeracos, A.: History and physics of the Klein paradox. Contemp. Phys. **40**, 313 (1999) (quant-ph/9905076)
20. Sauter, F.: Zum Kleinschen Paradoxon. Z. Phys. **73**, 547 (1932)

Chapter 2
From Classical to Quantum Fields

We have learned how the consistency of quantum mechanics with special relativity forces us to abandon the single-particle interpretation of the wave function. Instead we have to consider quantum fields whose elementary excitations are associated with particle states, as we will see below. In this chapter we study the basics of field quantization using both the canonical formalism and the path integral method.

2.1 Particles and Quantum Fields

In any scattering experiment the only information available to us is the set of quantum numbers associated with the set of free particles in the initial and final states. Ignoring for the moment other quantum numbers like spin and flavor, one-particle states are labelled by the three-momentum $\mathbf{p}$ and span the single-particle Hilbert space $\mathscr{H}_1$

$$|\mathbf{p}\rangle \in \mathscr{H}_1, \quad \langle \mathbf{p}|\mathbf{p}'\rangle = \delta(\mathbf{p}-\mathbf{p}'). \tag{2.1}$$

The states $\{|\mathbf{p}\rangle\}$ form a basis of $\mathscr{H}_1$ and therefore satisfy the closure relation

$$\int d^3p |\mathbf{p}\rangle\langle\mathbf{p}| = \mathbf{1}. \tag{2.2}$$

The group of spatial rotations acts unitarily on the states $|\mathbf{p}\rangle$. This means that for every rotation $R \in \mathrm{SO}(3)$ there is a unitary operator $\mathscr{U}(R)$ such that

$$\mathscr{U}(R)|\mathbf{p}\rangle = |R\mathbf{p}\rangle \tag{2.3}$$

where $R\mathbf{p}$ represents the action of the rotation on the vector $\mathbf{p}$, $(R\mathbf{p})^i = R^i_{\ j}p^j$. Using a spectral decomposition, the momentum operator can be written as

$$\hat{P}^i = \int d^3p |\mathbf{p}\rangle p^i \langle\mathbf{p}|. \tag{2.4}$$

L. Álvarez-Gaumé and M. Á. Vázquez-Mozo, *An Invitation to Quantum Field Theory*, Lecture Notes in Physics 839, DOI: 10.1007/978-3-642-23728-7_2,

With the help of Eq. (2.3) it is straightforward to check that the momentum operator transforms as a vector under rotations:

$$\mathscr{U}(R)^{-1}\hat{P}^i\mathscr{U}(R) = \int d^3p|R^{-1}\mathbf{p}\rangle p^i\langle R^{-1}\mathbf{p}| = R^i_{\ j}\hat{P}^j, \tag{2.5}$$

where we have used that the integration measure is invariant under SO(3).

Since, as argued above, we are forced to deal with multiparticle states, it is convenient to introduce creation-annihilation operators associated with a single-particle state of momentum $\mathbf{p}$

$$\left[\hat{a}(\mathbf{p}), \hat{a}^\dagger(\mathbf{p}')\right] = \delta(\mathbf{p}-\mathbf{p}'), \quad [\hat{a}(\mathbf{p}), \hat{a}(\mathbf{p}')] = \left[\hat{a}^\dagger(\mathbf{p}), \hat{a}^\dagger(\mathbf{p}')\right] = 0, \tag{2.6}$$

such that the state $|\mathbf{p}\rangle$ is created out of the Fock space vacuum $|0\rangle$ (normalized such that $\langle 0|0\rangle = 1$) by the action of a creation operator $\hat{a}^\dagger(\mathbf{p})$

$$|\mathbf{p}\rangle = \hat{a}^\dagger(\mathbf{p})|0\rangle, \quad \hat{a}(\mathbf{p})|0\rangle = 0 \quad \text{for all} \quad \mathbf{p}. \tag{2.7}$$

Covariance under spatial rotations is all we need if we are interested in a nonrelativistic theory. However in a relativistic quantum field theory we must preserve more that SO(3), we need the expressions to be covariant under the full Poincaré group ISO(1, 3) consisting of spatial rotations, boosts and space-time translations (see Sect. 3.1 and Appendix B). Therefore, in order to build the Fock space of the theory we need two key ingredients: first an invariant normalization for the states, since we want a normalized state in one reference frame to be normalized in any other inertial frame. And secondly a relativistic invariant integration measure in momentum space, so the spectral decomposition of operators is covariant under the full Poincaré group.

Let us begin with the invariant measure. Given an invariant function $f(p)$ of the four-momentum p^μ of a particle of mass m with positive energy $p^0 > 0$, there is an integration measure which is invariant under proper Lorentz transformations[1]

$$\int \frac{d^4p}{(2\pi)^4}(2\pi)\delta(p^2-m^2)\theta(p^0)f(p), \tag{2.8}$$

where the factors of 2π are introduced for later convenience, and $\theta(x)$ is the Heaviside step function

$$\theta(x) = \begin{cases} 0 & x < 0 \\ 1 & x > 0 \end{cases}. \tag{2.9}$$

The integration over p^0 can be easily done using the delta function identity

[1] The identity $p^2 = m^2$ satisfied by the four-momentum of a real particle will be referred to in the following as the *on-shell* condition.

$$\delta[g(x)] = \sum_{x_i = \text{zeros of } g} \frac{1}{|g'(x_i)|}\delta(x - x_i), \tag{2.10}$$

valid for any function $g(x)$ with simple zeroes. In our case this implies

$$\delta(p^2 - m^2) = \frac{1}{2p^0}\delta\left(p^0 - \sqrt{\mathbf{p}^2 + m^2}\right) + \frac{1}{2p^0}\delta\left(p^0 + \sqrt{\mathbf{p}^2 + m^2}\right). \tag{2.11}$$

The second term has support on states with negative energy and therefore does not contribute to the integral. We can write

$$\int \frac{d^4p}{(2\pi)^4}(2\pi)\delta(p^2 - m^2)\theta(p^0)f(p) = \int \frac{d^3p}{(2\pi)^3}\frac{1}{2\sqrt{\mathbf{p}^2 + m^2}}f\left(\sqrt{\mathbf{p}^2 + m^2}, \mathbf{p}\right). \tag{2.12}$$

Hence, the relativistic invariant measure is given by

$$\int \frac{d^3p}{(2\pi)^3}\frac{1}{2E_{\mathbf{p}}} \quad \text{with} \quad E_{\mathbf{p}} \equiv \sqrt{\mathbf{p}^2 + m^2}. \tag{2.13}$$

Once we have an invariant measure the next step is to find an invariant normalization for the states. We work with a basis $\{|p\rangle\}$ of eigenstates of the four-momentum operator $\hat{P}^\mu$

$$\hat{P}^0|p\rangle = E_{\mathbf{p}}|p\rangle, \quad \hat{P}^i|p\rangle = \mathbf{p}^i|p\rangle. \tag{2.14}$$

Since the states $|p\rangle$ are eigenstates of the three-momentum operator we can express them in terms of the non-relativistic states $|\mathbf{p}\rangle$ introduced in Eq. (2.1)

$$|p\rangle = N(\mathbf{p})|\mathbf{p}\rangle \tag{2.15}$$

with $N(\mathbf{p})$ a normalization to be determined now. The states $\{|p\rangle\}$ form a complete basis, so they should satisfy the Lorentz invariant closure relation

$$\int \frac{d^4p}{(2\pi)^4}(2\pi)\delta(p^2 - m^2)\theta(p^0)|p\rangle\langle p| = \mathbf{1}. \tag{2.16}$$

At the same time, this closure relation can be expressed, using Eq. (2.15), in terms of the nonrelativistic basis of states $\{|\mathbf{p}\rangle\}$ as

$$\int \frac{d^4p}{(2\pi)^4}(2\pi)\delta(p^2 - m^2)\theta(p^0)|p\rangle\langle p| = \int \frac{d^3p}{(2\pi)^3}\frac{1}{2E_{\mathbf{p}}}|N(\mathbf{p})|^2|\mathbf{p}\rangle\langle\mathbf{p}|. \tag{2.17}$$

Using Eq. (2.4) we get the expression (2.16) provided

$$|N(\mathbf{p})|^2 = (2\pi)^3(2E_{\mathbf{p}}). \tag{2.18}$$

Taking the overall phase in Eq. (2.15) so that $N(\mathbf{p})$ is real and positive, we define the Lorentz invariant states $|p\rangle$ as

$$|p\rangle = (2\pi)^{\frac{3}{2}}\sqrt{2E_{\mathbf{p}}}|\mathbf{p}\rangle, \tag{2.19}$$

and given the normalization of $|\mathbf{p}\rangle$ we find the one of the relativistic states to be

$$\langle p|p'\rangle = (2\pi)^3(2E_{\mathbf{p}})\delta(\mathbf{p}-\mathbf{p}'). \tag{2.20}$$

It might not be obvious at first sight, but the previous normalization is Lorentz invariant. Although it is not difficult to show this in general, here we consider the simpler case of 1+1 dimensions where the two components (p^0, p^1) of the on-shell momentum $p^2 = m^2$ can be parametrized in terms of a single hyperbolic angle λ as

$$p^0 = m\cosh\lambda, \quad p^1 = m\sinh\lambda. \tag{2.21}$$

Now, the combination $2E_{\mathbf{p}}\delta(p^1 - p'^1)$ can be written as

$$2E_{\mathbf{p}}\delta(p^1 - p'^1) = 2m\cosh\lambda\,\delta(m\sinh\lambda - m\sinh\lambda') = 2\delta(\lambda - \lambda'), \tag{2.22}$$

where we have made use of the property (2.10) of the delta function. Lorentz transformations in $1+1$ dimensions are labelled by a parameter $\xi \in \mathbb{R}$ and act on the momentum by shifting the hyperbolic angle $\lambda \to \lambda + \xi$. However, Eq. (2.22) is invariant under a common shift of λ and λ', so the whole expression is obviously invariant under Lorentz transformations.

To summarize what we did so far, we have succeeded in constructing a Lorentz covariant basis of states for the one-particle Hilbert space $\mathscr{H}_1$. The generators of space-time translations act on the basis states $|p\rangle$ as

$$\hat{P}^\mu|p\rangle = p^\mu|p\rangle, \tag{2.23}$$

whereas the action of Lorentz transformations is implemented by the unitary operator

$$\mathscr{U}(\Lambda)|p\rangle = |\Lambda^\mu{}_\nu p^\nu\rangle \equiv |\Lambda p\rangle \quad \text{with} \quad \Lambda \in \mathrm{SO}(1,3). \tag{2.24}$$

This transformation is compatible with the Lorentz invariant normalization (2.20),

$$\langle p|p'\rangle = \langle p|\mathscr{U}(\Lambda)^{-1}\mathscr{U}(\Lambda)|p'\rangle = \langle \Lambda p|\Lambda p'\rangle. \tag{2.25}$$

On $\mathscr{H}_1$ the operator $\hat{P}^\mu$ admits the following spectral representation

$$\hat{P}^\mu = \int \frac{d^3p}{(2\pi)^3}\frac{1}{2E_{\mathbf{p}}}|p\rangle p^\mu\langle p|. \tag{2.26}$$

Using (2.25) and the fact that the measure is invariant under Lorentz transformation, one can easily show that $\hat{P}^\mu$ transform covariantly under SO(1, 3)

$$\mathscr{U}(\Lambda)^{-1}\hat{P}^{\mu}\mathscr{U}(\Lambda) = \int \frac{d^3p}{(2\pi)^3}\frac{1}{2E_{\mathbf{p}}}|\Lambda^{-1}p\rangle p^{\mu}\langle\Lambda^{-1}p| = \Lambda^{\mu}{}_{\nu}\hat{P}^{\nu}. \tag{2.27}$$

A set of covariant creation-annihilation operators can be constructed now in terms of the operators $\hat{a}(\mathbf{p})$, $\hat{a}^{\dagger}(\mathbf{p})$ introduced above

$$\hat{\alpha}(\mathbf{p}) \equiv (2\pi)^{\frac{3}{2}}\sqrt{2E_{\mathbf{p}}}\hat{a}(\mathbf{p}), \quad \hat{\alpha}^{\dagger}(\mathbf{p}) \equiv (2\pi)^{\frac{3}{2}}\sqrt{2E_{\mathbf{p}}}\hat{a}^{\dagger}(\mathbf{p}). \tag{2.28}$$

with the Lorentz invariant commutation relations

$$\begin{aligned}\left[\hat{\alpha}(\mathbf{p}), \hat{\alpha}^{\dagger}(\mathbf{p}')\right] &= (2\pi)^3(2E_{\mathbf{p}})\delta(\mathbf{p}-\mathbf{p}'),\\ \left[\hat{\alpha}(\mathbf{p}), \hat{\alpha}(\mathbf{p}')\right] &= \left[\hat{\alpha}^{\dagger}(\mathbf{p}), \hat{\alpha}^{\dagger}(\mathbf{p}')\right] = 0.\end{aligned} \tag{2.29}$$

Particle states are created by acting with any number of creation operators $\alpha(\mathbf{p})$ on the Poincaré invariant vacuum state $|0\rangle$ satisfying

$$\begin{aligned}&\langle 0|0\rangle = 1,\\ &\hat{P}^{\mu}|0\rangle = 0,\\ &\mathscr{U}(\Lambda)|0\rangle = |0\rangle, \quad \text{for all } \Lambda \in \mathrm{SO}(1,3).\end{aligned} \tag{2.30}$$

A general one-particle state $|f\rangle \in \mathscr{H}_1$ can be written as

$$|f\rangle = \int \frac{d^3p}{(2\pi)^3}\frac{1}{2E_{\mathbf{p}}}f(\mathbf{p})\hat{\alpha}^{\dagger}(\mathbf{p})|0\rangle, \tag{2.31}$$

while a n-particle state $|f\rangle \in \mathscr{H}_1^{\otimes n}$ is

$$|f\rangle = \int \left[\prod_{i=1}^{n}\frac{d^3p_i}{(2\pi)^3}\frac{1}{2\omega_{p_i}}\right] f(\mathbf{p}_1,\ldots,\mathbf{p}_n)\hat{\alpha}^{\dagger}(\mathbf{p}_1)\ldots\hat{\alpha}^{\dagger}(\mathbf{p}_n)|0\rangle. \tag{2.32}$$

That these states are Lorentz invariant can be checked by noticing that from the definition of the creation-annihilation operators follows the transformation

$$\mathscr{U}(\Lambda)\hat{\alpha}(\mathbf{p})\mathscr{U}(\Lambda)^{\dagger} = \hat{\alpha}(\Lambda\mathbf{p}) \tag{2.33}$$

and the corresponding one for creation operators.

As we have argued above, the very fact that measurements have to be localized implies the necessity of introducing quantum fields. Here we will consider the simplest case of a quantum scalar field $\hat{\phi}(x)$ satisfying the following properties:

- *Hermiticity*

$$\hat{\phi}(x)^{\dagger} = \hat{\phi}(x). \tag{2.34}$$

- *Microcausality* Since measurements cannot interfere with each other when performed in causally disconnected points of space-time, the commutator of two fields has to vanish outside the relative light-cone

$$\left[\hat{\phi}(x), \hat{\phi}(y)\right] = 0, \quad (x-y)^2 < 0. \tag{2.35}$$

- *Translation invariance*

$$e^{i\hat{P}\cdot a}\hat{\phi}(x)e^{-i\hat{P}\cdot a} = \hat{\phi}(x-a). \tag{2.36}$$

- *Lorentz invariance*

$$\mathcal{U}(\Lambda)^\dagger\hat{\phi}(x)\mathcal{U}(\Lambda) = \hat{\phi}(\Lambda^{-1}x). \tag{2.37}$$

- *Linearity* To simplify matters we will also assume that $\phi(x)$ is linear in the creation-annihilation operators $\alpha(\mathbf{p})$, $\alpha^\dagger(\mathbf{p})$

$$\hat{\phi}(x) = \int \frac{d^3p}{(2\pi)^3}\frac{1}{2E_{\mathbf{p}}}\left[f(\mathbf{p},x)\hat{\alpha}(\mathbf{p}) + g(\mathbf{p},x)\hat{\alpha}^\dagger(\mathbf{p})\right]. \tag{2.38}$$

Since $\hat{\phi}(x)$ should be hermitian we are forced to take $g(\mathbf{p},x) = f(\mathbf{p},x)^*$. Moreover, $\phi(x)$ satisfies the equations of motion of a free scalar field, $(\partial_\mu\partial^\mu + m^2)\hat{\phi}(x) = 0$, only if $f(\mathbf{p},x)$ is a complete basis of solutions of the Klein–Gordon equation. These considerations leads to the expansion

$$\hat{\phi}(x) = \int \frac{d^3p}{(2\pi)^3}\frac{1}{2E_{\mathbf{p}}}\left[e^{-iE_{\mathbf{p}}t + i\mathbf{p}\cdot\mathbf{x}}\hat{\alpha}(\mathbf{p}) + e^{iE_{\mathbf{p}}t - i\mathbf{p}\cdot\mathbf{x}}\hat{\alpha}^\dagger(\mathbf{p})\right]. \tag{2.39}$$

It can be checked that $\hat{\phi}(x)$ and $\partial_t\hat{\phi}(x)$ satisfy the equal-time canonical commutation relations

$$\left[\hat{\phi}(t,\mathbf{x}), \partial_t\hat{\phi}(t,\mathbf{y})\right] = i\delta(\mathbf{x}-\mathbf{y}). \tag{2.40}$$

The general (non-equal time) commutator

$$\left[\hat{\phi}(x), \hat{\phi}(x')\right] = i\Delta(x-x') \tag{2.41}$$

can also be computed using the expression (2.39). The function $\Delta(x-y)$ is given by

$$\begin{aligned} i\Delta(x-y) &= -\mathrm{Im}\int \frac{d^3p}{(2\pi)^3}\frac{1}{2E_{\mathbf{p}}}e^{-iE_{\mathbf{p}}(t-t') + i\mathbf{p}\cdot(\mathbf{x}-\mathbf{x}')} \\ &= \int \frac{d^4p}{(2\pi)^4}(2\pi)\delta(p^2-m^2)\mathrm{sign}(p^0)e^{-ip\cdot(x-x')}, \end{aligned} \tag{2.42}$$

where the sign function is defined as

$$\text{sign}(x) \equiv \theta(x) - \theta(-x) = \begin{cases} 1 & x > 0 \\ -1 & x < 0 \end{cases}. \tag{2.43}$$

Using the last expression in Eq. (2.42) it is easy to show that $i\Delta(x - x')$ vanishes when x and x' are space-like separated. Indeed, if $(x - x')^2 < 0$ there is always a reference frame in which both events are simultaneous, and since $i\Delta(x - x')$ is Lorentz invariant we can compute it in this frame. In this case $t = t'$ and the exponential in the second line of (2.42) does not depend on p^0. Therefore, the integration over p^0 gives

$$\begin{aligned}
&\int_{-\infty}^{\infty} dp^0 \varepsilon(p^0)\delta(p^2 - m^2) \\
&= \int_{-\infty}^{\infty} dp^0 \left[\frac{1}{2E_{\mathbf{p}}}\varepsilon(p^0)\delta(p^0 - E_{\mathbf{p}}) + \frac{1}{2E_{\mathbf{p}}}\varepsilon(p^0)\delta(p^0 + E_{\mathbf{p}})\right] \\
&= \frac{1}{2E_{\mathbf{p}}} - \frac{1}{2E_{\mathbf{p}}} = 0.
\end{aligned} \tag{2.44}$$

So we have concluded that $i\Delta(x - x') = 0$ if $(x - x')^2 < 0$, as required by microcausality. Notice that the situation is completely different when $(x - x')^2 \geq 0$, since in this case the exponential depends on p^0 and the integration over this component of the momentum does not vanish.

2.2 Canonical Quantization

So far we have contented ourselves with requiring a number of properties from the quantum scalar field: existence of asymptotic states, locality, microcausality and relativistic invariance. With only these ingredients we have managed to go quite far. The previous results can also be obtained using canonical quantization. One starts with a classical free scalar field theory in the Hamiltonian formalism and obtains the quantum theory by replacing Poisson brackets by commutators. Since this quantization procedure is based on the use of the canonical formalism, which gives time a privileged role, it is important to check at the end of the calculation that the resulting quantum theory is Lorentz invariant. In the following we will briefly overview the canonical quantization of the Klein–Gordon scalar field.

The starting point is the action functional $S[\phi(x)]$ which, in the case of a free real scalar field of mass m is given by

$$S[\phi(x)] \equiv \int d^4x \mathscr{L}(\phi, \partial_\mu\phi) = \frac{1}{2}\int d^4x \left(\partial_\mu\phi\partial^\mu\phi - m^2\phi^2\right). \tag{2.45}$$

The equations of motion are obtained, as usual, from the Euler–Lagrange equations

$$\partial_\mu \left[\frac{\partial \mathcal{L}}{\partial(\partial_\mu \phi)} \right] - \frac{\partial \mathcal{L}}{\partial \phi} = 0 \implies (\partial_\mu \partial^\mu + m^2)\phi = 0. \tag{2.46}$$

In the Hamiltonian formalism the physical system is described in terms of the field $\phi(x)$, its spatial derivatives and its canonically conjugated momentum

$$\pi(x) \equiv \frac{\partial \mathcal{L}}{\partial(\partial_0 \phi)} = \frac{\partial \phi}{\partial t}. \tag{2.47}$$

The dynamics of the system is determined by the Hamiltonian functional

$$H = \int d^3x \left(\pi \frac{\partial \phi}{\partial t} - \mathcal{L} \right) = \frac{1}{2} \int d^3x \left[\pi^2 + (\nabla \phi)^2 + m^2 \right]. \tag{2.48}$$

The canonical equations of motion can be written in terms of Poisson brackets. Given two functionals $A[\phi, \pi]$, $B[\phi, \pi]$ of the canonical variables

$$A[\phi, \pi] = \int d^3x \mathcal{A}(\phi, \pi), \quad B[\phi, \pi] = \int d^3x \mathcal{B}(\phi, \pi), \tag{2.49}$$

their Poisson bracket is defined by

$$\{A, B\}_{\text{PB}} \equiv \int d^3x \left(\frac{\delta A}{\delta \phi} \frac{\delta B}{\delta \pi} - \frac{\delta A}{\delta \pi} \frac{\delta B}{\delta \phi} \right). \tag{2.50}$$

Here $\frac{\delta}{\delta \phi}$ denotes the functional derivative defined as

$$\frac{\delta A}{\delta \phi} \equiv \frac{\partial \mathcal{A}}{\partial \phi} - \partial_\mu \left[\frac{\partial \mathcal{A}}{\partial(\partial_\mu \phi)} \right]. \tag{2.51}$$

In particular, the canonically conjugated fields satisfy the following equal time Poisson brackets

$$\begin{aligned} \{\phi(t, \mathbf{x}), \phi(t, \mathbf{x}')\}_{\text{PB}} &= \{\pi(t, \mathbf{x}), \pi(t, \mathbf{x}')\}_{\text{PB}} = 0, \\ \{\phi(t, \mathbf{x}), \pi(t, \mathbf{x}')\}_{\text{PB}} &= \delta(\mathbf{x} - \mathbf{x}'), \end{aligned} \tag{2.52}$$

The canonical equations of motion are

$$\partial_0 \phi(x) = \{\phi(x), H\}_{\text{PB}}, \quad \partial_0 \pi(x) = \{\pi(x), H\}_{\text{PB}}, \tag{2.53}$$

where H is the Hamiltonian of the system.

In the case of the scalar field, a general solution of the classical field equations (2.46) can be obtained by working with the Fourier transform of the equation of motion

$$(\partial_\mu \partial^\mu + m^2)\phi(x) = 0 \implies (-p^2 + m^2)\tilde{\phi}(p) = 0, \tag{2.54}$$

whose general solution can be written as[2]

$$\begin{aligned}\phi(x) &= \int \frac{d^4p}{(2\pi)^4}(2\pi)\delta(p^2-m^2)\theta(p^0)\left[\alpha(p)e^{-ip\cdot x}+\alpha(p)^* e^{ip\cdot x}\right]\\ &= \int \frac{d^3p}{(2\pi)^3}\frac{1}{2E_{\mathbf{p}}}\left[\alpha(\mathbf{p})e^{-iE_{\mathbf{p}}t+\mathbf{p}\cdot\mathbf{x}}+\alpha(\mathbf{p})^* e^{iE_{\mathbf{p}}t-\mathbf{p}\cdot\mathbf{x}}\right]\end{aligned} \tag{2.55}$$

and we have required $\phi(x)$ to be real. The conjugate momentum is

$$\pi(x) = -\frac{i}{2}\int \frac{d^3p}{(2\pi)^3}\left[\alpha(\mathbf{p})e^{-iE_{\mathbf{p}}t+\mathbf{p}\cdot\mathbf{x}}-\alpha(\mathbf{p})^* e^{iE_{\mathbf{p}}t-\mathbf{p}\cdot\mathbf{x}}\right]. \tag{2.56}$$

Canonical quantization proceeds by replacing classical fields with operators and Poisson brackets with commutators according to the rule

$$i\{\cdot,\cdot\}_{\mathrm{PB}} \longrightarrow [\cdot,\cdot]. \tag{2.57}$$

Now $\phi(x)$ and $\pi(x)$ are promoted to operators by replacing the functions $\alpha(\mathbf{p})$, $\alpha(\mathbf{p})^*$ by the corresponding operators

$$\alpha(\mathbf{p}) \longrightarrow \hat{\alpha}(\mathbf{p}), \quad \alpha(\mathbf{p})^* \longrightarrow \hat{\alpha}^\dagger(\mathbf{p}). \tag{2.58}$$

Moreover, demanding $[\phi(t,\mathbf{x}),\pi(t,\mathbf{x}')] = i\delta(\mathbf{x}-\mathbf{x}')$ forces the operators $\hat{\alpha}(\mathbf{p})$, $\hat{\alpha}(\mathbf{p})^\dagger$ to have the commutation relations found in Eq. (2.29). Therefore they are identified as a set of creation-annihilation operators creating states with well-defined momentum $\mathbf{p}$ out of the vacuum $|0\rangle$. In the canonical quantization formalism the concept of particle appears as a result of the quantization of a classical field.

From the expressions of $\hat{\phi}$ and $\hat{\pi}$ in terms of the creation-annihilation operators we can evaluate the Hamiltonian operator. After a simple calculation one arrives at

$$\begin{aligned}\hat{H} &= \frac{1}{2}\int \frac{d^3p}{(2\pi)^3}\left[\hat{\alpha}^\dagger(\mathbf{p})\hat{\alpha}(\mathbf{p})+(2\pi)^3E_{\mathbf{p}}\delta(\mathbf{0})\right]\\ &= \int \frac{d^3p}{(2\pi)^3}\frac{1}{2E_{\mathbf{p}}}E_{\mathbf{p}}\hat{\alpha}^\dagger(\mathbf{p})\hat{\alpha}(\mathbf{p})+\frac{1}{2}\int d^3p E_{\mathbf{p}}\delta(\mathbf{0}).\end{aligned} \tag{2.59}$$

The first integral has a simple physical interpretation: the integrand is the number operator of particles with momentum $\mathbf{p}$, weighted by the energy $E_{\mathbf{p}}$ of the particle and integrated using the Lorentz-invariant measure. The second term diverges and it is equal to the expectation value of the Hamiltonian in the ground state, $\langle 0|\hat{H}|0\rangle$. It measures the energy stored in the vacuum.

We should make sense of the divergent vacuum energy in Eq. (2.59). It has two sources of divergence. One is of infrared origin and it is associated with the delta

---

[2] In momentum space, the general solution to this equation is $\tilde{\phi}(p) = f(p)\delta(p^2-m^2)$, with $f(p)$ a completely general function of p^μ. The solution in position space is obtained by inverse Fourier transform. The step function $\theta(p^0)$ enforces positivity of the energy.

function evaluated at $\mathbf{p} = \mathbf{0}$, reflecting the fact that we work in infinite volume. The second one comes from the integration of $E_{\mathbf{p}}$ at large values of the momentum and it is then an ultraviolet divergence. The infrared divergence can be regularized by putting the system in a box of finite but large volume and replacing $\delta(\mathbf{0}) \sim V$. Since now the momentum gets discretized, we have

$$E_{\text{vac}} \equiv \langle 0|\hat{H}|0\rangle = \sum_{\mathbf{p}} \frac{1}{2} E_{\mathbf{p}}. \tag{2.60}$$

Written in this form the interpretation of the vacuum energy is straightforward. A free scalar quantum field can be seen as a infinite collection of harmonic oscillators per unit volume, each one labelled by $\mathbf{p}$. Even if those oscillators are not excited, they contribute to the vacuum energy with their zero-point energy, given by $\frac{1}{2}E_{\mathbf{p}}$. Due to the ultraviolet divergence, the vacuum contribution to the energy adds up to infinity even working at finite volume: there are modes with arbitrary high momentum contributing to the sum, $p_i \sim \frac{n_i}{L_i}$, with L_i the sides of the box of volume V and n_i an integer.

For many practical purposes we can shift the origin of energies and subtract the vacuum energy. This is done by replacing $\hat{H}$ by the normal-ordered Hamiltonian

$$:\hat{H}:\equiv \hat{H} - \langle 0|\hat{H}|0\rangle = \frac{1}{2}\int \frac{d^3p}{(2\pi)^3}\hat{\alpha}^\dagger(\mathbf{p})\hat{\alpha}(\mathbf{p}). \tag{2.61}$$

In spite of this, in the next section we will see that under certain conditions the vacuum energy has observable effects. In addition, in general relativity the energy of the vacuum is a source of the gravitational field and contributes to the cosmological constant (see Chap. 12).

All relevant information about the free scalar field theory is encoded in the time-ordered correlation functions

$$G_n(x_1, \ldots, x_n) = \langle 0|T\Big[\hat{\phi}(x_1)\ldots\hat{\phi}(x_n)\Big]|0\rangle. \tag{2.62}$$

The symbol T indicates that we have a time-ordered product, i.e. the noncommuting field operators are multiplied in the order in which they occur in time. For example, for the time-ordered product of two scalar fields we have

$$T\Big[\hat{\phi}(x_1)\hat{\phi}(x_2)\Big] = \theta\left(x_1^0 - x_2^0\right)\hat{\phi}(x_1)\hat{\phi}(x_2) + \theta\left(x_2^0 - x_1^0\right)\hat{\phi}(x_2)\hat{\phi}(x_1). \tag{2.63}$$

The generalization to monomials with more than two operators is straightforward: operators evaluated at earlier times always appear to the right.

In the case of our free scalar field theory the only independent time-ordered correlation function is the Feynman propagator $G_2(x_1, x_2)$. After some manipulations, it can be written as

$$\langle 0|T\Big[\hat{\phi}(x_1)\hat{\phi}(x_2)\Big]|0\rangle = \int \frac{d^4p}{(2\pi)^4}\frac{ie^{-ip\cdot(x_1-x_2)}}{p^2 - m^2 + i\varepsilon}. \tag{2.64}$$

The term $i\varepsilon$ in the denominator is a reminder of how to surround the poles in the integration over p^0. This is crucial to reproduce correctly the step functions in the time-ordered product (2.63). To calculate higher order correlation functions one uses a mathematical result known as Wick's theorem that allows to write a time-ordered product as a combination of normal-ordered products with coefficients given by the Feynman propagator $G_2(x_1, x_2)$. We will not give a general proof but state it for the case of three fields

$$\begin{aligned} T\left[\hat{\phi}(x_1)\hat{\phi}(x_2)\hat{\phi}(x_3)\right] =& : \hat{\phi}(x_1)\hat{\phi}(x_2)\hat{\phi}(x_3) : + \wick{\c{\hat{\phi}}(x_1)\c{\hat{\phi}}(x_2)} : \hat{\phi}(x_3) : \\ & + \hat{\phi}(x_1) : \wick{\c{\hat{\phi}}(x_2)\c{\hat{\phi}}(x_3)} + \wick{\c{\hat{\phi}}(x_1) : \hat{\phi}(x_2) : \c{\hat{\phi}}(x_3)}. \end{aligned} \tag{2.65}$$

The pairs of operators connected by braces, called Wick contractions, have to be replaced by a Feynman propagator according to

$$\wick{\c{\hat{\phi}}(x_i)\c{\hat{\phi}}(x_j)} \longrightarrow G_2(x_i, x_j). \tag{2.66}$$

From this example we read the structure of the general case: the time-ordered product of n fields can be written as the sum of all monomials of n fields with any number of Wick contractions (from 0 to the integer part of $\frac{n}{2}$) done in all possible nonequivalent ways. In each of these monomial the product of those fields that are not Wick-contracted is always normal ordered.[3]

Using this result the correlation functions (2.62) can be easily computed. Since the vacuum expectation value of a normal ordered operator is zero, the only terms that contribute are those in which all fields are Wick-contracted among themselves. This automatically implies that all time-ordered correlation functions with an odd number of scalar fields are equal to zero. For correlation functions with an even number of insertion we illustrate how it works in the case of the four-point function, where there are three different contractions

$$\begin{aligned} G_4(x_1, \ldots, x_4) =& \langle 0|T\left[\wick{\c1{\hat{\phi}}(x_1)\c1{\hat{\phi}}(x_2)\c2{\hat{\phi}}(x_3)\c2{\hat{\phi}}(x_4)}\right]|0\rangle \\ & + \langle 0|T\left[\wick{\c1{\hat{\phi}}(x_1)\c2{\hat{\phi}}(x_2)\c1{\hat{\phi}}(x_3)\c2{\hat{\phi}}(x_4)}\right]|0\rangle + \langle 0|T\left[\wick{\c1{\hat{\phi}}(x_1)\c2{\hat{\phi}}(x_2)\c2{\hat{\phi}}(x_3)\c1{\hat{\phi}}(x_4)}\right]|0\rangle. \end{aligned} \tag{2.67}$$

[3] We remind the reader that in a normal-ordered product all annihilation operators appear to the right.

Replacing now each Wick contraction by the corresponding propagator according to (2.66), we find

$$G_4(x_1, \ldots, x_4) = G_2(x_1, x_2)G_2(x_3, x_4) + G_2(x_1, x_3)G_2(x_2, x_4) + G_2(x_1, x_4)G_2(x_2, x_3). \tag{2.68}$$

Any other correlation function is computed in terms of $G_2(x_1, x_2)$ following the same algorithm. In fact, this property is the defining feature of *any* free quantum field theory: the propagator completely determines all other correlation functions of the theory.

2.3 The Casimir Effect

The vacuum energy encountered in the quantization of the free scalar field is not exclusive of this theory. It is also present in other field theories and in particular in quantum electrodynamics. In 1948 Hendrik Casimir pointed out [1] that although a formally divergent vacuum energy would not be observable, any variation in this energy would be (see [2–4] for comprehensive reviews).

To show this he devised the following experiment. Consider a couple of infinite, perfectly conducting plates placed parallel to each other at a distance d (see Fig. 2.1). The plates fix the boundary condition of the vacuum modes of the electromagnetic field. These modes are discrete in between the plates (region II), while outside them they have a continuous spectrum (regions I and III). The vacuum energy of the electromagnetic field is equal to that of two massless scalar fields, corresponding to the two physical polarizations of the photon (see Sect. 4.2). Hence we can apply the formulae derived above.

A naive calculation of the vacuum energy in this system gives a divergent result. This infinity can be removed by subtracting the vacuum energy corresponding to the situation where the plates are removed

$$E(d)_{\text{reg}} = E(d)_{\text{vac}} - E(\infty)_{\text{vac}}. \tag{2.69}$$

This subtraction cancels the contribution of all the modes outside the plates. The boundary conditions of the electromagnetic field at the conducting plates dictate the quantization of the momentum modes perpendicular to them according to $p_\perp = \frac{n\pi}{d}$, with n a non-negative integer. When the size of the plates is much larger than their separation d, the momenta parallel to the plates $\mathbf{p}_\parallel$ can be treated as continuous. For $n > 0$ there are two polarizations for each vacuum mode of the electromagnetic field, each one contributing

$$\frac{1}{2}\sqrt{\mathbf{p}_\parallel^2 + p_\perp^2} \tag{2.70}$$

to the vacuum energy. When $p_\perp = 0$ (i.e., $n = 0$) the modes of the field are effectively (2+1)-dimensional and there is only one physical polarization. Taking all these

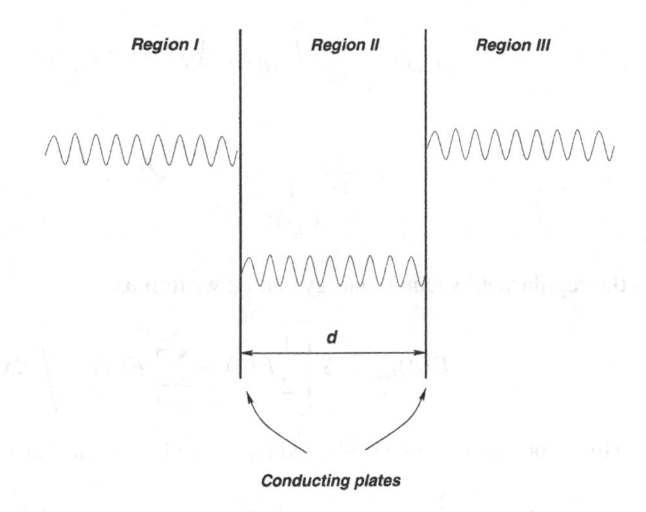

Fig. 2.1 Illustration of the Casimir effect. In regions I and II the spetrum of modes of the momentum $p_\perp$ is continuous, while in the space between the plates (region II) it is quantized in units of $\frac{\pi}{d}$

elements into account, we write

$$E(d)_{\text{reg}} = S\int \frac{d^2 p_\parallel}{(2\pi)^2}\frac{1}{2}|\mathbf{p}_\parallel| + 2S\int \frac{d^2 p_\parallel}{(2\pi)^2}\sum_{n=1}^{\infty}\frac{1}{2}\sqrt{\mathbf{p}_\parallel^2 + \left(\frac{n\pi}{d}\right)^2}$$
$$- 2Sd\int \frac{d^3 p}{(2\pi)^3}\frac{1}{2}|\mathbf{p}|, \tag{2.71}$$

where S is the area of the plates. The factors of two count the two propagating degrees of freedom of the electromagnetic field, as discussed above.

The integrals and the infinite sum in Eq. (2.71) are divergent. In order to define them we insert an exponential damping factor[4]

$$E(d)_{\text{reg}} = \frac{1}{2}S\int \frac{d^2 p_\perp}{(2\pi)^2}e^{-\frac{1}{\Lambda}|\mathbf{p}_\parallel|}|\mathbf{p}_\parallel| + S\sum_{n=1}^{\infty}\int \frac{d^2 p_\parallel}{(2\pi)^2}e^{-\frac{1}{\Lambda}\sqrt{\mathbf{p}_\parallel^2+\left(\frac{n\pi}{d}\right)^2}}\sqrt{\mathbf{p}_\parallel^2 + \left(\frac{n\pi}{d}\right)^2}$$
$$- Sd\int_{-\infty}^{\infty}\frac{dp_\perp}{2\pi}\int \frac{d^2 p_\parallel}{(2\pi)^2}e^{-\frac{1}{\Lambda}\sqrt{\mathbf{p}_\parallel^2+p_\perp^2}}\sqrt{\mathbf{p}_\parallel^2 + p_\perp^2}, \tag{2.72}$$

where Λ is an ultraviolet cutoff. It is now straightforward to see that in terms of the function

---

[4] Alternatively, one could introduce any cutoff function $f(p_\perp^2 + p_\parallel^2)$ going to zero fast enough as $p_\perp, p_\parallel \to \infty$. The result is independent of the particular function used in the calculation.

$$F(x) = \frac{1}{2\pi}\int_0^\infty y dy e^{-\frac{1}{\Lambda}\sqrt{y^2+(\frac{x\pi}{d})^2}}\sqrt{y^2+\left(\frac{x\pi}{d}\right)^2}$$
$$= \frac{1}{4\pi}\int_{(\frac{x\pi}{d})^2}^\infty dz e^{-\frac{\sqrt{z}}{\Lambda}}\sqrt{z} \tag{2.73}$$

the regularized vacuum energy can be written as

$$E(d)_{\text{reg}} = S\left[\frac{1}{2}F(0) + \sum_{n=1}^\infty F(n) - \int_0^\infty dx F(x)\right]. \tag{2.74}$$

This expression can be evaluated using the Euler-MacLaurin formula [5]

$$\sum_{n=1}^\infty F(n) - \int_0^\infty dx F(x) = -\frac{1}{2}[F(0)+F(\infty)] + \frac{1}{12}\left[F'(\infty)-F'(0)\right]$$
$$-\frac{1}{720}\left[F'''(\infty)-F'''(0)\right] + \cdots \tag{2.75}$$

Our function satisfies $F(\infty) = F'(\infty) = F'''(\infty) = 0$ and $F'(0) = 0$, whereas higher derivative terms give contributions that go to zero as the cutoff is sent to infinity. Hence the value of $E(d)_{\text{reg}}$ is determined by $F'''(0)$. Computing this term and taking the limit $\Lambda \to \infty$ we find the result

$$E(d)_{\text{reg}} = \frac{S}{720}F'''(0) = -\frac{\pi^2 S}{720 d^3}. \tag{2.76}$$

This shows that the vacuum energy between the two plates decreases when their separation is reduced. Therefore there should be a force per unit area between the plates given by

$$P_{\text{Casimir}} = -\frac{\pi^2}{240}\frac{1}{d^4}. \tag{2.77}$$

The minus sign indicates that the force is attractive. This is called the Casimir effect. It was experimentally measured for the first time in 1958 by Sparnaay [6] and since then the Casimir effect has been checked with better and better precision in a variety of situations [2–4].

2.4 Path Integrals

The canonical quantization formalism relies in the Hamiltonian formulation of the theory. It has the obvious disadvantage of singling out time from the spatial coordinates, making Lorentz covariance nonexplicit. This could be avoided if quantization could be carried out directly in the Lagrangian formalism, where Lorentz covariance is explicit. This is achieved by the path integral quantization method introduced by Feynman [7, 8]. In addition, when applied to the quantization of fields, path integral quantization presents many advantages over canonical quantization.

To describe the main ideas of path integral quantization we will not enter into technical details that can be found in many available textbooks [9–12]. Also, to make the discussion more transparent, we first illustrate the method in the case of nonrelativistic quantum mechanical system with a single degree of freedom denoted by q and Lagrangian $L(q, \dot{q})$. To quantize this theory we only need to know its propagator defined by

$$K(q, q'; \tau) = \langle q'; \tau | q; 0 \rangle. \tag{2.78}$$

Here we have used the Heisenberg representation where the time-independent eigenstates of the time-dependent operator $q(t)$ are denoted by $|q; t\rangle$. Physically, $K(q, q'; \tau)$ represents the amplitude for the system to "propagate" from q to q' in a time τ. That the knowledge of the propagator is enough to solve the quantum system can be seen by noticing that the Schrödinger wave function $\psi(t, q)$ at any time can be written in terms of the initial data as

$$\psi(t, q') = \int_{-\infty}^{\infty} dq K(q, q'; t) \psi(0, q). \tag{2.79}$$

This equation follows from the fact that $K(q, q'; \tau)$ is the Green function of the time-dependent Schrödinger wave equation.

Another physically meaningful quantity is the fixed-energy propagator, defined in terms of (2.78) by[5]

$$G(q, q'; E) = \int_{0}^{\infty} d\tau e^{\frac{i}{\hbar} E\tau} K(q, q'; \tau). \tag{2.80}$$

This propagator is the Green function for the time-independent Schrödinger problem $(\hat{H} - E)\psi(t, q) = 0$. In fact, $G(q, q'; E)$ contains all the information about the spectrum of the theory codified in the structure of its singularities in the complex E plane.

Both $K(q, q'; \tau)$ and $G(q, q'; E)$ can be calculated using canonical quantization. Here instead we would rather follow the Lagrangian formalism and use of the following observation due to Dirac [13]

$$\langle q + \delta q; t + \delta t | q; t \rangle \sim \exp\left[\frac{i}{\hbar} \delta t L\left(q, \frac{\delta q}{\delta t}\right)\right]. \tag{2.81}$$

---

[5] For the remaining of this chapter we restore the powers of $\hbar$.

It states that the amplitude for the propagation of the system between q and $q+\delta q$ in an infinitesimal time δt can be expressed in terms of the Lagrangian function of the system. We justify this equation in the case of a particle moving in one dimension in the presence of a potential $V(q)$,

$$\begin{aligned}\langle q+\delta q; t+\delta t|q; t\rangle &= \langle q+\delta q|e^{-\frac{i}{\hbar}\delta t\hat{H}}|q\rangle \\ &= \langle q+\delta q|e^{-\frac{i}{\hbar}\delta t\left[\frac{\hat{p}^2}{2m}+V(\hat{q})\right]}|q\rangle. \end{aligned} \tag{2.82}$$

The kinetic and potential energy in the exponent can be taken to commute up to terms of order $(\delta t)^2$. Hence, to linear order in δt, the exponential can be split into two terms depending respectively on $\hat{q}$ and $\hat{p}$. Inserting between them the completeness relation for the momentum eigenstates we find

$$\langle q+\delta q; t+\delta t|q; t\rangle = \int_{-\infty}^{\infty} dp e^{\frac{i}{\hbar}p\delta q-\frac{i}{\hbar}\delta t\left[\frac{p^2}{2m}-V(q)\right]}. \tag{2.83}$$

We complete now the square and perform the Gaussian integration over the momentum to arrive at

$$\langle q+\delta q; t+\delta t|q; t\rangle = \sqrt{\frac{m}{2\pi i\hbar\delta t}}e^{\frac{i}{\hbar}\left[\frac{1}{2}m\left(\frac{\delta q}{\delta t}\right)^2-V(q)\right]}. \tag{2.84}$$

This calculation shows that the proportionality constant omitted in Eq. (2.81) does not depend on the value of the coordinate q.

To compute the propagator (2.78) we split the time interval τ in $N+1$ subintervals of duration δt and insert the identity, $\int dq|q; n\delta t\rangle\langle q; n\delta t| = 1$ for $n = 1, \ldots, N$, at each intermediate time

$$K(q, q'; \tau) = \left(\int_{-\infty}^{\infty}\prod_{i=1}^{N} dq_i\right)\langle q'; \tau|q_N; N\delta t\rangle \ldots \langle q_1; \delta t|q; 0\rangle. \tag{2.85}$$

This representation of the propagator can be interpreted as a summation over continuous discretized paths defined by $q_n = q(n\delta t)$ and satisfying the boundary conditions $q(0) = q,\ q(\tau) = q$. We can go now to the continuous limit of the path by taking $\delta t \to 0$ and $N \to \infty$ while keeping $N\delta t = \tau$ fixed. Then, each overlap inside the integral can be evaluated using Eq. (2.81) and the result defines the path integral

$$\langle q'; \tau|q; 0\rangle = \mathcal{N}\int_{\substack{q(0)=q \\ q(\tau)=q'}} \mathcal{D}q(t)\exp\left[\frac{i}{\hbar}\int_0^{\tau} dt L(q, \dot{q})\right], \tag{2.86}$$

where $\mathcal{N}$ is a normalization constant. This equation is a shorthand to indicate that the quantum mechanical amplitude (2.78) is obtained by summing over *all* possible trajectories $q(t)$ joining the points q and q' in a time τ, each one weighted by a phase given by the action of the corresponding trajectory measured in units of $\hbar$.

From a purely technical point of view, the path integral formulation of quantum mechanics does not present any important advantage over other quantization methods, notably canonical quantization. It is however in quantum field theory that path integrals show their real power. Path integrals for quantum field theories can be constructed by looking at a quantum field $\phi(t, \mathbf{x})$ as a quantum mechanical system with one degree of freedom per point of space $\mathbf{x}$. In other words, $\mathbf{x}$ is treated as a (continuous) label counting the number of degrees of freedom of the system. Now, as in the quantum mechanical case, path integrals can be used to write the amplitude for the system to evolve from the field configuration $\phi_0(\mathbf{x})$ at $t = 0$ to $\phi_1(\mathbf{x})$ at $t = \tau$.

$$\langle\phi_1(\mathbf{x}); \tau|\phi_0(\mathbf{x}); 0\rangle = \mathcal{N} \int_{\substack{\phi(0,\mathbf{x})=\phi_0(\mathbf{x}) \\ \phi(\tau,\mathbf{x})=\phi_1(\mathbf{x})}} \mathscr{D}\phi(t, \mathbf{x}) e^{\frac{i}{\hbar}S[\phi(t,\mathbf{x})]}, \tag{2.87}$$

where $S[\phi(t, \mathbf{x})]$ is the action functional of the theory. As above, this expression states that the amplitude is obtained by summing over all field configurations interpolating between the boundary values at $t = 0$, and $t = \tau$, each one multiplied by the phase factor $\exp\{\frac{i}{\hbar}S[\phi(t, \mathbf{x})]\}$.

Far more interesting, however, than the amplitudes (2.87) are the time-ordered correlation functions of fields that we already studied in the case of a free scalar field. For an interacting theory they are generalized to

$$G_n(x_1, \ldots, x_n) = \langle\Omega|T\Big[\hat{\phi}(x_1)\ldots\hat{\phi}(x_n)\Big]|\Omega\rangle. \tag{2.88}$$

where $|\Omega\rangle$ is the ground state of the theory. In Chap. 6 we will explain how the correlation functions (2.88) are related to scattering amplitudes. Here we only want to point out that they admit the following path integral representation[6]

$$\langle\Omega|T\Big[\hat{\phi}(x_1)\ldots\hat{\phi}(x_n)\Big]|\Omega\rangle = \frac{\int \mathscr{D}\phi(x)\phi(x_1)\ldots\phi(x_n)e^{\frac{i}{\hbar}S[\phi(x)]}}{\int \mathscr{D}\phi(x)e^{\frac{i}{\hbar}S[\phi(x)]}}. \tag{2.89}$$

In the left-hand side of this expression $\hat{\phi}(x)$ is the field operator, while in the right-hand side $\phi(x)$ represents a commuting function of the space-time coordinates. Unlike Eq. (2.87), here the functional integration is performed over *all* field configurations irrespective of any boundary conditions. Moreover, given that (2.89) contains the quotient of two path integrals the overall numerical normalization $\mathcal{N}$ cancels

---

[6] Here we focus on bosonic fields. Path integrals for fermions will be discussed in Chap. 3 (see page 43).

out. A salient feature of path integrals to be noticed here is that they automatically implement time ordering.

Path integrals are not easy to evaluate. In fact they cannot be computed exactly in most cases. This notwithstanding, path integrals provide an extremely useful tool in quantum field theory. They can be formally manipulated to obtain results whose derivation using canonical quantization methods would be much harder. The only path integrals that can be computed exactly are the so-called *Gaussian integrals* where the action functional of the theory is at most quadratic in the fields. This is the case, for example, of the free scalar field theory whose canonical quantization we studied in Sect. 2.2.

2.5 The Semiclassical Limit

One of the most interesting aspects of the application of the path integral formalism to quantum mechanics is that it clarifies how the classical laws of motion emerge from quantum dynamics. In the limit $\hbar \to 0$ the phase $\exp\left(\frac{i}{\hbar}S\right)$ varies wildly when going from a path to a neighboring one. The consequence is that the contributions from these paths to the propagator tend to cancel each other. There is however one important exception to this that are those trajectories making the action stationary. Since the linear perturbation of the action around these paths vanishes, these are the only ones contributing to the functional integral (2.86) in the classical limit. This is the way in which the classical laws of mechanics are recovered.

It is kind of remarkable how the principle of least action can be seen in this light as a residual effect of quantum physics. What from the point of view of classical mechanics is just an elegant principle to derive the Newtonian equations of motion is in fact hinting at the existence of an underlying theory.

We can make this qualitative discussion more precise by looking at the case of a nonrelativistic quantum particle moving in one dimension in the presence of a potential $V(q)$. The Lagrangian function of the system is

$$L(q, \dot{q}) = \frac{1}{2}m\dot{q}^2 - V(q). \tag{2.90}$$

We have argued that in the limit $\hbar \to 0$ the path integral (2.86) is dominated by paths around the classical solution $q_{\mathrm{cl}}(t)$ that solves the equations of motion with the appropriate boundary conditions. Thus, the propagator $K(q, q'; \tau)$ can be computed in the semiclassical limit by considering only the contribution to the path integral coming from paths that are "close" to the classical one. In technical terms this means that we write

$$q(t) = q_{\mathrm{cl}}(t) + \sqrt{\hbar}\delta q(t) \tag{2.91}$$

and integrate over the perturbation $\delta q(t)$ keeping in the action only terms that are at most linear in $\hbar$

$$S[q] = S_{\rm cl}[q, q'; \tau] + \frac{\hbar}{2} \int_0^\tau dt \Big[m(\delta\dot{q})^2 - V''(q_{\rm cl})(\delta q)^2\Big] + \mathscr{O}(\hbar^3). \tag{2.92}$$

Here $S_{\rm cl}[q, q'; \tau]$ is the action evaluated on the classical trajectory with the boundary conditions $q(0) = q$ and $q(\tau) = q'$. Since $q_{\rm cl}(t)$ satisfies the classical equations of motion, the term linear in the perturbation vanishes.

With this prescription, the semiclassical propagator is given by

$$\begin{aligned} K(q, q'; \tau) &\overset{\hbar\to 0}{\approx} \mathscr{N} e^{\frac{i}{\hbar} S_{\rm cl}[q,q';\tau]} \int\limits_{\substack{\delta q(0)=0 \\ \delta q(\tau)=0}} \mathscr{D}(\delta q)\, e^{\frac{i}{2}\int_0^\tau dt [m(\delta\dot{q})^2 - V^n(q_{cl}(\delta q)^2)]} \\ &= \mathscr{N} \sqrt{\frac{i}{2\pi\hbar} \frac{\partial^2 S_{\rm cl}[q, q'; \tau]}{\partial q \partial q'}}\, e^{\frac{i}{\hbar} S_{\rm cl}[q,q';\tau]}. \end{aligned} \tag{2.93}$$

Since the classical solution $q_{\rm cl}(t)$ satisfies the boundary conditions, the perturbation $\delta q(t)$ has to vanish at both $t = 0$ and $t = \tau$. The path integral over the fluctuations is Gaussian and can be computed exactly for any potential $V(q)$. Its evaluation is however nontrivial. The details of the calculation can be found in the literature [9–12].

A similar analysis can be applied to the computation of the semiclassical limit of the fixed-energy propagator $G(q, q'; E)$ defined in Eq. (2.80). Using the semiclassical expression for the full propagator (2.93) we are left with the integral

$$G(q, q'; E) \overset{\hbar\to 0}{\approx} \mathscr{N} \int_0^\infty d\tau \sqrt{\frac{i}{2\pi\hbar} \frac{\partial^2 S_{\rm cl}[q, q'; \tau]}{\partial q \partial q'}}\, e^{\frac{i}{\hbar}\{E\tau + S_{\rm cl}[q,q';\tau]\}}, \tag{2.94}$$

that has to be evaluated using the stationary phase method. The value $\tau = \tau_c$ that makes the phase stationary is the one solving the equation

$$E + \frac{\partial}{\partial\tau} S_{\rm cl}[q, q'; \tau]\bigg|_{\tau=\tau_c} = 0. \tag{2.95}$$

In this expression we recognize the Hamilton-Jacobi equation for a particle with constant energy E. Hence, the path dominating the path integral in the semiclassical computation of $G(q, q'; E)$ is the one solving the classical equation of motion

$$\dot{q}_{\rm cl}(\tau)^2 = \frac{2}{m}\Big[E - V(q_{\rm cl})\Big]. \tag{2.96}$$

and connecting the points q and q'. The calculation of τ_c reduces to the following quadrature

$$\tau_c = \sqrt{\frac{m}{2}} \int_q^{q'} \frac{dz}{\sqrt{E - V(z)}}. \tag{2.97}$$

The calculation of the semiclassical propagators requires some extra care in situations where quantum tunneling can occur. This is the case, for example, of a particle propagating in a barrier potential where the points q and q' are on different sides of the barrier. The calculation of $K(q, q'; \tau)$ in the limit $\hbar \to 0$ can be done in this case following the steps we have described, since there is always an above-the-barrier classical trajectory joining the points q and q' in a time τ that dominates the path integral.

The problem comes in the computation of $G(q, q'; E)$ when E is lower that the maximum of the barrier, $E < \max[V(q)]$. In this case there are no classical trajectories going through the classically forbidden region and therefore no saddle point value for the integral (2.94) is found in the domain of integration.

The key to solving the problem lies in performing an analytic continuation on the integrand of (2.94) and deforming the integration contour to capture the saddle points that occur for complex values of τ. For simplicity we concentrate on the case shown on the left panel of Fig. 2.2 where q and q' correspond to the classical turning points of a trajectory with energy E. Now, to compute the saddle point values of τ we have to continue the integrand of (2.97) and deform the contour of integration to surround the branch cut joining q and q'. As it happens, there is an infinite number of critical values given by[7]

$$\tau_n = -i(2n+1)\sqrt{\frac{m}{2}} \int_q^{q'} \frac{dx}{\sqrt{V(x) - E}}, \qquad n = 0, 1, \ldots \tag{2.98}$$

Here n is the number of times the contour surrounds the branch cut before reaching the endpoint. As a matter of fact we only need to consider the saddle point with $n = 0$, since the remaining ones give contributions to the semiclassical limit of $G(q, q'; E)$ that are exponentially suppressed with respect to it.

This analysis shows that the quantum tunneling under a barrier proceeds semiclassically *as if* the particle is propagating in imaginary time. A look at Eq. (2.94) leads to a nice interpretation of this fact. This expression says that there are infinitely many classical trajectories connecting the points q and q' that contribute to the fixed-energy propagator describing the tunneling process. As we explained above, all these real-time trajectories have energies above the height of the barrier. However, when $E < \max\{V(q)\}$ and in the limit $\hbar \to 0$, the coherent effect of all these paths is "resummed" into a single imaginary-time trajectory with an energy below the maximum of the barrier. This is called an *instanton*.

---

[7] There is a global sign ambiguity associated with the sense in which the integration contour surrounds the branch cut. Here we take it clockwise.

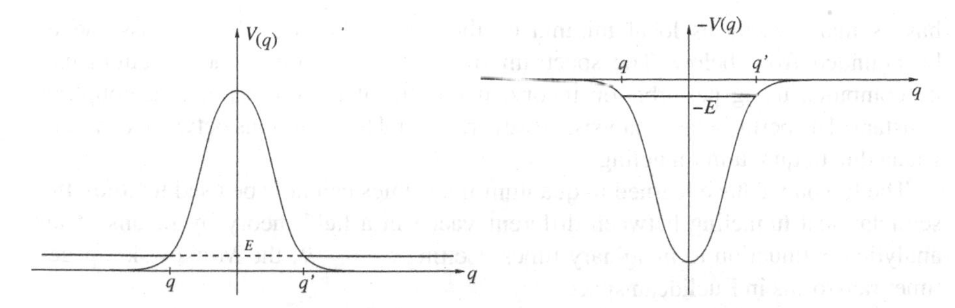

Fig. 2.2 On the left picture the tunneling of a particle with energy E from q to q' through the potential barrier is represented. In the right panel we have depicted the Euclidean trajectory describing this semiclassical tunneling

In fact, the imaginary-time trajectory that is found to dominate the path integral computation of $G(q, q'; E)$ in the semiclassical limit is a solution of the equations of motion derived from the Euclidean action $S_E[q]$, obtained by analytically continuing the action of the system to imaginary times

$$S[q] \overset{\tau \to -i\tau}{\longrightarrow} iS_E[q] = i \int_0^\tau dt \left[\frac{1}{2}m\dot{q}^2 + V(q)\right] \tag{2.99}$$

and the equations of motion of the Euclidean trajectories of energy E are

$$\dot{q}_{\rm cl}^2 = \frac{2}{m}\Big[-E + V(q_{\rm cl})\Big]. \tag{2.100}$$

Heuristically these equations can be interpreted as those of a "real" particle with energy $-E$ in the inverted potential $-V(q)$ (see the right panel of Fig. 2.2). From this point of view, the infinite number of saddle points trajectories found above correspond to this particle bouncing n times in the inverted potential before reaching the endpoint at q'. It should be clear, however, that the time parameter in Eq. (2.100) does not have any meaning as a physical time.

The previous discussion carries over to field theory. By the same arguments used above, the path integrals in (2.87) and (2.89) are dominated in the limit $\hbar \to 0$ by those field configurations making the action stationary, that is, satisfying the Euler–Lagrange equations. The semiclassical approximation is obtained by expanding around these classical field solutions to second order in the perturbations and carrying out the resulting Gaussian integral.

Field theories can have many vacua separated by energy barriers. For example, a scalar field theory

$$S[\phi] = \int d^4x \left[\frac{1}{2}\eta^{\mu\nu}\partial_\mu\phi\partial_\nu\phi - V(\phi)\right]. \tag{2.101}$$

has as many vacua as local minima of the potential $V(\phi)$, that we assume to be bounded from below. The spectrum of excitations around each vacuum can be computed using perturbation theory in powers of the corresponding coupling constant. The perturbative analysis, however, is blind to transitions between different vacua due to quantum tunneling.

The lesson we have learned in quantum mechanics can now be used to study the semiclassical tunneling between different vacua in a field theory by means of an analytic continuation to imaginary times. Letting $t \to -it$, the Minkowski space-time transforms in Euclidean space

$$ds^2 = \eta_{\mu\nu} dx^\mu dx^\nu \quad \Longrightarrow \quad ds^2 = -\delta_{\mu\nu} dx^\mu dx^\nu. \tag{2.102}$$

In the example of the scalar field theory discussed above, this analytical continuation to imaginary time leads to the Euclidean action

$$S_E[\phi] = \int d^4x \left[\frac{1}{2} \delta^{\mu\nu} \partial_\mu \phi \partial_\nu \phi + V(\phi) \right]. \tag{2.103}$$

Instantons, i.e. the solution to the field equations derived from this action, are interpreted in quantum field theory as representing semiclassical tunneling between the different vacua of the theory. In Chap. 4 we will study field theory instantons in some more detail in the case of nonabelian gauge theories, where their existence has important physical consequences.

References

1. Casimir, H.B.G.: On the attraction between two perfectly conducting plates. Proc. Kon. Ned. Akad. Wet. **60**, 793 (1948)
2. Plunien, G., Müller, B., Greiner, W.: The Casimir effect. Phys. Rept. **134**, 87 (1986)
3. Milton, K.A.: The Casimir effect: recent controversies and progress. J. Phys. A **37**, R209 (2004)
4. Lamoreaux, S.K.: The Casimir force: background, experiments, and applications. Rep. Prog. Phys. **68**, 201 (2005)
5. Abramowitz, M., Stegun, I.A.: Handbook of Mathematical Functions. Dover, New York (1972)
6. Sparnaay, M.J.: Measurement of attractive forces between flat plates. Physica **24**, 751 (1958)
7. Feynman, R.P.: The principle of least action in quantum mechanics. PhD Thesis (1942)
8. Feynman, R.P.: Space-time approach to non-relativistic quantum mechanics. Rev. Mod. Phys. **20**, 367 (1948)
9. Feynman, R.P., Hibbs, A.R.: Quantum Mechanics and Path Integrals. (edition emended by Steyer, D.F.) Dover, New York (2010)
10. Schulman, L.S.: Techniques and Applications of Path Integration. Dover, New York (2005)
11. Kleinert, H.: Path Integrals in Quantum Mechanics, Statistical & Polymer Physics & Financial Markets. World Scientific, Singapore (2004)
12. Zinn-Justin, J.: Path Integrals in Quantum Mechanics. Oxford University Press, Oxford (2009)
13. Dirac, P.A.M.: The Lagrangian in Quantum Mechanics. Phys. Z. Sowjetunion **3**, 64 (1933)

Chapter 3
Theories and Lagrangians I: Matter Fields

Up to this point we have used a scalar field to illustrate our discussion of the quantization procedure. However, Nature is richer than that and it is necessary to consider other fields with more complicated behavior under Lorentz transformations. Before considering these other fields we pause and study the properties of the Lorentz group.

3.1 Representations of the Lorentz Group

The Lorentz group is the group of linear coordinate transformations that leave invariant the Minkowskian line element. It has a very rich mathematical structure that we review in Appendix B. Here our interest is focused on its representations.

In four dimensions the Lorentz group has six generators. Three of them are the generators J_i of the group of rotations in three dimensions SO(3). A finite rotation of angle φ with respect to the axis determined by a unitary vector $\mathbf{e}$ can be written as

$$R(\mathbf{e}, \varphi) = e^{-i\varphi\,\mathbf{e}\cdot\mathbf{J}}, \quad \mathbf{J} = \begin{pmatrix} J_1 \\ J_2 \\ J_3 \end{pmatrix}. \tag{3.1}$$

The other three generators of the Lorentz group are associated with boosts M_i along the three spatial directions. A boost with rapidity λ along a direction $\mathbf{u}$ is given by

$$B(\mathbf{u}, \lambda) = e^{-i\lambda\,\mathbf{u}\cdot\mathbf{M}}, \quad \mathbf{M} = \begin{pmatrix} M_1 \\ M_2 \\ M_3 \end{pmatrix}. \tag{3.2}$$

The six generators J_i, M_i satisfy the algebra

$$\begin{aligned} [J_i, J_j] &= i\varepsilon_{ijk} J_k, \\ [J_i, M_k] &= i\varepsilon_{ijk} M_k, \\ [M_i, M_j] &= -i\varepsilon_{ijk} J_k, \end{aligned} \tag{3.3}$$

L. Álvarez-Gaumé and M. Á. Vázquez-Mozo, *An Invitation to Quantum Field Theory*,
Lecture Notes in Physics 839, DOI: 10.1007/978-3-642-23728-7_3,

The first line are the commutation relations of SO(3), while the second one implies that the generators of the boosts transform like a vector under rotations. The six generators of the Lorentz group can be collected into the six independent components of an antisymmetric rank-two tensor $\mathscr{J}_{\mu\nu}$ according to

$$\mathscr{J}_{0i} = M_i, \quad \mathscr{J}_{ij} = \varepsilon_{ijk} J_k. \tag{3.4}$$

They satisfy

$$\left[\mathscr{J}_{\mu\nu}, \mathscr{J}_{\sigma\lambda}\right] = i\eta_{\mu\sigma}\mathscr{J}_{\nu\lambda} - i\eta_{\mu\lambda}\mathscr{J}_{\nu\sigma} + i\eta_{\nu\lambda}\mathscr{J}_{\mu\sigma} - i\eta_{\nu\sigma}\mathscr{J}_{\mu\lambda}. \tag{3.5}$$

The Lorentz algebra in terms of $\mathscr{J}_{\mu\nu}$ has the same form in any space-time dimension.

The task of finding representations of the algebra (3.3) [or (3.5)] might seem difficult at first sight. In four dimensions the problem is greatly simplified by combining the generators in the following way

$$J_k^{\pm} = \frac{1}{2}(J_k \pm i M_k). \tag{3.6}$$

Using (3.3), the new generators $J_k^{\pm}$ are found to satisfy

$$\begin{aligned} \left[J_i^{\pm}, J_j^{\pm}\right] &= i\varepsilon_{ijk} J_k^{\pm}, \\ \left[J_i^{+}, J_j^{-}\right] &= 0. \end{aligned} \tag{3.7}$$

Thus, the four-dimensional Lorentz algebra is equivalent to two copies of the algebra of $SU(2) \approx SO(3)$. Their irreducible representations are identified by $(\mathbf{s}_+, \mathbf{s}_-)$, where $\mathbf{s}_\pm = k_\pm$ or $k_\pm + \frac{1}{2}$ (with $k_\pm \in \mathbb{N}$) are the spins of the representations of the two copies of SU(2).

To get familiar with this way of labeling the representations of the Lorentz group we study some particular examples. Let us start with the simplest one $(\mathbf{s}_+, \mathbf{s}_-) = (\mathbf{0}, \mathbf{0})$. This state is a singlet under $J_i^{\pm}$ and therefore also under rotations and boosts. Therefore we have a scalar.

The next interesting cases are $(\mathbf{\frac{1}{2}}, \mathbf{0})$ and $(\mathbf{0}, \mathbf{\frac{1}{2}})$. States transforming in these representations are respectively right and left-handed Weyl spinors. Their properties will be studied in more detail below. Next we deal with $(\mathbf{\frac{1}{2}}, \mathbf{\frac{1}{2}})$. Equation (3.6) shows that $J_i = J_i^+ + J_i^-$. Applying the rules of addition of angular momenta we find that the states transforming in this representations decompose into a vector and a scalar under three-dimensional rotations. A more detailed analysis shows that the singlet state is identified with the time component of a four-vector, combining with the triplet to form a vector under the Lorentz group.

We can consider more "exotic" representations. For example the $(\mathbf{1},\mathbf{0})$ and $(\mathbf{0},\mathbf{1})$ representations correspond respectively to selfdual and anti-selfdual rank-two antisymmetric tensors $T^{\mu\nu} = -T^{\nu\mu}$,

$$T_{\mu\nu} = \pm\frac{1}{2}\varepsilon_{\mu\nu\sigma\lambda}T^{\sigma\lambda} \quad (+\,\text{selfdual}, -\,\text{anti-selfdual}), \tag{3.8}$$

Table 3.1 Representations of the Lorentz group in terms of the representations of SU(2)× SU(2)

Representation	Type of field
$(\mathbf{0},\mathbf{0})$	Scalar
$(\frac{1}{2},\mathbf{0})$	Right-handed spinor
$(\mathbf{0},\frac{1}{2})$	Left-handed spinor
$(\frac{1}{2},\frac{1}{2})$	Vector
$(\mathbf{1},\mathbf{0})$	Selfdual antisymmetric 2-tensor
$(\mathbf{0},\mathbf{1})$	Anti-selfdual antisymmetric 2-tensor

where $\varepsilon_{\mu\nu\sigma\lambda}$ is the Levi-Civita symbol with four indices. Table 3.1 summarizes the previous discussion.

To conclude our analysis of the representations of the Lorentz group we notice that under parity the generators of SO(1,3) transform as[1]

$$P\colon J_i \longrightarrow J_i, \quad P : M_i \longrightarrow -M_i. \tag{3.9}$$

This implies that $P\colon J_i^{\pm} \longrightarrow J_i^{\mp}$ and therefore a representation $(\mathbf{s}_1, \mathbf{s}_2)$ is transformed into $(\mathbf{s}_2, \mathbf{s}_1)$. As a consequence a vector $(\frac{1}{2}, \frac{1}{2})$ is invariant under parity, whereas a left-handed Weyl spinor $(\frac{1}{2}, \mathbf{0})$ transforms into a right-handed one $(\mathbf{0}, \frac{1}{2})$ and vice versa.

It is instructive to see how the representations of the Lorentz group differ from those of SO(4), the isometry group of four-dimensional Euclidean space. Like the Lorentz group, it is generated by a set of six generators $\mathscr{J}_{\mu\nu}$ whose algebra can be obtained from Eq. (3.5) by replacing $\eta_{\mu\nu} \to -\delta_{\mu\nu}$. The Lie algebra of SO(4) is isomorphic to that of SU(2)× SU(2). This can be seen by introducing the generators

$$N^a = \eta^a_{\mu\nu} J^{\mu\nu}, \quad \overline{N}^a = \overline{\eta}^a_{\mu\nu} J^{\mu\nu}. \tag{3.10}$$

The numerical coefficients $\eta^a_{\mu\nu}$ and $\overline{\eta}^a_{\mu\nu}$ (with $a = 1, 2, 3$ and $\mu, \nu = 0, \ldots, 3$) are called *'t Hooft symbols* and are given by

$$\begin{aligned} \eta^a_{\mu\nu} &= \varepsilon_{a\mu\nu} + \delta_{a\mu}\delta_{\nu 0} - \delta_{a\nu}\delta_{\mu 0}, \\ \overline{\eta}^a_{\mu\nu} &= \varepsilon_{a\mu\nu} - \delta_{a\mu}\delta_{\nu 0} + \delta_{a\nu}\delta_{\mu 0}. \end{aligned} \tag{3.11}$$

Here $\varepsilon_{a\mu\nu}$ represents the Levi-Civita antisymmetric symbol with three indices and it is taken to be zero whenever μ or ν are equal to zero. Now it is not difficult to check that the generators (3.10) satisfy the Lie algebra of SU(2)× SU(2)

$$\left[N^a, N^b\right] = i\varepsilon^{abc} N^c, \quad \left[\overline{N}^a, \overline{N}^b\right] = i\varepsilon^{abc}\overline{N}^c, \quad \left[N^a, \overline{N}^b\right] = 0. \tag{3.12}$$

This shows that the representations of SO(4) can also be labelled in terms of the irreducible representations of SU(2).

---

[1] Parity and other discrete symmetries are studied in detail in Chap. 11.

3.2 Weyl Spinors

A Weyl spinor $u_\pm$ is a complex two-component object that transforms in the representations $(\frac{1}{2}, \mathbf{0})$ and $(\mathbf{0}, \frac{1}{2})$ respectively. The generators $J_i^\pm$ can be explicitly constructed using the Pauli matrices as

$$\begin{aligned} J_i^+ &= \frac{1}{2}\sigma_i, \quad J_i^- = 0 \quad \text{for} \quad (\tfrac{1}{2}, \mathbf{0}), \\ J_i^+ &= 0, \qquad J_i^- = \frac{1}{2}\sigma_i \quad \text{for} \quad (\mathbf{0}, \tfrac{1}{2}). \end{aligned} \tag{3.13}$$

Going back to J^i and K^i, we find that under a rotation of angle θ and axis $\mathbf{n}$ and a boost of rapidity $\boldsymbol{\beta} = (\beta_1, \beta_2, \beta_3)$ the spinors $u_\pm$ transform as

$$u_\pm \longrightarrow e^{-\frac{i}{2}(\theta\mathbf{n} \mp i\boldsymbol{\beta})\cdot\boldsymbol{\sigma}} u_\pm. \tag{3.14}$$

To construct a free Lagrangian for the fields $u_\pm$ we have to look for quadratic combinations of the fields that are Lorentz scalars. Defining $\sigma_\pm^\mu = (\mathbf{1}, \pm\sigma_i)$, we can construct the following quantities

$$u_+^\dagger \sigma_+^\mu u_+, \qquad u_-^\dagger \sigma_-^\mu u_-. \tag{3.15}$$

The first thing to point out is that, since $(J_i^\pm)^\dagger = J_i^\mp$, the hermitian conjugate fields $u_\pm^\dagger$ are in the $(\mathbf{0}, \frac{1}{2})$ and $(\frac{1}{2}, \mathbf{0})$ representation respectively. The combinations (3.15) transform as a four-vector under (3.14), due to the property

$$e^{\frac{i}{2}(\theta\mathbf{n} \pm i\boldsymbol{\beta})\cdot\boldsymbol{\sigma}} \sigma_\pm^\mu e^{-\frac{i}{2}(\theta\mathbf{n} \mp i\boldsymbol{\beta})\cdot\boldsymbol{\sigma}} = \Lambda^\mu_\nu(\theta\mathbf{n}, \boldsymbol{\beta})\sigma_\pm^\nu, \tag{3.16}$$

where $\Lambda^\mu_\nu(\theta\mathbf{n}, \boldsymbol{\beta})$ gives the transformation of the coordinates x^μ.

Once the transformation properties of (3.15) are known we can start building invariants. If, in addition, we also demand that the Lagrangian be invariant under global phase rotations

$$u_\pm \longrightarrow e^{i\theta} u_\pm \tag{3.17}$$

we are left with just one possibility up to a sign, namely

$$\mathscr{L}^\pm_{\text{Weyl}} = i u_\pm^\dagger \left(\partial_t \pm \boldsymbol{\sigma}\cdot\boldsymbol{\nabla}\right) u_\pm = i u_\pm^\dagger \sigma_\pm^\mu \partial_\mu u_\pm. \tag{3.18}$$

This is the Weyl Lagrangian. In order to get a more clear idea of the physical meaning of the spinors $u_\pm$ we write the equations of motion

$$\left(\partial_0 \pm \boldsymbol{\sigma}\cdot\boldsymbol{\nabla}\right) u_\pm = 0. \tag{3.19}$$

Multiplying this equation on the left by $(\partial_0 \mp \boldsymbol{\sigma}\cdot\boldsymbol{\nabla})$ and applying the algebraic properties of the Pauli matrices, we conclude that $u_\pm$ satisfy the massless Klein-Gordon equation

$$\partial_\mu \partial^\mu u_\pm = 0, \tag{3.20}$$

whose solutions are

$$u_\pm(x) = u_\pm(k)e^{-ik\cdot x}, \quad \text{with} \quad k^0 = |\mathbf{k}|. \tag{3.21}$$

Plugging them back into the equations of motion (3.19) we find

$$(|\mathbf{k}| \mp \mathbf{k}\cdot\boldsymbol{\sigma})\, u_\pm = 0, \tag{3.22}$$

implying the following conditions

$$\begin{aligned} u_+ &: \quad \frac{\boldsymbol{\sigma}\cdot\mathbf{k}}{|\mathbf{k}|} = 1, \\ u_- &: \quad \frac{\boldsymbol{\sigma}\cdot\mathbf{k}}{|\mathbf{k}|} = -1. \end{aligned} \tag{3.23}$$

Since the spin operator is $\mathbf{s} = \frac{1}{2}\boldsymbol{\sigma}$, the previous expressions give the helicity of the states with wave function $u_\pm$, i.e. the projection of the spin along the momentum of the particle

$$\lambda = \mathbf{s}\cdot\frac{\mathbf{k}}{|\mathbf{k}|}. \tag{3.24}$$

We conclude that u_+ is a Weyl spinor of positive helicity $\lambda = \frac{1}{2}$, while u_- has negative helicity $\lambda = -\frac{1}{2}$. This agrees with our assertion in the previous section that the representation $(\frac{1}{2}, \mathbf{0})$ corresponds to a right-handed Weyl fermion (positive helicity) whereas $(\mathbf{0}, \frac{1}{2})$ is a left-handed Weyl fermion (negative helicity). For example, the standard model neutrinos are left-handed Weyl spinors and therefore transform in the representation $(\mathbf{0}, \frac{1}{2})$ of the Lorentz group.

Nevertheless, it is possible that we were too restrictive in constructing the Weyl Lagrangian (3.18). There we constructed the invariants from the vector bilinears (3.15) corresponding to the product representations

$$\left(\tfrac{1}{2}, \tfrac{1}{2}\right) = \left(\tfrac{1}{2}, \mathbf{0}\right) \otimes \left(\mathbf{0}, \tfrac{1}{2}\right) \quad \text{and} \quad \left(\tfrac{1}{2}, \tfrac{1}{2}\right) = \left(\mathbf{0}, \tfrac{1}{2}\right) \otimes \left(\tfrac{1}{2}, \mathbf{0}\right). \tag{3.25}$$

In particular our insistence in demanding the Lagrangian to be invariant under the global symmetry $u_\pm \to e^{i\theta}u_\pm$ rules out the scalar term that appears in the product representations

$$\begin{aligned} \left(\tfrac{1}{2}, \mathbf{0}\right) \otimes \left(\tfrac{1}{2}, \mathbf{0}\right) &= (\mathbf{1}, \mathbf{0}) \oplus (\mathbf{0}, \mathbf{0}), \\ \left(\mathbf{0}, \tfrac{1}{2}\right) \otimes \left(\mathbf{0}, \tfrac{1}{2}\right) &= (\mathbf{0}, \mathbf{1}) \oplus (\mathbf{0}, \mathbf{0}). \end{aligned} \tag{3.26}$$

The singlet representations corresponds to the antisymmetric combinations

$$\varepsilon_{ab} u_\pm^a u_\pm^b, \tag{3.27}$$

where ε_{ab} is the antisymmetric symbol $\varepsilon_{12} = -\varepsilon_{21} = 1$.

At first sight it might seem that the term (3.27) vanishes identically due to the antisymmetry of the ε-symbol. However we should keep in mind that the spin-statistics theorem (more on this later) demands that fields with half-integer spin have to satisfy the Fermi-Dirac statistics and therefore satisfy anticommutation relations, whereas fields of integer spin follow the statistic of Bose-Einstein and, as a consequence, quantization replaces Poisson brackets by commutators. This implies that the components of the Weyl fermions $u_\pm$ are anticommuting Grassmann fields

$$u_\pm^a u_\pm^b + u_\pm^b u_\pm^a = 0. \tag{3.28}$$

It is important to realize that, strictly speaking, fermions (i.e., objects that satisfy the Fermi-Dirac statistics) do not exist classically. The reason is that they satisfy the Pauli exclusion principle and therefore each quantum state can be occupied, at most, by one fermion. Therefore the naive definition of the classical limit as a limit of large occupation numbers cannot be applied. Fermion fields do not really make sense classically.

Since the combination (3.27) does not vanish, we can construct a new Lagrangian

$$\mathscr{L}_{\text{Weyl}}^{\pm} = i u_\pm^\dagger \sigma_\pm^\mu \partial_\mu u_\pm - \frac{m}{2}\left(\varepsilon_{ab} u_\pm^a u_\pm^b + \text{h.c.}\right) \tag{3.29}$$

This mass term, called of Majorana type, is allowed if we do not worry about breaking the global U(1) symmetry $u_\pm \to e^{i\theta} u_\pm$. This is not the case, for example, of charged chiral fermions, since the Majorana mass violates the conservation of electric charge or any other gauge U(1) charge. In the standard model, however, there is no such a problem if we introduce Majorana masses for right-handed neutrinos, since they are singlets under all standard model gauge groups. Such a term will break, however, the global U(1) lepton number charge, the operator $\varepsilon_{ab} \nu_R^a \nu_R^b$ changes the lepton number by two units. We will have more to say about this later.

3.3 Dirac Spinors

We have seen that parity interchanges the representations $(\mathbf{\frac{1}{2}}, \mathbf{0})$ and $(\mathbf{0}, \mathbf{\frac{1}{2}})$, i.e. it changes right-handed with left-handed fermions

$$P : u_\pm \longrightarrow u_\mp. \tag{3.30}$$

An obvious way to build a parity invariant theory is to combine a pair or Weyl fermions u_+ and u_- of opposite helicity in a single four-component spinor

$$\psi = \begin{pmatrix} u_+ \\ u_- \end{pmatrix} \tag{3.31}$$

transforming in the reducible representation $(\frac{1}{2}, \mathbf{0}) \oplus (\mathbf{0}, \frac{1}{2})$.

Since now we have both u_+ and u_- simultaneously at our disposal, the equations of motion for $u_\pm$, $i\sigma_\pm^\mu \partial_\mu u_\pm = 0$ can be modified, while keeping them linear, to introduce a mass term

$$\left.\begin{aligned} i\sigma_+^\mu \partial_\mu u_+ &= m u_- \\ i\sigma_-^\mu \partial_\mu u_- &= m u_+ \end{aligned}\right\} \quad \Longrightarrow \quad i\begin{pmatrix} \sigma_+^\mu & 0 \\ 0 & \sigma_-^\mu \end{pmatrix} \partial_\mu \psi = m \begin{pmatrix} 0 & 1 \\ 1 & 0 \end{pmatrix} \psi. \tag{3.32}$$

These equations of motion can be derived from the Lagrangian density

$$\mathscr{L}_{\text{Dirac}} = i\psi^\dagger \begin{pmatrix} \sigma_+^\mu & 0 \\ 0 & \sigma_-^\mu \end{pmatrix} \partial_\mu \psi - m\psi^\dagger \begin{pmatrix} 0 & 1 \\ 1 & 0 \end{pmatrix} \psi. \tag{3.33}$$

To simplify the notation it is useful to define the Dirac γ-matrices as

$$\gamma^\mu = \begin{pmatrix} 0 & \sigma_-^\mu \\ \sigma_+^\mu & 0 \end{pmatrix}. \tag{3.34}$$

and the Dirac conjugate spinor $\overline{\psi}$

$$\overline{\psi} \equiv \psi^\dagger \gamma^0 = \psi^\dagger \begin{pmatrix} 0 & 1 \\ 1 & 0 \end{pmatrix}. \tag{3.35}$$

The Lagrangian (3.33) can be written in the more compact form

$$\mathscr{L}_{\text{Dirac}} = \overline{\psi} \left(i\gamma^\mu \partial_\mu - m \right) \psi, \tag{3.36}$$

whose equations of motion give the Dirac equation (1.10) with the identifications

$$\gamma^0 = \beta, \qquad \gamma^i = i\alpha^i. \tag{3.37}$$

The γ-matrices defined in (3.34) satisfy the Dirac algebra

$$\{\gamma^\mu, \gamma^\nu\} = 2\eta^{\mu\nu}. \tag{3.38}$$

In d dimensions this algebra admits representations of dimension $2^{\left[\frac{d}{2}\right]}$. Equation (3.34) gives the chiral representation of the algebra (3.38). Other equivalent representations can be constructed exploiting the invariance of (3.38) under unitary transformations $\gamma^\mu \to U\gamma^\mu U^\dagger$.

A representation of the Lorentz algebra SO$(1, d-1)$ can be constructed using the γ-matrices as

$$\mathscr{J}^{\mu\nu} = -\frac{i}{4}\left[\gamma^\mu, \gamma^\nu\right] \equiv \sigma^{\mu\nu}. \tag{3.39}$$

By definition, Dirac fermions ψ in d dimensions transform under the Lorentz group in this representation.

When d is even the representation (3.39) is reducible. In the case of interest $d = 4$ this result is easy to prove by defining the chirality matrix

$$\gamma_5 = -i\gamma^0\gamma^1\gamma^2\gamma^3 = \begin{pmatrix} \mathbf{1} & 0 \\ 0 & -\mathbf{1} \end{pmatrix}. \tag{3.40}$$

The matrix γ_5 anticommutes with all other γ-matrices and as a consequence

$$[\gamma_5, \sigma^{\mu\nu}] = 0. \tag{3.41}$$

Using Schur's lemma (see Appendix B) this implies that the representation of the Lorentz group provided by $\sigma^{\mu\nu}$ is reducible into subspaces spanned by the eigenvectors of γ_5 with the same eigenvalue. Introducing the projectors $P_{\pm} = \frac{1}{2}(1 \pm \gamma_5)$ these subspaces correspond to

$$P_+\psi = \begin{pmatrix} u_+ \\ 0 \end{pmatrix}, \quad P_-\psi = \begin{pmatrix} 0 \\ u_- \end{pmatrix}, \tag{3.42}$$

which are precisely the Weyl spinors introduced above.

Our next task is to quantize the Dirac Lagrangian. This will be done along the lines followed for the free real scalar field, starting with a general solution to the Dirac equation and introducing the corresponding set of creation–annihilation operators. Therefore we start by looking for a complete basis of solutions to the Dirac equation. In the case of the scalar field the elements of the basis were labelled by their four-momentum k^μ. Now, however, the field has several components so we have to add an extra label. Equation (3.23) suggest the following definition of the helicity operator of a Dirac spinor

$$\lambda = \begin{pmatrix} \frac{1}{2}\boldsymbol{\sigma} \cdot \frac{\mathbf{k}}{|\mathbf{k}|} & 0 \\ 0 & \frac{1}{2}\boldsymbol{\sigma} \cdot \frac{\mathbf{k}}{|\mathbf{k}|} \end{pmatrix}. \tag{3.43}$$

Each element of the basis of functions is labelled by its four-momentum k^μ and the corresponding eigenvalue s of the helicity operator.

For positive energy solutions of the Dirac equation we take

$$u(k,s)e^{-ik\cdot x}, \qquad s = \pm\frac{1}{2}, \tag{3.44}$$

where $u_\alpha(k,s)$ ($\alpha = 1, \ldots, 4$) is a four-component spinor. Substituting in the Dirac equation we obtain[2]

$$(\not{k} - m)u(k,s) = 0. \tag{3.45}$$

In the same way, for negative energy solutions we have

---

[2] From now on we will frequently use the Feynman slash notation, $\not{a} \equiv \gamma^\mu a_\mu$.

$$v(k,s)e^{ik\cdot x}, \qquad s = \pm\frac{1}{2}, \tag{3.46}$$

where $v_\alpha(k,s)$ has to satisfy

$$(\not{k} + m)v(k,s) = 0. \tag{3.47}$$

Multiplying Eqs. (3.45) and (3.47) on the left respectively by $(\not{k} \mp m)$ we find that the momentum is on the mass shell, $k^2 = m^2$. Hence, the wave function for both positive- and negative-energy solutions is labelled by the three-momentum $\mathbf{k}$ of the particle, $u(\mathbf{k},s)$, $v(\mathbf{k},s)$.

Before proceeding any further we consider the case of a massless Dirac fermion. Using the equation $\not{k}u(\mathbf{k},s) = 0$ it is not difficult to show that the helicity operator (3.43) satisfies

$$\lambda u(\mathbf{k},s) = \frac{1}{2}\gamma_5 u(\mathbf{k},s), \tag{3.48}$$

and similarly for $v(\mathbf{k},s)$. This means that when $m = 0$ helicity (i.e., the projection of the spin along the direction of motion) and chirality (the eigenvalue of the γ_5 matrix) are equivalent concepts. In this case the helicity of the spinor is a relativistic invariant. This is no longer true when $m \neq 0$ because when the particle moves with a speed smaller than the speed of light the sign of λ can be changed by a boost reversing the direction of $\mathbf{k}$. Hence, the helicity of a massive Dirac spinor has no invariant meaning and moreover it is not equivalent to its chirality.

The spinors $u(\mathbf{k},s)$, $v(\mathbf{k},s)$ can be normalized according to

$$\begin{aligned} \overline{u}(\mathbf{k},s)u(\mathbf{k},s) &= 2m, \\ \overline{v}(\mathbf{k},s)v(\mathbf{k},s) &= -2m. \end{aligned} \tag{3.49}$$

Given this normalization, the following identities can be obtained

$$\begin{aligned} \overline{u}(\mathbf{k},s)\gamma^\mu u(\mathbf{k},s) &= 2k^\mu, \\ \overline{v}(\mathbf{k},s)\gamma^\mu v(\mathbf{k},s) &= 2k^\mu, \end{aligned} \tag{3.50}$$

as well as the completeness relations

$$\begin{aligned} \sum_{s=\pm\frac{1}{2}} u_\alpha(\mathbf{k},s)\overline{u}_\beta(\mathbf{k},s) &= (\not{k} + m)_{\alpha\beta}, \\ \sum_{s=\pm\frac{1}{2}} v_\alpha(\mathbf{k},s)\overline{v}_\beta(\mathbf{k},s) &= (\not{k} - m)_{\alpha\beta}, \end{aligned} \tag{3.51}$$

with $k^0 = E_{\mathbf{k}} = \sqrt{\mathbf{k}^2 + m^2}$. A general solution to the Dirac equation including creation and annihilation operators can be written as

$$\hat{\psi}_\alpha(t, \mathbf{x}) = \sum_{s=\pm\frac{1}{2}} \int \frac{d^3k}{(2\pi)^3} \frac{1}{2\omega_\mathbf{k}} \Big[u_\alpha(\mathbf{k}, s)\hat{b}(\mathbf{k}, s) e^{-iE_\mathbf{k}t + i\mathbf{k}\cdot\mathbf{x}}$$
$$+ v_\alpha(\mathbf{k}, s)\hat{d}^\dagger(\mathbf{k}, s) e^{iE_\mathbf{k}t - i\mathbf{k}\cdot\mathbf{x}} \Big]. \tag{3.52}$$

Unlike the real scalar field studied in the previous chapter, the Dirac field is not hermitian. As a consequence, the operators $\hat{b}(\mathbf{k}, s)$ and $\hat{d}(\mathbf{k}, s)$ are independent and not related by Hermitian conjugation.

Since we are dealing with half-integer spin fields, the spin-statistics theorem forces a modification of the canonical quantization prescription (2.57). In the case of the Dirac field the canonical Poisson brackets are replaced by *anticommutators*

$$i\{\cdot, \cdot\}_{\text{PB}} \longrightarrow \{\cdot, \cdot\}. \tag{3.53}$$

Thus we arrive to the following canonical anticommutation relations for $\hat{\psi}(t, \mathbf{x})$

$$\{\hat{\psi}_\alpha(t, \mathbf{x}), \hat{\psi}_\beta^\dagger(t, \mathbf{y})\} = \delta(\mathbf{x} - \mathbf{y})\delta_{\alpha\beta}, \tag{3.54}$$

with the other anticommutators vanishing. From Eq. (3.52) we find that the operators $\hat{b}^\dagger(\mathbf{k}, s)$, $\hat{b}(\mathbf{k}, s)$ satisfy the algebra[3]

$$\begin{aligned} \{b(\mathbf{k}, s), b^\dagger(\mathbf{k}', s')\} &= (2\pi)^3(2E_\mathbf{k})\delta(\mathbf{k} - \mathbf{k}')\delta_{ss'}, \\ \{b(\mathbf{k}, s), b(\mathbf{k}', s')\} &= \{b^\dagger(\mathbf{k}, s), b^\dagger(\mathbf{k}', s')\} = 0. \end{aligned} \tag{3.55}$$

They respectively create and annihilate a spin-$\frac{1}{2}$ particle (for example, an electron) out of the vacuum with momentum $\mathbf{k}$ and helicity s.

In the case of $d(\mathbf{k}, s)$, $d^\dagger(\mathbf{k}, s)$, they satisfy the fermionic algebra

$$\begin{aligned} \{d(\mathbf{k}, s), d^\dagger(\mathbf{k}', s')\} &= (2\pi)^3(2E_\mathbf{k})\delta(\mathbf{k} - \mathbf{k}')\delta_{ss'}, \\ \{d(\mathbf{k}, s), d(\mathbf{k}', s')\} &= \{d^\dagger(\mathbf{k}, s), d^\dagger(\mathbf{k}', s')\} = 0. \end{aligned} \tag{3.56}$$

Hence we have a set of creation–annihilation operators for the corresponding antiparticles (for example positrons). This is clear if we notice that $d^\dagger(\mathbf{k}, s)$ can be seen as the annihilation operator of a negative energy state of the Dirac equation with wave function $v_a(\mathbf{k}, s)$. In the Dirac picture this corresponds to the creation of an antiparticle out of the vacuum (see Fig. 1.2). Finally, all other anticommutators between $b(\mathbf{k}, s)$, $b^\dagger(\mathbf{k}, s)$ and $d(\mathbf{k}, s)$, $d^\dagger(\mathbf{k}, s)$ vanish.

The Hamiltonian operator for the Dirac field is

$$\hat{H} = \frac{1}{2} \sum_{s=\pm\frac{1}{2}} \int \frac{d^3k}{(2\pi)^3} \Big[b^\dagger(\mathbf{k}, s)b(\mathbf{k}, s) - d(\mathbf{k}, s)d^\dagger(\mathbf{k}, s) \Big]. \tag{3.57}$$

---

[3] To simplify notation, and since there is no risk of confusion, we drop from now on the hats to indicate operators.

At this point we realize again the necessity of quantizing the theory using anti-commutators instead of commutators. Had we used canonical *commutation* relations, the second term inside the integral in (3.57) would give the number operator $d^\dagger(\mathbf{k},s)d(\mathbf{k},s)$ with a minus sign in front. As a consequence, the Hamiltonian would be unbounded from below and we would be facing again the instability of the theory already noticed in the context of relativistic quantum mechanics. However, using the *anticommutation* relations (3.56), the Hamiltonian (3.57) takes the form

$$\hat{H} = \sum_{s=\pm\frac{1}{2}} \int \frac{d^3k}{(2\pi)^3} \frac{1}{2E_\mathbf{k}} \Big[E_\mathbf{k} b^\dagger(\mathbf{k},s) b(\mathbf{k},s) + E_\mathbf{k} d^\dagger(\mathbf{k},s) d(\mathbf{k},s) \Big] \\ - 2 \int d^3k E_\mathbf{k} \delta(\mathbf{0}). \tag{3.58}$$

As with the scalar field, we find a divergent vacuum energy contribution due to the zero-point energy of an infinite number of harmonic oscillators. Unlike the case of the scalar field, the vacuum energy here is negative. This is interesting because, as it will be explaned in Chap. 13, there is a certain type of theories called supersymmetric where the number of bosonic and fermionic degrees of freedom is the same. For this kind of theories the contribution of the vacuum energy of the bosonic field exactly cancels that of the fermions. The divergent contribution in the Hamiltonian (3.58) can be removed by the normal order prescription

$$:\hat{H}:= \sum_{s=\pm\frac{1}{2}} \int \frac{d^3k}{(2\pi)^3} \frac{1}{2E_\mathbf{k}} \Big[E_\mathbf{k} b^\dagger(\mathbf{k},s) b(\mathbf{k},s) + E_\mathbf{k} d^\dagger(\mathbf{k},s) d(\mathbf{k},s) \Big]. \tag{3.59}$$

Finally, let us mention that using the Dirac equation it is easy to prove the conservation of the four-current

$$j^\mu = \overline{\psi}\gamma^\mu\psi, \quad \partial_\mu j^\mu = 0. \tag{3.60}$$

As we will explain further in Chap. 7, this current is associated to the invariance of the Dirac Lagrangian under the global phase shift $\psi \to e^{i\theta}\psi$. In electrodynamics the associated conserved charge

$$Q = q \int d^3x j^0 \tag{3.61}$$

is identified with the electric charge, with q the charge of the particle created by $b^\dagger(\mathbf{k},s)$ acting on the vacuum.

Since we are dealing with a free theory, all correlation functions can be written in terms of those with two fields. The Feynman propagator is given by

$$S_{\alpha\beta}(x_1,x_2) = \langle 0|T\Big[\psi_\alpha(x_1)\overline{\psi}_\beta(x_2)\Big]|0\rangle \\ = \int \frac{d^4p}{(2\pi)^4} \left(\frac{i}{\not{p} - m + i\varepsilon}\right)_{\alpha\beta} e^{-ip\cdot(x_1-x_2)}, \tag{3.62}$$

while the other two-point correlation functions are zero

$$\langle 0|T\left[\psi_\alpha(x_1)\psi_\beta(x_2)\right]|0\rangle = \langle 0|T\left[\overline{\psi}_\alpha(x_1)\overline{\psi}_\beta(x_2)\right]|0\rangle = 0, \tag{3.63}$$

as can be seen by direct computation using the field expansion in terms of creation-annihilation operators. Due to the fermionic character of the Dirac field, the definition of the time-ordered product includes a number of minus signs associated with the permutation of the two fields. For the particular case of a Dirac spinor and its conjugate we have

$$T\Big[\psi_\alpha(x)\overline{\psi}_\beta(y)\Big] = \theta(x^0-y^0)\psi_\alpha(x)\overline{\psi}_\beta(y) - \theta(y^0-x^0)\overline{\psi}_\beta(y)\psi_\alpha(x). \tag{3.64}$$

The rule for higher order point functions is the same as in the bosonic case ("earlier" fields always to the right) apart from the fact that each term comes now multiplied by the sign needed to bring the original expression into the time order.

The computation of the vacuum expectation value of the time-ordered product of a number of ψ and $\overline{\psi}$ fields can be done using an extension of Wick's theorem introduced in Sect. 2.2 for a real scalar field. The main difference is that now the Wick contractions only occur between a Dirac field $\psi(x)$ and its conjugate $\overline{\psi}(x)$

$$\overbrace{\psi_\alpha(x_1)\overline{\psi}_\beta}(x_2) \longrightarrow S_{\alpha\beta}(x_1,x_2). \tag{3.65}$$

In addition, since the fields anticommute, there are extra signs associated with the permutations required to bring together in the correct order the fields that are Wick-contracted. The details can be found in the standard texts (see for example Ref. [1–15] in Chap. 1).

The Dirac field can also be quantized using the path integral formalism introduced in Chap. 2. The propagator (3.62) can be written as

$$S_{\alpha\beta}(x_1,x_2) = \frac{\displaystyle\int \mathscr{D}\overline{\psi}\mathscr{D}\psi\,\psi_\alpha(x_1)\overline{\psi}_\beta(x_2)e^{iS[\psi,\overline{\psi}]}}{\displaystyle\int \mathscr{D}\overline{\psi}\mathscr{D}\psi\, e^{iS[\psi,\overline{\psi}]}}. \tag{3.66}$$

This expression has, however, a very important difference with its bosonic counterpart shown in Eq. (2.89). Whereas in both cases all fields inside the path integral are functions and not operators, here ψ and $\overline{\psi}$ are *anticommuting* functions. This fact is crucial in performing the functional integration. Anticommuting objects have to be integrated using the so-called Berezin rules

$$\int d\theta_i 1 = 0, \quad \int d\theta_i \theta_j = \delta_{ij}, \tag{3.67}$$

where θ_i are anticommuting variables satisfying $\theta_i\theta_j = -\theta_j\theta_i$. The details of the computation of path integrals with fermionic fields can be found in Ref. [9–12] of Chap. 2.

Chapter 4
Theories and Lagrangians II: Introducing Gauge Fields

Gauge theories play a central role in our current understanding of the fundamental interactions. The weak, electromagnetic and strong interactions are well described by gauge theories. We introduce them in this chapter for the first time. Although we often talk about gauge invariance, or gauge symmetry, these terms are a bit misleading. The gauge symmetry is more a redundancy in the description of the physical degrees of freedom than a symmetry, as will be shown later on. The redundancy is of course very useful because it makes Lorentz invariance and locality explicit, but it is not a symmetry in the same sense as rotations or translations. Gauge theories have incredible richness and complexity. Many aspects of their dynamics are still poorly understood. In our presentation we just scratch the surface of a deep subject.

4.1 Classical Gauge Fields

In classical electrodynamics the basic physical quantities are the electric and magnetic fields **E** and **B**. They can be expressed in terms of the scalar and vector potentials φ and **A** as

$$\begin{aligned}\mathbf{E} &= -\nabla\varphi - \frac{\partial \mathbf{A}}{\partial t},\\ \mathbf{B} &= \nabla \times \mathbf{A}.\end{aligned} \tag{4.1}$$

From these equations we see that specifying **E** and **B** does not uniquely determine the potentials, since the former do not change under the gauge transformations

$$\varphi(t,\mathbf{x}) \to \varphi(t,\mathbf{x}) + \frac{\partial}{\partial t}\varepsilon(t,\mathbf{x}), \quad \mathbf{A}(t,\mathbf{x}) \to \mathbf{A}(t,\mathbf{x}) - \nabla\varepsilon(t,\mathbf{x}). \tag{4.2}$$

From a classical point of view the introduction of φ and **A** is seen as a technicality that helps solving the Maxwell equations, but without physical relevance.

L. Álvarez-Gaumé and M. Á. Vázquez-Mozo, *An Invitation to Quantum Field Theory*,
Lecture Notes in Physics 839, DOI: 10.1007/978-3-642-23728-7_4,

The equations of electrodynamics can be recast in a manifestly Lorentz invariant form using the four-vector gauge potential $A^{\mu} = (\varphi, \mathbf{A})$ and the antisymmetric field strength tensor defined by

$$F_{\mu\nu} = \partial_{\mu} A_{\nu} - \partial_{\nu} A_{\mu}. \tag{4.3}$$

The four Maxwell equations

$$\begin{aligned} \nabla \cdot \mathbf{E} &= \rho, \\ \nabla \cdot \mathbf{B} &= 0, \\ \nabla \times \mathbf{E} &= -\frac{\partial}{\partial t}\mathbf{B}, \\ \nabla \times \mathbf{B} &= \mathbf{j} + \frac{\partial}{\partial t}\mathbf{E}, \end{aligned} \tag{4.4}$$

are written in the form

$$\begin{aligned} \partial_{\mu} F^{\mu\nu} &= j^{\mu}, \\ \varepsilon^{\mu\nu\sigma\eta} \partial_{\nu} F_{\sigma\eta} &= 0, \end{aligned} \tag{4.5}$$

where the four-current $j^{\mu} = (\rho, \mathbf{j})$ contains the charge density and the electric current. The second set of equations are called the Bianchi identities and are satisfied by any field strength (4.3). Notice that $F_{\mu\nu}$, and therefore the Maxwell equations, are invariant under the gauge transformations (4.2), which in covariant form read

$$A_{\mu} \longrightarrow A_{\mu} + \partial_{\mu}\varepsilon. \tag{4.6}$$

Finally, the equations of motion of a particle with mass m and charge q

$$m\ddot{\mathbf{x}} = q\left(\mathbf{E} + \dot{\mathbf{x}} \times \mathbf{B}\right) \tag{4.7}$$

take the form

$$m\frac{du^{\mu}}{d\tau} = qF^{\mu\nu}u_{\nu}, \tag{4.8}$$

where $u^{\mu}(\tau)$ is the particle four-velocity as a function of the proper time τ. These equations of motion, depending only on the field strength $F_{\mu\nu}$, are also gauge invariant.

The physical role of the vector potential becomes manifest only in quantum mechanics. Using the prescription of minimal substitution $\mathbf{p} \rightarrow \mathbf{p} - q\mathbf{A}$, the Schrödinger equation describing a particle with charge q moving in an electromagnetic field is

$$i\frac{\partial}{\partial t}\Psi = \left[-\frac{1}{2m}(\nabla - iq\mathbf{A})^2 + q\varphi\right]\Psi. \tag{4.9}$$

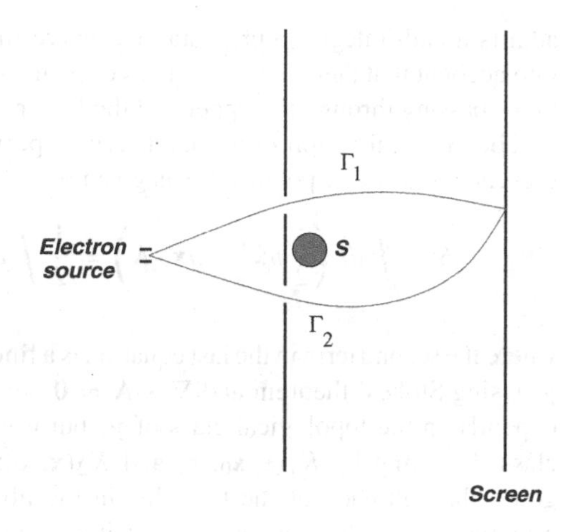

Fig. 4.1 Illustration of an interference experiment to show the Aharonov–Bohm effect. *S* represents the solenoid where the magnetic field is confined

Due to the explicit dependence on the electromagnetic potentials φ and $\mathbf{A}$, this equation seems to change under the gauge transformations (4.2). This is physically acceptable only if the ambiguity does not affect the probability density given by $|\Psi(t,\mathbf{x})|^2$. Therefore, a gauge transformation of the electromagnetic potential should amount to a change in the (unobservable) global phase of the wave function. This is indeed what happens: the Schrödinger equation (4.9) is invariant under the gauge transformations (4.2) provided the phase of the wave function is transformed at the same time according to

$$\Psi(t,\mathbf{x}) \longrightarrow e^{-iq\varepsilon(t,\mathbf{x})}\Psi(t,\mathbf{x}). \tag{4.10}$$

The Aharonov–Bohm Effect

This interplay between gauge transformations and the phase of the wave function gives rise to surprising phenomena. A first evidence of the role played by the electromagnetic potentials at the quantum level was pointed out by Yakir Aharonov and David Bohm [1]. Let us consider a double slit experiment as shown in Fig. 4.1, where we have placed a shielded solenoid just behind the first screen. Although the magnetic field is confined to the interior of the solenoid, the vector potential is nonvanishing also outside. The value of $\mathbf{A}$ outside the solenoid is locally a pure gauge, i.e., $\nabla\times\mathbf{A}=\mathbf{0}$, however since the region outside the solenoid is not simply connected the vector potential cannot be gauged to zero everywhere.

The dependence of the interference pattern with the magnetic field inside the solenoid can be calculated very easily using the path integral formalism introduced in Sect. 2.4. The probability amplitude for an electron emitted at $t=0$ to be detected at some given position $\mathbf{x}$ on the screen at a later time τ is given by the propagator $K(\mathbf{x},\mathbf{x}_0;\tau)$, where $\mathbf{x}_0$ is the point where the electron is emitted. This propagator admits a path integral representation, where the integration has to be done taking into account that there are two classes of paths that are topologically non-equivalent: those passing through the upper and the lower slits.

The classical action of a nonrelativistic particle of mass m and charge q in the presence of a vector potencial $\mathbf{A}$ is given by

$$S = \int dt \left(\frac{1}{2} m \dot{\mathbf{x}}^2 + q \dot{\mathbf{x}} \cdot \mathbf{A} \right) = \frac{1}{2} \int dt m \dot{\mathbf{x}}^2 + q \int_{\gamma} d\mathbf{x} \cdot \mathbf{A}, \tag{4.11}$$

where the second term in the last equation is a line integral along the particle trajectory γ. Using Stokes' theorem and $\nabla \times \mathbf{A} = \mathbf{0}$ we find that the value of this term only depends on the topological class of γ, but not in the particular curve within each class. Denoting by $K_1(\mathbf{x}, \mathbf{x}_0; \tau)$ and $K_2(\mathbf{x}, \mathbf{x}_0; \tau)$ the propagators of the electron going through each of the two slits in the absence of a magnetic field, the total propagator with the magnetic field switched on can be written as

$$\begin{aligned} K(\mathbf{x}, \mathbf{x}_0; \tau) &= e^{iq \int_{\Gamma_1} \mathbf{A} \cdot d\mathbf{x}} K_1(\mathbf{x}, \mathbf{x}_0; \tau) + e^{iq \int_{\Gamma_2} \mathbf{A} \cdot d\mathbf{x}} K_2(\mathbf{x}, \mathbf{x}_0; \tau) \\ &= e^{iq \int_{\Gamma_1} \mathbf{A} \cdot d\mathbf{x}} \left[K_1(\mathbf{x}, \mathbf{x}_0; \tau) + e^{iq \oint_{\Gamma} \mathbf{A} \cdot d\mathbf{x}} K_2(\mathbf{x}, \mathbf{x}_0; \tau) \right]. \end{aligned} \tag{4.12}$$

Here Γ_1 and Γ_2 are two arbitrary curves going through each of the two slits and joining $\mathbf{x}_0$ with $\mathbf{x}$ (see Fig. 4.1). Γ is the closed curve surrounding the solenoid defined by the union of Γ_1^{-1} and Γ_2.

The interference pattern on the screen is determined by the relative phase between the two terms in (4.12). The presence of the magnetic field confined to the solenoid introduces an extra term depending on the value of the vector potential outside the solenoid

$$U = \exp\left(iq \oint_{\Gamma} \mathbf{A} \cdot d\mathbf{x} \right). \tag{4.13}$$

Due again to Stokes' theorem and $\nabla \times \mathbf{A} = \mathbf{0}$ the value of the phase does not depend on the particular curve Γ chosen, so far as it surrounds the solenoid. The conclusion of this analysis is that the presence of the vector potential becomes observable even if the electrons do not feel the magnetic field directly. Performing the double-slit experiment when the magnetic field inside the solenoid is switched off we will observe the usual interference pattern on the second screen. Switching on the magnetic field a change in the interference pattern will appear due to the phase (4.13). This is the Aharonov–Bohm effect (see also [2] for an early prediction of the effect).

The first question that comes up is what happens with gauge invariance. Since $\mathbf{A}$ can be changed by a gauge transformation it seems that the resulting interference patters might depend on the gauge used. In fact the phase factor (4.13) is gauge invariant: the gauge variation of $\mathbf{A}$ is $-\nabla \varepsilon$ that, being a total derivative, gives zero upon integration over the close contour Γ.

The lesson we have learned is that in the quantum theory there are, apart from the electric and magnetic fields, other gauge invariant quantities giving observable effects. An important difference with respect to **E** and **B** is that these gauge invariant observables are non-local, as can be seen from the definition of the phase U.

Magnetic Monopoles

It is very easy to check that the vacuum Maxwell equations

$$\begin{aligned} \nabla\cdot\mathbf{E} &= 0 \\ \nabla\cdot\mathbf{B} &= 0 \\ \nabla\times\mathbf{E} &= -\frac{\partial}{\partial t}\mathbf{B} \\ \nabla\times\mathbf{B} &= \frac{\partial}{\partial t}\mathbf{E} \end{aligned} \tag{4.14}$$

remain invariant under the transformation

$$\mathbf{E}-i\mathbf{B}\longrightarrow e^{i\theta}(\mathbf{E}-i\mathbf{B}),\quad \theta\in[0,2\pi] \tag{4.15}$$

that for $\theta=\frac{\pi}{2}$ interchanges the electric and magnetic fields: $\mathbf{E}\to\mathbf{B}$, $\mathbf{B}\to-\mathbf{E}$. This duality symmetry is however broken in the presence of electric sources $(\rho,\mathbf{j})$. Nevertheless the Maxwell equations can be "completed" by introducing sources for the magnetic field $(\rho_m,\mathbf{j}_m)$ in such a way that the duality (4.15) is restored when supplemented by the transformation

$$\rho-i\rho_m\longrightarrow e^{i\theta}(\rho-i\rho_m),\quad \mathbf{j}-i\mathbf{j}_m\longrightarrow e^{i\theta}(\mathbf{j}-i\mathbf{j}_m). \tag{4.16}$$

In covariant language, this modification of the Maxwell equations implies adding sources on the right-hand side of the Bianchi identities

$$\partial_\mu\widetilde{F}^{\mu\nu}=j^\mu_m, \tag{4.17}$$

where $j^\mu_m=(\rho_m,\mathbf{j}_m)$ and

$$\widetilde{F}^{\mu\nu}=\frac{1}{2}\varepsilon^{\mu\nu\sigma\lambda}F_{\sigma\lambda} \tag{4.18}$$

is the dual electromagnetic tensor field. This means that, while electric charges act as sources for $F_{\mu\nu}$, magnetic charges are sources for $\widetilde{F}^{\mu\nu}$. The duality transformation (4.15, 4.16) is written now as

$$\begin{aligned} F_{\mu\nu}+i\widetilde{F}_{\mu\nu} &\longrightarrow e^{i\theta}\left(F_{\mu\nu}+i\widetilde{F}_{\mu\nu}\right), \\ j^\mu+ij^\mu_m &\longrightarrow e^{i\theta}\left(j^\mu+ij^\mu_m\right), \end{aligned} \tag{4.19}$$

keeping the extended Maxwell equations invariant. For $\theta = \frac{\pi}{2}$ electric and magnetic sources get interchanged and the field strength is replaced by its dual.

In 1931 Dirac [3] studied the possibility of finding solutions of the completed Maxwell equations with a magnetic monopoles of charge g as a source

$$\nabla \cdot \mathbf{B} = g\delta(\mathbf{x}). \tag{4.20}$$

Away from the position of the monopole $\nabla \cdot \mathbf{B} = 0$ and the magnetic field can still be derived locally from a vector potential $\mathbf{A}$ according to $\mathbf{B} = \nabla \times \mathbf{A}$. However, this potential cannot be regular everywhere since otherwise Gauss' theorem would imply that the magnetic flux threading a closed surface around the monopole should vanish, contradicting (4.20).

A solution to Eq. (4.20) in spherical coordinates is given by

$$B_r = \frac{1}{4\pi}\frac{g}{|\mathbf{x}|^2}, \quad B_\varphi = B_\theta = 0, \tag{4.21}$$

that for $\mathbf{x} \neq \mathbf{0}$ can be derived from the vector potential

$$A_\varphi = \frac{1}{4\pi}\frac{g}{|\mathbf{x}|}\tan\frac{\theta}{2}, \quad A_r = A_\theta = 0. \tag{4.22}$$

As expected, we find that this vector potential is singular at the half-line $\theta = \pi$ (see Fig. 4.2). This singular line starting at the position of the monopole is called the Dirac string and its position changes with a change of gauge but cannot be eliminated by any gauge transformation. Physically, we can see it as an infinitely thin solenoid confining a magnetic flux entering into the magnetic monopole from infinity that equals the outgoing magnetic flux from the monopole.

Since the position of the Dirac string depends on the gauge chosen it seems that we are facing a physical ambiguity. This would be rather strange since the Maxwell equations are gauge invariant also in the presence of magnetic sources. The solution to this apparent riddle lies in the fact that the presence of the Dirac string does not pose any consistency problem as far as it does not produce any physical effect, i.e., if its presence turns out to be undetectable. From our discussion of the Aharonov–Bohm effect we know that the wave function of charged particles picks up a phase (4.13) when surrounding a region where a magnetic flux is confined (such as the solenoid in the Aharonov–Bohm experiment). Since the Dirac string is like an infinitely thin solenoid, it will be unobservable if the phase picked up by the wave function of a charged particle surrounding it is equal to one. An evaluation of (4.13) in the field of the monopole shows that

$$e^{iqg} = 1 \quad \Longrightarrow \quad qg = 2\pi n \quad \text{with } n \in \mathbb{Z}. \tag{4.23}$$

Interestingly, we are led to the conclusion that the presence of a single magnetic monopole somewhere in the universe implies for consistency the quantization of the

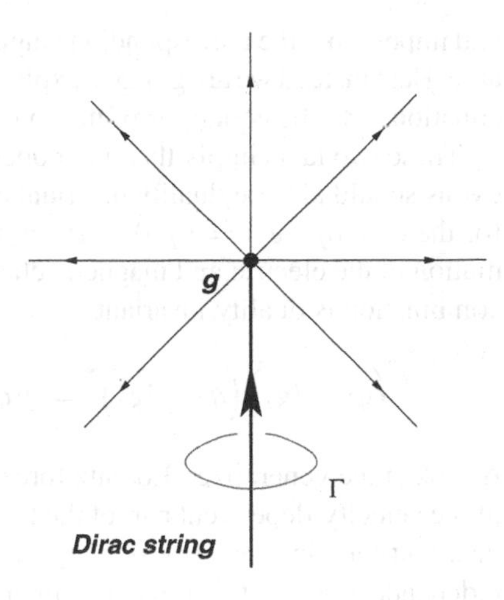

Fig. 4.2 The Dirac monopole

electric charge in units of $2\pi/g$, where g is the magnetic charge of the monopole.[1] This is called the Dirac charge quantization condition.

The idea of the magnetic monopole can be extended to dyons, particles having both electric and magnetic charge (q, g). The equations of motion for such particles in an electromagnetic field can be written remembering that magnetic charges couple to the dual field strength and requiring invariance under duality. This leads to

$$m\ddot{x}^{\mu} = \left(qF^{\mu\nu} + g\widetilde{F}^{\mu\nu}\right)\dot{x}_{\nu}, \tag{4.24}$$

where m is the mass of the dyon and the dot indicates differentiation with respect to the proper time. Writing the right-hand side of this equation in components in the nonrelativistic limit, we get the generalization of the Lorentz force acting on a dyon with charges (q, g)

$$\mathbf{F} = q\left(\mathbf{E} + \mathbf{v}\times\mathbf{B}\right) + g\left(\mathbf{B} - \mathbf{v}\times\mathbf{E}\right). \tag{4.25}$$

The invariance under duality is obvious noticing that the parentheses in the right-hand side of (4.24) can be written as $\text{Im}[(q - ig)(F_{\mu\nu} - i\widetilde{F}_{\mu\nu})^{*}]$, which is manifestly invariant.

The Dirac quantization condition, valid for an electrically charged particle and a magnetic monopole, can be extended to two dyons with charges (q_1, g_1) and (q_2, g_2). To obtain this new condition one could proceed as in the case of the Dirac monopole

---

[1] The quantization of the electric charge has another consequence, which is that the gauge transformation of the wave function (4.10) is periodic. Using technical jargon one says that the U(1) gauge group gets compactified (see Appendix B). Although this might seem just a technical point, it has important physical consequences for the production of monopoles in gauge theories.

and impose that the corresponding singularities of the gauge potentials are unobservable. Here instead we are going to exploit the invariance of both the extended Maxwell equations and the equations of motion of the dyons under duality transformations.

These two facts imply that the proper quantization condition for the charge of the dyons should also be duality invariant and, moreover, reduce to the Dirac condition for the case $(q_1, g_1) = (q, 0)$, $(q_2, g_2) = (0, g)$. Taking into account the transformation of the electric and magnetic charges it is immediate to see that the following combination is duality invariant

$$\left(q_1 - ig_1\right)\left(q_2 - ig_2\right)^* = q_1 q_2 + g_1 g_2 + i\left(q_1 g_2 - q_2 g_1\right). \tag{4.26}$$

A look at the generalized Lorentz force shows $q_1 g_2 - q_2 g_1$ is the coupling constant of the velocity-dependent part of the force between the two dyons. The other duality-invariant combination, $q_1 q_2 + g_1 g_2$, gives the strength of the coupling of the velocity-independent part of this force, i.e., their "Coulomb" interaction. Since the imaginary part of Eq. (4.26) reduces to the Dirac quantization condition in the appropriate limit, we arrive at

$$q_1 g_2 - q_2 g_1 = 2\pi n, \quad \text{where } n \in \mathbb{Z}, \tag{4.27}$$

called the Dirac–Schwinger–Zwanziger quantization condition [4, 5].

There are some difficulties in considering quantum theories with *fundamental* magnetic monopoles. One of them is that they cannot be handled in perturbation theory, since the Dirac quantization condition implies that electric and magnetic coupling constants are inverse of each other and cannot be simultaneously small. This problem is avoided if monopoles are not fundamental objects but field configurations with finite size and energy. It was proved by 't Hooft and Polyakov [6, 7] that many gauge theories contain such monopoles as solitonic solutions. The 't Hooft-Polyakov monopoles have masses that scale with the inverse of the coupling constant, and therefore they are very heavy when the theory is weakly coupled. Only at large gauge couplings this objects become light and can be counted among the low-lying excitations of the system.

Monopoles are believed to have been produced copiously in the very early Universe. It is a generic prediction of grand unified theories that monopoles occur when a semisimple gauge group is spontaneously broken leaving a U(1) factor (spontaneous symmetry breaking will be explained in Chap. 7). The reason is that this U(1) is compact in the sense explained in the footnote of page 53, and therefore can "accommodate" monopole solutions. The fact that these monopoles are not observed today is believed to be the result of the dilution they underwent during the inflationary era that presumably followed their production.

4.2 Quantization of the Electromagnetic Field

We now proceed to the quantization of the electromagnetic field in the absence of sources $\rho = 0$, $\mathbf{j} = \mathbf{0}$. In this case the Maxwell equations (4.14) can be derived from the Lagrangian density

$$\mathscr{L}_{\text{Maxwell}} = -\frac{1}{4}F_{\mu\nu}F^{\mu\nu} = \frac{1}{2}\left(\mathbf{E}^2 - \mathbf{B}^2\right). \tag{4.28}$$

Although in general the procedure to quantize the Maxwell Lagrangian is not very different from the one used for the Klein–Gordon or the Dirac field, here we need to deal with a new ingredient: gauge invariance. Unlike the cases studied so far, here the photon field A_μ is not unambiguously defined because the action and the equations of motion are insensitive to the gauge transformations $A_\mu \rightarrow A_\mu + \partial_\mu\varepsilon$. A first consequence of this symmetry is that the theory has less physical degrees of freedom than what would be expected for a vector field.

The way to tackle the problem of gauge invariance is to fix the freedom in choosing the electromagnetic potential before quantization. This can be done in several ways, for example by imposing the Lorentz gauge fixing condition

$$\partial_\mu A^\mu = 0. \tag{4.29}$$

Notice that this condition does not fix completely the gauge freedom since Eq. (4.29) is left invariant by gauge transformations satisfying $\partial_\mu\partial^\mu\varepsilon = 0$. One of the advantages of the Lorentz gauge is that it is covariant and therefore does not pose any danger to the Lorentz invariance of the quantum theory. Besides, applying it to the Maxwell equation $\partial_\mu F^{\mu\nu} = 0$ one finds

$$0 = \partial_\mu\partial^\mu A^\nu - \partial_\nu\left(\partial_\mu A^\mu\right) = \partial_\mu\partial^\mu A^\nu. \tag{4.30}$$

Since A_μ satisfies the massless Klein-Gordon equation the photon, the quantum of the electromagnetic interaction, has zero mass.

Once gauge invariance is fixed, $A_\mu(t, \mathbf{x})$ can be expanded in a complete basis of plane-wave solutions to Eq. (4.30)

$$\varepsilon_\mu(\mathbf{k}, \lambda)e^{-i|\mathbf{k}|t+i\mathbf{k}\cdot\mathbf{x}}, \tag{4.31}$$

where $\varepsilon_\mu(\mathbf{k}, \lambda)$ are the polarization vectors. In principle there are four independent polarizations for the photon, labelled by λ. The Lorentz gauge condition (4.29), however, forces the polarization vectors to be transverse

$$k^\mu\varepsilon_\mu(\mathbf{k}, \lambda) = k^\mu\varepsilon_\mu(\mathbf{k}, \lambda)^* = 0. \tag{4.32}$$

This condition can be used to eliminate one polarization. We can get rid of another one by using the on-shell condition $k^2 = 0$ and the residual gauge transformations mentioned after Eq. (4.29). Finally we are left with just two physical independent transverse polarizations $\lambda = \pm 1$. They correspond to right and left circularly polarized photons.

Now, upon quantization, the gauge field operator $\hat{A}_\mu(t, \mathbf{x})$ can be written as the following expansion

$$\hat{A}_\mu(t, \mathbf{x}) = \sum_{\lambda=\pm 1} \int \frac{d^3k}{(2\pi)^3} \frac{1}{2|\mathbf{k}|} \Big[\varepsilon_\mu(\mathbf{k}, \lambda) \hat{a}(\mathbf{k}, \lambda) e^{-i|\mathbf{k}|t + i\mathbf{k}\cdot\mathbf{x}} + \varepsilon_\mu(\mathbf{k}, \lambda)^* \hat{a}^\dagger(\mathbf{k}, \lambda) e^{i|\mathbf{k}|t - i\mathbf{k}\cdot\mathbf{x}} \Big], \tag{4.33}$$

where the canonical commutation relations imply that

$$\begin{aligned} \Big[\hat{a}(\mathbf{k}, \lambda), \hat{a}^\dagger(\mathbf{k}', \lambda')\Big] &= (2\pi)^3 (2|\mathbf{k}|) \delta(\mathbf{k} - \mathbf{k}') \delta_{\lambda\lambda'} \\ \big[\hat{a}(\mathbf{k}, \lambda), \hat{a}(\mathbf{k}', \lambda')\big] &= \Big[\hat{a}^\dagger(\mathbf{k}, \lambda), \hat{a}^\dagger(\mathbf{k}', \lambda')\Big] = 0. \end{aligned} \tag{4.34}$$

Therefore $\hat{a}(\mathbf{k}, \lambda)$, $\hat{a}^\dagger(\mathbf{k}, \lambda)$ form a set of creation-annihilation operators for photons with momentum $\mathbf{k}$ and helicity λ.

Had we kept the unphysical degrees of freedom removed by the residual gauge transformations, the spectrum would contain states with negative norm. To decouple these states with negative probability is one of the main concerns in quantizing theories with gauge invariance. In these theories there is a redundancy in the way physical states are represented by rays in the Hilbert space $\mathscr{H}$: a physical state is represented by infinitely many rays in $\mathscr{H}$. Here we have dealt with this problem by eliminating this redundancy explicitly, i.e., keeping only those polarizations that are physical. Other strategies to handle this problem can be found in standard textbooks (see Ref. [1–15] in Chap. 1). In Sect. 4.6 we will return to the problem of fixing the gauge redundancy, this time using the path integral formalism.

From the previous discussion the reader might think that we have worked too hard unnecessarily. If the photon has only two physical degrees of freedom, perhaps we could describe it using two scalar degrees of freedom, instead of introducing a redundant four-component gauge field. The obstacle is Lorentz invariance: the only known way of describing the two photon polarizations in a Lorentz invariant way is through the gauge field A_μ. The gauge redundancy is the prize we pay for a Lorentz invariant and local description of massless photons.

4.3 Coupling Gauge Fields to Matter

Once we know how to quantize the electromagnetic field we can consider interacting theories containing electrically charged particles, for example electrons. To couple the Dirac Lagrangian to electromagnetism we use the analysis of the Schrödinger equation for a charged particle presented in pages 48–49. There we learned that the gauge ambiguity of the electromagnetic potential is compensated by a U(1) phase shift in the wave function. The Lagrangian (3.36) is invariant under $\psi \to e^{-iq\varepsilon}\psi$, with ε a constant. This invariance is broken as soon as one identifies ε with the position-dependent gauge transformation parameter of the electromagnetic field.

To promote this global U(1) symmetry of the Dirac Lagrangian to a local one $\psi \to \psi' = e^{-iq\varepsilon(x)}\psi$ it is enough to replace ∂_μ by a covariant derivative D_μ, also transforming under a gauge transformation $D_\mu \to D'_\mu$, and satisfying

$$D'_\mu \psi' = D'_\mu \left[e^{-iq\varepsilon(x)}\psi \right] = e^{-iq\varepsilon(x)} D_\mu \psi. \tag{4.35}$$

Such a covariant derivative can be constructed in terms of the gauge potential A_μ as

$$D_\mu = \partial_\mu + iqA_\mu. \tag{4.36}$$

The gauge transformation of A_μ absorbs the derivative of the gauge parameter and Eq. (4.35) is satisfied. The electromagnetic field strength can be written in terms of the commutator of two covariant derivatives as

$$[D_\mu, D_\nu] = iqF_{\mu\nu}. \tag{4.37}$$

This identity will be useful in the construction of nonabelian gauge theories in the next section.

The Lagrangian of quantum electrodynamics (QED), i.e., a spin-$\frac{1}{2}$ field coupled to electromagnetism,

$$\mathscr{L}_{\text{QED}} = -\frac{1}{4}F_{\mu\nu}F^{\mu\nu} + \overline{\psi}(i\not{D} - m)\psi, \tag{4.38}$$

is invariant under the U(1) gauge transformations

$$\psi \longrightarrow e^{-iq\varepsilon(x)}\psi, \quad A_\mu \longrightarrow A_\mu + \partial_\mu \varepsilon(x). \tag{4.39}$$

Unlike the theories we encountered so far, QED is an interacting theory. By plugging (4.36) into the Lagrangian we find that the interaction term between fermions and photons has the form

$$\mathscr{L}^{(\text{int})}_{\text{QED}} = -\mathscr{H}^{(\text{int})}_{\text{QED}} = -qA_\mu \overline{\psi}\gamma^\mu \psi. \tag{4.40}$$

This shows that, as anticipated in the previous chapter (see page 43), the electric current four-vector is given by $j^\mu = q\overline{\psi}\gamma^\mu\psi$. In the following we stick to the general convention and denote the charge by e. In the case of electrons or muons, for example, e is negative and equal to the elementary charge.

The quantization of interacting field theories like QED poses new problems that we did not meet in the case of the free theories. In particular in most cases it is not possible to solve the theory exactly. When this happens the physical observables have to be computed in perturbation theory in powers of the coupling constant. An added problem appears in the computation of quantum corrections to the classical result, which is plagued with infinities that should be taken care of. All these issues will be addressed in Chaps. 6 and 8.

Here we can connect with the comments made at the beginning of the chapter. The end result of our quantization procedure is to write the gauge field in terms of the two physical degrees of freedom appearing in (4.33). Out of the four components of A_μ only two represent physical degrees of freedom. It is clear that if we wrote the theory (after including interactions) only in terms of the transverse degrees of freedom the result would be a theory without explicit Lorentz symmetry and also with non-local interactions. The inclusion of longitudinal- and timelike photons makes these apparently lost, but fundamental properties, explicit. The basic problem in the quantization of gauge theories is to make sure that at the quantum level the additional components continue to be irrelevant. Unfortunately this is not always possible, in some cases there are quantum anomalies making the theory inconsistent (see Chap. 9).

4.4 Nonabelian Gauge Theories

QED is the simplest example of a gauge theory coupled to matter based on the abelian gauge symmetry of local U(1) phase rotations. Gauge theories based on nonabelian groups can also be constructed. Our knowledge of the strong and weak interactions is in fact based on the use of the nonabelian generalizations of QED, called *Yang–Mills theories*.

Let us consider a gauge group G with hermitian generators T^A, $A = 1, \ldots, \dim G$ satisfying the Lie algebra[2]

$$\left[T^A, T^B\right] = if^{ABC}T^C. \tag{4.41}$$

We introduce a vector field $A_\mu \equiv A_\mu^A T^A$ taking values on the Lie algebra $\mathfrak{g}$ of the group G. Its gauge transformation is given by

$$A_\mu \longrightarrow A'_\mu = -\frac{1}{ig_{\rm YM}} U\partial_\mu U^{-1} + UA_\mu U^{-1}, \quad U = e^{i\chi(x)}, \tag{4.42}$$

where $\chi(x) = \chi^A(x)T^A$ and $g_{\rm YM}$ is the coupling constant. These gauge transformations are non-linear in the gauge function $\chi(x)$. Infinitesimally, the matrix-valued field A_μ transforms according to

$$\delta A_\mu = \frac{1}{g_{\rm YM}}\partial_\mu\chi - i[A_\mu, \chi], \tag{4.43}$$

which in components reads

---

[2] Some basics facts about Lie groups have been summarized in Appendix B.

$$\delta A_\mu^A = \frac{1}{g_{\rm YM}} \partial_\mu \chi^A + f^{ABC} A_\mu^B \chi^C. \tag{4.44}$$

As in the abelian case, the coupling of matter to a nonabelian gauge field is done by introducing a covariant derivative. Let Φ be a field (scalar or spinor) transforming in a representation $\mathbf{R}$ of the gauge group G

$$\Phi \longrightarrow \Phi' = U_{\mathbf{R}} \Phi. \tag{4.45}$$

The covariant derivative satisfying $D'_\mu \Phi' = U_{\mathbf{R}} D_\mu \Phi$ is defined by

$$D_\mu \Phi = \partial_\mu \Phi - i g_{\rm YM} A_\mu \Phi, \tag{4.46}$$

where $A_\mu = A_\mu^A T_{\mathbf{R}}^A$, with $T_{\mathbf{R}}^A$ the generators in the representation $\mathbf{R}$. In the particular case of the adjoint representation the generators can be written in terms of the structure constants

$$\left(T_{\rm adj}^A\right)^B{}_C = -i f^{ABC}, \tag{4.47}$$

and the covariant derivative takes the form

$$D_\mu \Phi = \partial_\mu \Phi - i g_{\rm YM} \left[A_\mu, \Phi\right] \qquad \text{(adjoint representation).} \tag{4.48}$$

Comparing this expression with (4.43) we find that the infinitesimal transformation of the gauge field can be expressed as

$$\delta A_\mu = \frac{1}{g_{\rm YM}} D_\mu \chi. \tag{4.49}$$

Our last task is to find the kinetic term for the nonabelian gauge fields. Generalizing Eq. (4.37), we write

$$\left[D_\mu, D_\nu\right] = -i g_{\rm YM} F_{\mu\nu}, \tag{4.50}$$

where $F_{\mu\nu}$ is the nonabelian field strength

$$F_{\mu\nu} = \partial_\mu A_\nu - \partial_\nu A_\mu - i g_{\rm YM} \left[A_\mu, A_\nu\right] \tag{4.51}$$

This expression reduces to (4.3) for abelian gauge groups, when the commutator of the gauge fields vanishes. The field strength tensor takes values in the Lie algebra, $F_{\mu\nu} = F_{\mu\nu}^A T^A$, where

$$F_{\mu\nu}^A = \partial_\mu A_\nu^A - \partial_\nu A_\mu^A + g_{\rm YM} f^{ABC} A_\mu^B A_\nu^C. \tag{4.52}$$

Unlike the case of the Maxwell theory the field strength for nonabelian gauge fields is not gauge invariant. Using (4.50) and the transformation of the covariant derivative it is easy to show that it transforms as

$$F_{\mu\nu} \longrightarrow U F_{\mu\nu} U^{-1}. \tag{4.53}$$

This gives the clue to constructing a gauge invariant Lagrangian for the nonabelian gauge field A_μ as

$$\mathscr{L} = -\frac{1}{2}\mathrm{Tr}\Big(F_{\mu\nu}F^{\mu\nu}\Big) = -\frac{1}{4}F^{A}_{\mu\nu}F^{A\mu\nu}, \tag{4.54}$$

where the normalization $\mathrm{Tr}(T^A T^B) = \frac{1}{2}\delta^{AB}$ has been used. A crucial difference between this and the Lagrangian of electromagnetism is the presence of cubic and quartic terms in the gauge field A_μ. This means that, unlike the photon, the nonabelian gauge bosons act themselves as sources of the field. The equations of motion derived from the Lagrangian (4.54) can be written as

$$D_\mu F^{\mu\nu} = 0, \tag{4.55}$$

where D_μ is the covariant derivative in the adjoint representation shown in Eq. (4.48).

Just as in the Maxwell theory, the components of the nonabelian field strength tensor $F^A_{\mu\nu}$ in four dimensions can be decomposed into electric and magnetic fields $\mathbf{E}^A$ and $\mathbf{B}^A$

$$E^A_i = F^A_{0i}, \quad B^A_i = -\frac{1}{2}\varepsilon_{ijk}F^A_{jk}. \tag{4.56}$$

From (4.53) it follows that the nonabelian electric and magnetic fields are gauge dependent. In terms of them the Lagrangian (4.54) becomes

$$\mathscr{L} = \frac{1}{2}\Big(\mathbf{E}^A \cdot \mathbf{E}^A - \mathbf{B}^A \cdot \mathbf{B}^A\Big). \tag{4.57}$$

In QCD $\mathbf{E}^A$ and $\mathbf{B}^A$ are respectively known as chromoelectric and chromomagnetic fields.

With all this information we can write a generic Lagrangian for a nonabelian gauge field coupled to scalars ϕ and spinors ψ as

$$\begin{aligned}\mathscr{L} = &-\frac{1}{2}\mathrm{Tr}\Big(F_{\mu\nu}F^{\mu\nu}\Big) + i\overline{\psi}\not{D}\psi + (D_\mu\phi)^\dagger D^\mu\phi \\ &- \overline{\psi}\Big[M_1(\phi) + i\gamma_5 M_2(\phi)\Big]\psi - V(\phi),\end{aligned} \tag{4.58}$$

where the covariant derivatives are in the representation of the field involved. The Lagrangian of the standard model is of this form, with $M_1(\phi)$ and $M_2(\phi)$ linear in ϕ and $V(\phi)$ of quartic order. This particular form of the functions appearing in (4.58) is related to the good properties of the standad model at high energies.

4.5 Understanding Gauge Symmetry

In classical mechanics the application of the Hamiltonian formalism starts with the replacement of generalized velocities by momenta

$$p_i \equiv \frac{\partial L}{\partial \dot{q}_i} \quad \Longrightarrow \quad \dot{q}_i = \dot{q}_i(q, p). \tag{4.59}$$

Most of the time there is no problem in inverting the relations $p_i = p_i(q, \dot{q})$. However in some systems these relations might not be invertible and result in a number of constraints of the type

$$f_a(q, p) = 0, \quad a = 1, \ldots, N_1. \tag{4.60}$$

These systems are called degenerate or constrained [8, 9].

The presence of constraints of the type (4.60) makes the formulation of the Hamiltonian formalism more involved. The first problem is related to the ambiguity in defining the Hamiltonian, since the addition of any linear combination of the constraints does not modify its value. Secondly, one has to make sure that the constraints are consistent with the time evolution in the system. In the language of Poisson brackets this means that further constraints have to be imposed in the form

$$\{f_a, H\}_{PB} \approx 0. \tag{4.61}$$

Following [8], we use the symbol $\approx$ to indicate a "weak" equality holding when the constraints $f_a(q, p) = 0$ are satisfied. Notice however that since the computation of the Poisson brackets involves derivatives, the constraints can be used only after the bracket is computed. In principle, the conditions (4.61) can give rise to a new set of constraints $g_b(q, p) = 0$, $b = 1, \ldots, N_2$. Again these constraints have to be consistent with time evolution and we have to repeat the procedure. Eventually this finishes when a set of constraints is found that do not require any further constraint to be preserved in time.[3]

Once all the constraints of a degenerate system have been found we consider the so-called first class constraints $\phi_a(q, p) = 0, a = 1, \ldots, M$, those whose mutual Poisson bracket vanishes weakly

$$\{\phi_a, \phi_b\}_{\text{PB}} = c_{abc}\phi_c \approx 0. \tag{4.62}$$

The constraints that do not satisfy this condition, called second class constraints, can be eliminated by modifying the Poisson bracket [8], so for all practical purposes we can forget about them. The total Hamiltonian of the theory is defined as the canonical Hamiltonian plus a linear combination of all first-class constraints with arbitrary coefficients

---

[3] In principle it is also possible that the procedure finishes because some kind of inconsistent identity is found. In this case the system itself is inconsistent as it happens with the Lagrangian $L(q, \dot{q}) = q$.

$$H_T = p_i \dot{q}_i - L + \sum_{a=1}^{M} \lambda_a(t)\phi_a. \tag{4.63}$$

The total Hamiltonian and the canonical one coincide on the submanifold of phase space defined by the first class constraints, where the dynamical evolution of the system takes place.

What is the relation with gauge invariance? The answer lies in the fact that for a singular system the first class constraints ϕ_a generate gauge transformations. Indeed, the time evolution generated by the Hamiltonian (4.63) is ambiguous due to the presence of the arbitrary functions $\lambda_a(t)$. Specifying the state of the system by the values of the canonical variables at some reference time t_0, the ambiguity in the time evolution translates into a redundancy in the description of the state of the system in terms of the values of the canonical variables at a later time t: the phase space trajectories related by the infinitesimal transformations

$$\begin{aligned} q_i &\longrightarrow q_i + \sum_{a=1}^{M} \varepsilon_a(t)\{q_i, \phi_a\}_{\text{PB}}, \\ p_i &\longrightarrow p_i + \sum_{a=1}^{M} \varepsilon_a(t)\{p_i, \phi_a\}_{\text{PB}} \end{aligned} \tag{4.64}$$

describe one and the same state.

This ambiguity in the description of the system in terms of the generalized coordinates and momenta can be traced back to the equations of motion in Lagrangian language. Writing them in the form

$$\frac{\partial^2 L}{\partial \dot{q}_i \partial \dot{q}_j}\ddot{q}_j = -\frac{\partial^2 L}{\partial \dot{q}_i \partial q_j}\dot{q}_j + \frac{\partial L}{\partial q_i}, \tag{4.65}$$

we find that in order to determine the accelerations in terms of the positions and velocities, the matrix $\frac{\partial^2 L}{\partial \dot{q}_i \partial \dot{q}_j}$ has to be invertible. However, the existence of constraints (4.60) precisely implies that the determinant of this matrix vanishes and therefore the time evolution is not uniquely determined in terms of the initial conditions.

Applications to Electrodynamics

After a general discussion we particularize the analysis to the Maxwell Lagrangian

$$L = -\frac{1}{4}\int d^3x\, F_{\mu\nu}F^{\mu\nu}. \tag{4.66}$$

The generalized momenta conjugate to A_μ is defined by

$$\pi^\mu = \frac{\delta L}{\delta(\partial_0 A_\mu)} = F^{\mu 0}, \tag{4.67}$$

hence, $\pi^0 = 0$ and $\pi^i = E^i$. The Hamiltonian is given by

$$H = \int d^3x \left(\pi^\mu \partial_0 A_\mu - \mathscr{L}\right) = \int d^3x \left[\frac{1}{2}\left(\mathbf{E}^2 + \mathbf{B}^2\right) + A_0 \nabla\cdot\mathbf{E}\right], \tag{4.68}$$

where we have used $\partial_0 \mathbf{A} = \nabla A_0 - \pi = \nabla A_0 - \mathbf{E}$ and integrated by parts the second term in the last integral.

The Hamiltonian (4.68) shows that $A_0(x)$ plays the role of a Lagrange multiplier implementing Gauss' law $\nabla\cdot\mathbf{E} = 0$ as a constraint.[4] Thus $\pi^0 = 0$ and $\nabla\cdot\pi = 0$ form a set of two first class constraints generating gauge transformations. The ones generated by π^0 can be used to fix the value of $A_0(x)$, thus defining a temporal gauge. This does not completely fix the gauge freedom, since there are the gauge transformations generated by Gauss' law. Using the canonical Poisson brackets

$$\{A_i(t,\mathbf{x}), E_j(t,\mathbf{x}')\}_{\mathrm{PB}} = \delta_{ij}\delta(\mathbf{x}-\mathbf{x}') \tag{4.69}$$

we find these to be

$$\delta A_i(t,\mathbf{x}) = \{A_i(t,\mathbf{x}), \int d^3x' \varepsilon(t,\mathbf{x}')\nabla\cdot\mathbf{E}(t,\mathbf{x}')\}_{\mathrm{PB}} = \partial_i \varepsilon(t,\mathbf{x}), \tag{4.70}$$

while $A_0(t,\mathbf{x})$ is left invariant. This is equivalent to a general gauge transformation generated by a time-independent gauge function $\varepsilon(\mathbf{x})$. Thus, for consistency, we take $\varepsilon(t,\mathbf{x})$ in (4.70) to depend only on the spatial coordinates. The constraint $\nabla\cdot\mathbf{E} = 0$ can be implemented by demanding $\nabla\cdot\mathbf{A} = 0$, reducing the three degrees of freedom of $\mathbf{A}$ to the two physical degrees of freedom of the photon.

So much for the classical analysis. In the quantum theory the constraint $\nabla\cdot\mathbf{E} = 0$ has to be imposed on the physical states $|\mathrm{phys}\rangle$. This is done by defining the following unitary operator in the Hilbert space

$$\mathscr{U}(\varepsilon) \equiv \exp\left[i\int d^3x \varepsilon(\mathbf{x})\nabla\cdot\mathbf{E}\right]. \tag{4.71}$$

By definition, physical states should not change when a gauge transformations is performed. This is implemented by requiring the operator $\mathscr{U}(\varepsilon)$ to act trivially on them

$$\mathscr{U}(\varepsilon)|\mathrm{phys}\rangle = |\mathrm{phys}\rangle \quad \Longrightarrow \quad (\nabla\cdot\mathbf{E})|\mathrm{phys}\rangle = 0. \tag{4.72}$$

[4] This constraint can also be obtained from the requirement that $\pi^0 = 0$ be preserved by the time evolution, $\{\pi^0, H\}_{\mathrm{PB}} = 0$. A detailed analysis of Maxwell electrodynamics using the general formalism for constrained systems can be found in [9].

In the presence of a charge density ρ, this condition becomes $(\nabla \cdot \mathbf{E} - \rho)|\text{phys}\rangle = 0$.

The action of the gauge transformations in the quantum theory is very illuminating in understanding the real role of gauge invariance [10–12]. We have learned that the presence of a gauge symmetry in a theory reflects a degree of redundancy in the description of physical states in terms of the degrees of freedom appearing in the Lagrangian. In classical mechanics, for example, the state of a system is determined by the value of the canonical coordinates (q_i, p_i). We know, however, that this is not the case for constrained Hamiltonian systems, where the transformations generated by the first class constraints change the value of q_i and p_i without actually changing the physical state. Physical (i.e., measurable) quantities have to be free from such ambiguity and therefore be represented by gauge invariant objects. The same happens in classical field theory: in the Maxwell theory for every physical configuration determined by the gauge invariant quantities $\mathbf{E}$ and $\mathbf{B}$ there is an infinite number of possible values of A_μ related by gauge transformations $\delta A_\mu = \partial_\mu \varepsilon$.

In the quantum theory this means that one should identify into a single physical state all rays in the Hilbert space related by the operator $\mathscr{U}(\varepsilon)$ with any gauge function $\varepsilon(x)$. In other words, each physical state corresponds to a whole orbit of states transforming among themselves by gauge transformations.

This explains the necessity of gauge fixing. In order to avoid the redundancy in the states a further condition should be given selecting one single state on each orbit. Once again, we connect with the opening comments in this chapter. In the Hamiltonian quantization we see very clearly described how the gauge symmetry is more a redundancy than a symmetry. In going to the timelike gauge, i.e., imposing $A_0 = 0$, we eliminate one of the components of the gauge field. In the initial value surface we need to impose Gauss' law (by requiring for example $\nabla \cdot \mathbf{A} = 0$) to eliminate yet one more degree of freedom, reducing the number of physical degrees of freedom to two per gauge group generator.

4.6 Gauge Fields and Path Integrals

The redundancy in the Hilbert space is a source of complications when quantizing gauge theories. This we have seen already in Sect. 4.2: the photon had two unphysical polarizations removed using the Lorentz gauge fixing condition and the residual gauge invariance.

In the path integral formalism the problem of gauge invariance reflects in the necessity of carrying out the integration over gauge fields in a way that avoids over-counting. This means that two field configurations related by a gauge transformation should be considered as physically equivalent and included only once. For example, a naive evaluation of the vacuum-to-vacuum amplitude (partition function)

$$\mathscr{Z} = \int \mathscr{D}A_\mu e^{-\frac{i}{2}\int d^4x \mathrm{Tr}(F_{\mu\nu}F^{\mu\nu})} \tag{4.73}$$

would include together with each gauge field configuration A_μ all others obtained from it by an arbitrary gauge transformation, thus overcounting the result by an infinite factor equal to the volume of the gauge group.

The correct evaluation of the integral (4.73) requires restricting the integration to fields not related by gauge transformations. A practical way to do this is to notice that the computation of observables in quantum field theory generically involves quotients of path integrals [see Chap. 6 and in particular Eq. (6.35)]. Then it suffices to cancel the (infinite) volume factor in the numerator and denominator.

To carry out this program we follow ideas due to Faddeev and Popov [13] and begin by imposing a set of gauge fixing conditions of the form

$$\mathscr{F}^A(A_\mu) = 0. \tag{4.74}$$

They can be visualized as a "slice" in the space of all gauge field configurations. Each A_μ falls into a gauge orbit generated by the gauge transformations acting on it. Two gauge field configurations are nonequivalent if they lie on different orbits. The condition (4.74) selects a representative on each orbit and has to satisfy a number of requirements: it has to be reachable from any A_μ, i.e., each gauge orbit should have a representative satisfying (4.74), and this representative should be unique. To keep expressions simple in the following we drop the group theory index in Eq. (4.74) and denote the gauge conditions collectively by $\mathscr{F}(A_\mu) = 0$.

The next step is to split the functional integral (4.73) into an integration over the orbit representatives and an integral over each gauge orbit. This last integration results in a common factor equal to the volume of the gauge group. This is done by introducing the functional $\Delta_{\text{FP}}[A_\mu]$ through the following definition

$$1 = \Delta_{\text{FP}}[A_\mu] \int \mathscr{D}U \delta\Big[\mathscr{F}(A_\mu^U)\Big], \tag{4.75}$$

where we are integrating over all gauge transformations and by A_μ^U we denote the gauge potential transformed by U. For reasons that will be explained soon, $\Delta_{\text{FP}}[A_\mu]$ is called the Faddeev–Popov determinant. It is not difficult to show that it is gauge invariant. Indeed, for any gauge transformation U' we have

$$\begin{aligned}\Delta_{\text{FP}}[A_\mu^{U'}]^{-1} &= \int \mathscr{D}U \delta\Big[\mathscr{F}\Big(A_\mu^{UU'}\Big)\Big] \\ &= \int \mathscr{D}U'' \delta\Big[\mathscr{F}\Big(A_\mu^{U''}\Big)\Big] = \Delta_{\text{FP}}[A_\mu]^{-1},\end{aligned} \tag{4.76}$$

where we have made the change of variables $U'' = UU'$ and used the gauge invariance of the integration measure over the gauge group, $\mathscr{D}U'' = \mathscr{D}U$.

We insert now the identity (4.75) into the function integral (4.73)

$$\mathscr{Z} = \int \mathscr{D}A_\mu \mathscr{D}U \, \Delta_{\text{FP}}[A_\mu] \delta\Big[\mathscr{F}\Big(A_\mu^U\Big)\Big] e^{-\frac{i}{2}\int d^4x \operatorname{Tr}(F_{\mu\nu}F^{\mu\nu})}. \tag{4.77}$$

Doing the change of variables $A_\mu \to A_\mu^{U^{-1}}$ and using the gauge invariance of both the action and $\Delta_{\rm FP}[A_\mu]$, we remove all dependence on U from the integrand. If the integration measure over the gauge fields $\mathscr{D}A_\mu$ is gauge invariant, this change of variables does not induce any Jacobian and the integration over the gauge group can be factored out

$$\mathscr{Z} = \left(\int \mathscr{D}U\right) \int \mathscr{D}A_\mu \Delta_{\rm FP}[A_\mu]\delta\Big[\mathscr{F}(A_\mu)\Big] e^{-\frac{i}{2}\int d^4x {\rm Tr}(F_{\mu\nu}F^{\mu\nu})}. \tag{4.78}$$

We can ignore the divergent prefactor and replace (4.73) by the gauge-fixed functional integral

$$\mathscr{Z} = \int \mathscr{D}A_\mu \Delta_{\rm FP}[A_\mu]\delta\Big[\mathscr{F}(A_\mu)\Big] e^{-\frac{i}{2}\int d^4x {\rm Tr}(F_{\mu\nu}F^{\mu\nu})}. \tag{4.79}$$

The delta function restricts the integration to gauge configurations lying on the slice $\mathscr{F}(A_\mu) = 0$, i.e., the integral only includes the contributions of the representatives of each gauge orbit.

To find an explicit expression for $\Delta_{\rm FP}[A_\mu]$ we use a functional version of the delta-function identity (2.10), namely

$$\delta\Big[\mathscr{F}\Big(A_\mu^U\Big)\Big] = \left|\det\left[\frac{\delta\mathscr{F}(A_\mu^U)}{\delta U}\bigg|_{U=U'}\right]\right|^{-1} \delta(U - U'), \tag{4.80}$$

where U' is a gauge transformation such that $\mathscr{F}\Big(A_\mu^{U'}\Big) = 0$ for a given A_μ. Going back to Eq. (4.75) and integrating over U using the delta function, we find that $\Delta_{\rm FP}\big[A_\mu\big]$ can be expressed as the following functional determinant

$$\Delta_{\rm FP}[A_\mu] = \det\left[\frac{\delta\mathscr{F}(A_\mu^U)}{\delta U}\right]\bigg|_{U=1}. \tag{4.81}$$

In writing this expression we have used that $\Delta_{\rm FP}[A_\mu] = \Delta_{\rm FP}[A_\mu^{U'^{-1}}]$. This means that in the computation of the Faddeev–Popov determinant we have to impose that the gauge field lies on the gauge slice $\mathscr{F}(A_\mu) = 0$.

It should be clear that the value of the path integral (4.79) is not modified by changing the position of the slice defined by (4.74). That is, the value of $\mathscr{Z}$ does not change if we replace $\mathscr{F}(A_\mu)$ by $\mathscr{F}(A_\mu) = f(x)$, where $f(x)$ is an arbitrary Lie algebra valued function of the coordinates,

$$\mathscr{Z} = \int \mathscr{D}A_\mu \Delta_{\rm FP}[A_\mu]\delta\Big[\mathscr{F}(A_\mu) - f(x)\Big] e^{-\frac{i}{2}\int d^4x {\rm Tr}(F_{\mu\nu}F^{\mu\nu})}. \tag{4.82}$$

Since the previous expression is independent of $f(x)$ we can insert the constant term

$$\int \mathscr{D}f e^{-\frac{i}{\xi}\int d^4x \mathrm{Tr}[f(x)^2]} = \text{constant}, \tag{4.83}$$

and carry out the integration over $f(x)$ using the delta function. Modulo a global normalization, this gives

$$\mathscr{Z} = \int \mathscr{D}A_\mu \Delta_{\mathrm{FP}}[A_\mu] e^{i\int d^4x \mathrm{Tr}\left[-\frac{1}{2}F_{\mu\nu}F^{\mu\nu} - \frac{1}{\xi}\mathscr{F}(A_\mu)^2\right]}, \tag{4.84}$$

where ξ is an arbitrary real parameter. The new term added to the action is called the gauge fixing term.

We illustrate the previous discussion with two examples. We begin with QED and impose the Lorentz gauge $\mathscr{F}(A_\mu) = \partial_\mu A^\mu$. Using $U(x) = e^{ie\varepsilon(x)}$ we find

$$\mathscr{F}(A_\mu^U) = \partial_\mu A^\mu + \partial_\mu \partial^\mu \varepsilon \quad \Longrightarrow \quad \left.\frac{\delta \mathscr{F}(A_\mu^U)}{\delta U}\right|_{U=\mathbf{1}} = -\frac{1}{ie}\partial_\mu \partial^\mu. \tag{4.85}$$

Hence $\Delta_{\mathrm{FP}}[A_\mu] = |\det(-\frac{1}{ie}\partial_\mu\partial^\mu)|$ is independent of the gauge field. This means that we do not have to bother computing the determinant because it goes out of the path integral as an irrelevant global normalization constant. The typical functional integral for QED can be written as

$$\mathscr{Z}_{\mathrm{QED}} = \int \mathscr{D}\overline{\psi}\mathscr{D}\psi\mathscr{D}A_\mu e^{i(S_{\mathrm{QED}}+S_{\mathrm{gf}})}, \tag{4.86}$$

where the action and the gauge-fixing term read

$$S_{\mathrm{QED}} + S_{\mathrm{gf}} = \int d^4x \left[\overline{\psi}(i\not{D} - m)\psi - \frac{1}{4}F_{\mu\nu}F^{\mu\nu} - \frac{1}{2\xi}\left(\partial_\mu A^\mu\right)^2\right]. \tag{4.87}$$

The conclusion is that the problem of gauge invariance in the path integral quantization of QED is handled in a Lorentz-invariant way by adding a gauge fixing term to the action. The constant ξ is arbitrary and can be chosen to make some expressions simpler. In Chap. 6 we will learn how to compute observables in QED.

The case of nonabelian Yang–Mills theories is more complicated and here we only outline the procedure. Using the Lorentz condition $\mathscr{F}(A_\mu) = \partial_\mu A^\mu$ and the gauge transformation $\delta A_\mu = \frac{1}{g_{\mathrm{YM}}}D_\mu\chi$ we find

$$\left.\frac{\delta \mathscr{F}(A_\mu^U)}{\delta U}\right|_{U=\mathbf{1}} = \frac{1}{ig_{\mathrm{YM}}}\partial_\mu D^\mu, \tag{4.88}$$

where D_μ is the covariant derivative in the adjoint representation, given by (4.48). Unlike the case of QED, now the Faddeev–Popov determinant depends on the gauge field, even after imposing the Lorentz condition $\partial_\mu A^\mu = 0$. This has to be taken into account when carrying out the integration over A_μ. The standard way to proceed now is to write $\Delta_{\text{FP}}[A_\mu]$ as a path integral over some unphysical fields called the Faddeev–Popov ghosts. The details can be found in most of the textbooks listed in Ref. [1–15] of Chap. 1.

The use of Faddeev–Popov ghosts in nonabelian gauge theories can be avoided, for example, in the axial gauge $n^\mu A_\mu = 0$, with $n_\mu n^\mu < 0$. In this case

$$\left.\frac{\delta \mathscr{F}(A_\mu^U)}{\delta U}\right|_{U=1} = \frac{1}{i g_{\text{YM}}} n^\mu D_\mu. \tag{4.89}$$

Imposing the gauge condition $n^\mu A_\mu = 0$, we find that $n^\mu D_\mu = n^\mu \partial_\mu$ and $\Delta_{\text{FP}}[A_\mu]$ is independent of the gauge field. It can be absorbed in the global normalization of the path integral, and the partition function (4.79) becomes

$$\begin{aligned}
\mathscr{Z} &= \int \mathscr{D}A_\mu \delta[n^\nu A_\nu] e^{-\frac{i}{2}\int d^4x \operatorname{Tr}(F_{\mu\nu}F^{\mu\nu})} \\
&= \int \mathscr{D}A_\mu e^{i\int d^4x \operatorname{Tr}(-\frac{1}{2}F_{\mu\nu}F^{\mu\nu} - \frac{1}{\xi} n^\mu n^\nu A_\mu A_\nu)}.
\end{aligned} \tag{4.90}$$

4.7 The Structure of the Gauge Theory Vacuum

The topology of the gauge group plays an important physical role in Yang–Mills theories. To illustrate the issue, we first look at a toy model: a U(1) gauge theory in 1+1 dimensions. Later we will be more general. We will also point out a number of subtleties involved in the definition of the topology of the gauge field making the arguments presented more semiclassical rather than nonperturbative.

In the Hamiltonian formalism, gauge transformations $g(\mathbf{x})$ are functions defined on $\mathbb{R}$ with values on the gauge group U(1)

$$g : \mathbb{R} \longrightarrow U(1). \tag{4.91}$$

We assume that $g(x)$ is regular at infinity. In this case we can add to the real line $\mathbb{R}$ the point at infinity and compactify it to the circle S^1 (see Fig. 4.3). Once this is done, the $g(x)$'s are functions defined on S^1 with values on $\text{U}(1) = S^1$ that can be parametrized as

$$g : S^1 \longrightarrow U(1), \quad g(x) = e^{i\alpha(x)}, \tag{4.92}$$

with $x \in [0, 2\pi]$.

Since S^1 does have a nontrivial topology, $g(x)$ is divided into topological sectors. They are labelled by an integer number $n \in \mathbb{Z}$ and defined by

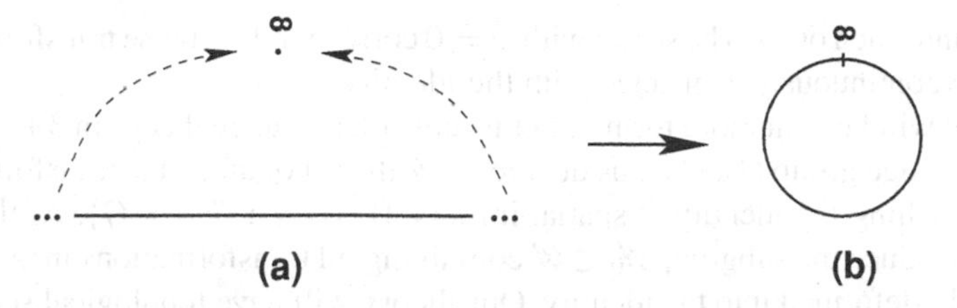

Fig. 4.3 Compactification of the real line (a) into the circumference S^1 (b) by adding the point at infinity

$$\alpha(2\pi) = \alpha(0) + 2\pi n. \tag{4.93}$$

Geometrically, n is the number of times that the spatial S^1 winds around the gauge group U(1). This winding number can be written equivalently as

$$\oint_{S^1} g(x)^{-1} dg(x) = 2\pi n, \tag{4.94}$$

where the integral is along the spatial S^1.

Something similar happens in the case of a SU(2) gauge theory in 3+1 dimensions.[5] Demanding $g(\mathbf{x}) \in SU(2)$ to be regular at spatial infinity, $|\mathbf{x}| \to \infty$, we can compactify $\mathbb{R}^3$ into a three-dimensional sphere S^3, exactly as we did in 1+1 dimensions. The matrices $g(\mathbf{x})$ can be parameterized as

$$g(\mathbf{x}) = a^0(\mathbf{x})\mathbf{1} + i\mathbf{a}(\mathbf{x}) \cdot \sigma, \tag{4.95}$$

with σ_i the Pauli matrices. The conditions $g(\mathbf{x})^\dagger g(\mathbf{x}) = \mathbf{1}$, $\det g = 1$ imply $(a^0)^2 + \mathbf{a}^2 = 1$. Hence SU(2) is a three-dimensional sphere and $g(\mathbf{x})$ defines a map from the spatial S^3 to the S^3 defined by the gauge group

$$g : S^3 \longrightarrow S^3. \tag{4.96}$$

As in the (1+1)-dimensional case, the gauge transformations $g(\mathbf{x})$ are divided into topological sectors labelled this time by the integer winding number

$$n = \frac{1}{24\pi^2} \int_{S^3} d^3x \varepsilon_{ijk} \mathrm{Tr}\Big[\left(g^{-1}\partial_i g\right) \left(g^{-1}\partial_i g\right) \left(g^{-1}\partial_i g\right) \Big]. \tag{4.97}$$

In U(1) and SU(2), gauge transformations split into different sectors labelled by an integer. Since this winding number is a continuous function of the gauge transformation $g(\mathbf{x})$, two transformations with different values of n cannot be smoothly

---

[5] Although we present for simplicity only the case of SU(2), similar arguments apply to any simple group.

deformed into each other. The sector with $n = 0$ corresponds to those transformations that can be continuously connected with the identity.

Now we will be a bit more formal. Let us consider a gauge theory in 3+1 dimensions with gauge group G and let us denote by $\mathscr{G}$ the set of all gauge transformations $g(\mathbf{x})$ approaching the identity at spatial infinity, $\mathscr{G} = \{g : S^3 \to G\}$. At the same time we introduce the subgroup $\mathscr{G}_0 \subset \mathscr{G}$ containing all transformations in $\mathscr{G}$ that can be smoothly deformed into the identity. Our theory will have topological sectors if

$$\mathscr{G}/\mathscr{G}_0 \neq \mathbf{1}. \tag{4.98}$$

The existence of these topological sectors in (3+1)-dimensional gauge theories is controlled by a mathematical object called the third homotopy group of the gauge group that is denoted by $\pi_3(G)$. For example, it can be proved [14] that $\pi_3(S^1) = \mathbf{1}$, i.e., the third homotopy group of U(1) is trivial and therefore no topological sectors appear in (3+1)-dimensional electrodynamics. On the other hand, $\pi_3(S^3) = \mathbb{Z}$ and as a consequence the topological sectors of the SU(2) gauge theory are labelled by a single integer, the winding number[6] (4.97).

In the case of electromagnetism, we have seen that Gauss' law annihilates physical states. For a nonabelian theory the analysis is similar and leads to the condition

$$\mathscr{U}(g_0)|\text{phys}\rangle \equiv \exp\left[i\int d^3x\,\chi^A(\mathbf{x})(\mathbf{D}\cdot\mathbf{E})^A\right]|\text{phys}\rangle = |\text{phys}\rangle, \tag{4.99}$$

where $g_0(\mathbf{x}) = e^{i\chi^A(\mathbf{x})T^A}$ is in the connected component of the identity $\mathscr{G}_0$, and D_i is the covariant derivative in the adjoint representation. The important point here is that only the elements of $\mathscr{G}_0$ can be written as exponentials of the infinitesimal generators. Since these generators annihilate the physical states, this implies $\mathscr{U}(g_0)|\text{phys}\rangle = |\text{phys}\rangle$ only when $g_0(\mathbf{x}) \in \mathscr{G}_0$.

What happens with gauge transformations in the other topological sectors? If $g \in \mathscr{G}/\mathscr{G}_0$ there is still a unitary operator $\mathscr{U}(g)$ implementing gauge transformations on the Hilbert space of the theory. However since g is not in the connected component of the identity, it cannot be written as the exponential of Gauss' law. Still, gauge invariance is preserved if $\mathscr{U}(g)$ only changes the overall global phase of the physical states. For example, if $g_1(\mathbf{x})$ is a gauge transformation with winding number $n = 1$

$$\mathscr{U}(g_1)|\text{phys}\rangle = e^{i\theta}|\text{phys}\rangle. \tag{4.100}$$

It is easy to convince oneself that all transformations with winding number $n=1$ have the same value of θ modulo 2π. This can be shown by noticing that if $g(\mathbf{x})$ has $n=1$ then $g(\mathbf{x})^{-1}$ has opposite winding number $n = -1$. It is a simple exercise to prove that the winding number is additive: given two transformations g_1, g_2 with winding number 1, $g_1^{-1}g_2$ has winding number $n=0$. This leads to

---

[6] The existence of topological sectors in (1+1)-dimensional electrodynamics is a consequence of the nontrivial character of the *first* homotopy group of S^1, namely $\pi_1(S^1) = \mathbb{Z}$.

$$\begin{aligned} |\text{phys}\rangle &= \mathscr{U}(g_1^{-1}g_2)|\text{phys}\rangle = \mathscr{U}(g_1)^\dagger \mathscr{U}(g_2)|\text{phys}\rangle \\ &= e^{i(\theta_2-\theta_1)}|\text{phys}\rangle, \end{aligned} \tag{4.101}$$

thus $\theta_1 = \theta_2 \bmod 2\pi$. Therefore a gauge transformation $g_n(\mathbf{x})$ with winding number n acts on physical states according to

$$\mathscr{U}(g_n)|\text{phys}\rangle = e^{in\theta}|\text{phys}\rangle, \quad n \in \mathbb{Z}. \tag{4.102}$$

To find a physical interpretation of this result, we look for a similar situation in a more familiar setup, for example the quantum states of electrons in the periodic potential produced by the ion lattice in a solid. For simplicity, we discuss the one-dimensional case where the minima of the potential are separated by a distance a. When the barrier between consecutive degenerate vacua is high enough, we can neglect tunneling between different vacua and consider the ground states $|na\rangle$ of the potential near the minimum located at $x = na$ $(n \in \mathbb{Z})$ as possible vacua of the theory. These ground states are not invariant under lattice translations

$$e^{ia\hat{P}}|na\rangle = |(n+1)a\rangle. \tag{4.103}$$

It is nevertheless possible to define a new vacuum state

$$|k\rangle = \sum_{n\in\mathbb{Z}} e^{-ikna}|na\rangle, \tag{4.104}$$

which under $e^{ia\hat{P}}$ transforms just by a global phase

$$e^{ia\hat{P}}|k\rangle = \sum_{n\in\mathbb{Z}} e^{-ikna}|(n+1)a\rangle = e^{ika}|k\rangle. \tag{4.105}$$

This ground state is labelled by the momentum k and corresponds to the Bloch wave function.

This is very similar to what we found for nonabelian gauge theories. The vacuum state labelled by θ plays a role similar to the Bloch wave function for the periodic potential with the identification of θ with the momentum k. To make this analogy more precise, let us write the Hamiltonian for nonabelian gauge theories

$$H = \frac{1}{2}\int d^3x\left(\pi^A\cdot\pi^A + \mathbf{B}^A\cdot\mathbf{B}^A\right) = \frac{1}{2}\int d^3x\left(\mathbf{E}^A\cdot\mathbf{E}^A + \mathbf{B}^A\cdot\mathbf{B}^A\right), \tag{4.106}$$

where we have used the expression of the canonical momenta π^{iA}. Moreover, we work in the gauge $A_0 = 0$ and assume that the Gauss law constraint is satisfied. The first term in the integral is the kinetic energy, $T = \frac{1}{2}\pi^A\cdot\pi^A$, and the second the potential energy, $V = \frac{1}{2}\mathbf{B}^A\cdot\mathbf{B}^A$. Since $V \geqslant 0$, the vacua of the theory can be identified with those gauge field configurations for which $V = 0$, modulo gauge transformations. This happens when $\mathbf{A}(0,\mathbf{x})$ is a pure gauge. Since gauge transformations are classified by their winding number, there are infinitely many ground states. Indeed, taking a representative gauge transformation $g_n(\mathbf{x})$ in the sector with winding number n, these vacua will be associated with the gauge potentials

$$\mathbf{A}(0,\mathbf{x}) = -\frac{1}{ig_{\mathrm{YM}}} g_n(\mathbf{x})\nabla g_n(\mathbf{x})^{-1}, \tag{4.107}$$

modulo topologically trivial gauge transformations. Thus the theory is characterized by an infinite number of ground states $|n\rangle$ labelled by the winding number.

These vacua are not gauge invariant. A gauge transformation with $n=1$ changes the winding number of the vacuum by one unit

$$\mathscr{U}(g_1)|n\rangle = |n+1\rangle. \tag{4.108}$$

As with Bloch waves, a gauge invariant vacuum can be defined

$$|\theta\rangle = \sum_{n\in\mathbb{Z}} e^{-in\theta}|n\rangle \quad \text{with } \theta \in \mathbb{R}, \tag{4.109}$$

transforming under a gauge transformation by a global phase

$$\mathscr{U}(g_1)|\theta\rangle = e^{i\theta}|\theta\rangle. \tag{4.110}$$

We have concluded that the nontrivial topology of the gauge group has very important physical consequences for the quantum theory. In particular, it implies an ambiguity in the definition of the vacuum. This can also be seen in a Lagrangian analysis. In constructing the Lagrangian for the nonabelian version of the Maxwell theory we only considered the term $F^A_{\mu\nu}F^{\mu\nu A}$. However this is not the only Lorentz and gauge invariant term containing just two derivatives. We can write the more general action

$$S = -\frac{1}{2}\int d^4x \mathrm{Tr}\Big(F_{\mu\nu}F^{\mu\nu}\Big) - \frac{\theta g^2_{\mathrm{YM}}}{16\pi^2}\int d^4x \mathrm{Tr}\Big(F_{\mu\nu}\widetilde{F}^{\mu\nu}\Big), \tag{4.111}$$

where $\widetilde{F}_{\mu\nu}$ is the dual of the field strength defined by

$$\widetilde{F}_{\mu\nu} = \frac{1}{2}\varepsilon_{\mu\nu\sigma\lambda}F^{\sigma\lambda}. \tag{4.112}$$

The constant θ is dimensionless in natural units. The extra term in (4.111), proportional to $\mathbf{E}^A\cdot\mathbf{B}^A$, is a total derivative and does not change the equations of motion or the quantum perturbation theory.

This, however, does not mean that the addition of the second piece in the action (4.111) does not change the physics. It can be directly checked that

$$\frac{g^2_{\mathrm{YM}}}{16\pi^2}\mathrm{Tr}\Big(F_{\mu\nu}\widetilde{F}^{\mu\nu}\Big) = \partial_\mu \mathscr{J}^\mu \tag{4.113}$$

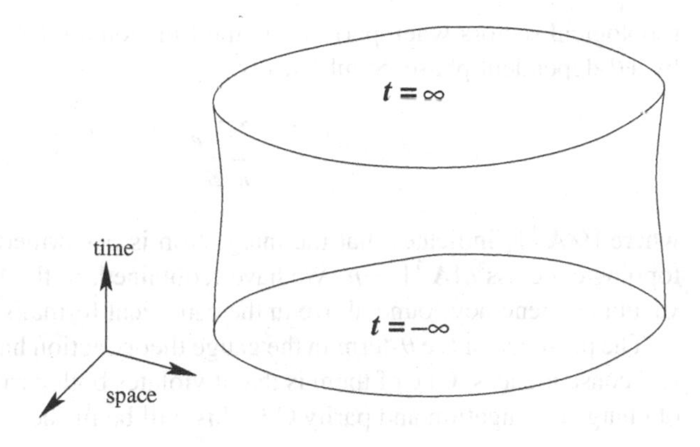

Fig. 4.4 Region of integration to compute the contribution of the θ-term to the gauge theory action. The gauge field $\mathbf{A}(t,\mathbf{x})$ tends to pure gauge configurations both at early and late times $t \to \pm\infty$ and at spatial infinity $|\mathbf{x}| \to \infty$ (the side of the cylinder)

with

$$\mathscr{J}^{\mu} = \frac{g_{\rm YM}^2}{16\pi^2}\varepsilon^{\mu\nu\sigma\lambda}\mathrm{Tr}\left(F_{\nu\sigma}A_{\lambda} - \frac{2ig_{\rm YM}}{3}A_{\nu}A_{\sigma}A_{\lambda}\right). \tag{4.114}$$

Thus, the contribution of the second term in (4.111) can be computed using Gauss' theorem. To ensure the convergence of the integral we assume that $\mathbf{A}(t,\mathbf{x})$ approaches a pure gauge configuration both at spatial infinity and at late and early times $t \to \pm\infty$. To be more precise we assume that

$$\mathbf{A}(t\to\infty,\mathbf{x}) \longrightarrow -\frac{1}{ig_{\rm YM}}g(\mathbf{x})\nabla g(\mathbf{x})^{-1}, \tag{4.115}$$

while $\mathbf{A}(t,\mathbf{x})$ is taken to vanish at $t \to -\infty$. This last condition implies no loss of generality, since it can always be achieved by an appropriate gauge transformation.

In the gauge $A^0 = 0$ it is easy to check that $\mathscr{J}^i \to 0$ at spatial infinity. Hence, the integral of the topological term in the action only receives contributions from the boundaries at $t \to \pm\infty$ (see Fig. 4.4). This yields

$$\begin{aligned}&\frac{g_{\rm YM}^2}{16\pi^2}\int d^4x\,\mathrm{Tr}\left(F_{\mu\nu}\widetilde{F}^{\mu\nu}\right)\\ &\quad= \frac{1}{24\pi^2}\int d^3x\,\varepsilon_{ijk}\mathrm{Tr}\Big[\left(g\partial_i g^{-1}\right)\left(g\partial_j g^{-1}\right)\left(g\partial_k g^{-1}\right)\Big].\end{aligned} \tag{4.116}$$

Comparing this expression with Eq. (4.97) we obtain

$$\frac{\theta g_{\rm YM}^2}{16\pi^2}\int d^4x\,\mathrm{Tr}\left(F_{\mu\nu}\widetilde{F}^{\mu\nu}\right) = \theta n[g] \equiv \theta n[\mathbf{A}^A]. \tag{4.117}$$

This term distinguishes gauge fields according to topological sectors: two gauge fields are in the same sector if the corresponding gauge transformations g giving their asymptotic behavior at late times have the same winding numbers. This is very important in the quantum theory, because it means that one must sum over all topological sectors when performing the functional integration, each one weighted by a θ-dependent phase. Symbolically,

$$\mathscr{D}\mathbf{A}_{\mu}^{A} = \sum_{n \in \mathbb{Z}} e^{-i\theta n} [\mathscr{D}\mathbf{A}^{A}]_{n} \tag{4.118}$$

where $[\mathscr{D}\mathbf{A}^{A}]_{n}$ indicates that the integration is performed over gauge fields in the topological class $n[\mathbf{A}^{A}] = n$. We have reobtained, in the Lagrangian language, the vacuum degeneracy found above in the canonical formalism.

The presence of the θ-term in the gauge theory action has several important physical consequences. One of them is that it violates both parity P and the combination of charge conjugation and parity CP. This will be further studied in Chap. 11.

Subtleties and Technicalities

Before closing this section we would like to mention a number of subtleties in the arguments presented concerning the structure of the gauge group. We have used the fact that $\pi_3(S^3) = \mathbb{Z}$ to characterize the number of components of the gauge group SU(2). In the argument it is crucial that the spatial topology is S^3. If this is not the case, the treatment should be refined. For instance, in *noncompact* three-dimensional Euclidean space the type of gauge transformations described by the elements of $\mathscr{G} = \{g : S^3 \to G\}$ are those approaching the identity at infinity fast enough and in a way that does not depend on angles. An equivalent way to describe them is to consider those gauge transformations that, outside a compact set surrounding the origin of coordinates, go to the identity very fast. The classes generated by these gauge transformations can be characterized by an integer number, but this may not exhaust the topological characterization of all possible nontrivial transformations.

Working on a three-dimensional box with periodic boundary conditions results in a spatial topology that is that of a three-dimensional torus T^3, and the topological structure of the mappings $\mathscr{G} = \{g : T^3 \to G\}$ is in general richer than the one described by the single winding number appearing in S^3. In this case we also have other gauge transformations not included in $\mathscr{G}_0$ and associated to the fact that the space is not simply connected. These additional transformations are physically relevant, and play an important role in 't Hooft's theory of confinement in nonabelian gauge theories. From this point of view, the topology of the space of gauge transformation often depends on the type of physical questions asked. Hence, apart from the θ angle, there may be other angles or quantum number characterizing the physical states (or the vacuum) of the theory (see for instance [15] and references therein).

To summarize, the implementation of the Gauss' law constrain and the set of physical parameter that characterize it depends on the physics and topology of the problem at hand. In the case of the θ-angle, we can introduce it by either refining our arguments on the structure of the space of nontrivial gauge transformations, or simply by arguing that the second term in (4.111) should be included because it is local, gauge and Lorentz invariant and with the same canonical dimension as the kinetic term. Extracting the dependence of physical quantities on the vacuum angle is in general a highly non-trivial problem that is not fully understood.

4.8 Instantons in Gauge Theories

The existence of multiple vacua in nonabelian gauge theories makes natural to study the possibility of tunneling between them. As explained in Sect. 2.5, in semiclassical tunneling this is described by solutions to the Euclidean field equations with finite action. For nonabelian gauge theories, the analytical continuation to imaginary times $t \to -it$, $A_0 \to iA_0$, leads to the Euclidean action

$$S_E[A_\mu] = \frac{1}{2}\int d^4x \mathrm{Tr}\Big(F_{\mu\nu}F^{\mu\nu}\Big), \tag{4.119}$$

where the indices now are lowered and raised using $\delta_{\mu\nu}$. Since we are interested in solutions to the Euclidean field equations with finite action, the gauge field $A_\mu(t, \mathbf{x})$ has to approach a pure gauge configuration both at spatial infinity $|\mathbf{x}| \to \infty$ as well as at "early" and "late" Euclidean times, $t \to \pm\infty$.

In the semiclassical evaluation of the path integral, the contribution of each saddle point comes weighted by the exponential factor $\exp\{-S_E[A_\mu]\}$, so the leading contribution is the one with the lowest value of the Euclidean action. To identify the dominant field configurations we use the following inequality valid in Euclidean space

$$0 \leqslant \mathrm{Tr}\Big[\left(F_{\mu\nu} \mp \widetilde{F}_{\mu\nu}\right)\left(F^{\mu\nu} \mp \widetilde{F}^{\mu\nu}\right)\Big] = 2\mathrm{Tr}\left(F_{\mu\nu}F^{\mu\nu}\right) \mp 2\mathrm{Tr}\left(F_{\mu\nu}\widetilde{F}^{\mu\nu}\right). \tag{4.120}$$

The combination of the inequalities for the two signs leads to the bound

$$S_E[A_\mu] \geqslant \frac{1}{2}\left|\int d^4x \mathrm{Tr}\Big(F_{\mu\nu}\widetilde{F}^{\mu\nu}\Big)\right|. \tag{4.121}$$

The right-hand side of this expression we already encountered in its Minkowskian version in Eq. (4.113). In the present setup, it is defined in four-dimensional Euclidean space and, being a total derivative, it can be written as an integral over the three-dimensional sphere at infinity, $|x| \equiv \sqrt{x^\mu x_\mu} \to \infty$. Notice that this term is independent of the metric and therefore it does not change when continued to Euclidean space, unlike the Yang–Mills action that picks up an imaginary unit in front, $S[A_\mu] \to iS_E[A_\mu]$.

Since the gauge field approaches a pure gauge when $|x| \to \infty$

$$A_\mu(x) \stackrel{|x|\to\infty}{\longrightarrow} -\frac{1}{ig_{\mathrm{YM}}} g\partial_\mu g^{-1}, \tag{4.122}$$

the integral in (4.121) is given in terms of the instanton charge Q defined by

$$Q = \frac{1}{24\pi^2} \int_{S^3_\infty} d\sigma_\mu \varepsilon^{\mu\nu\sigma\lambda} \mathrm{Tr}\Big(g\partial_\nu g^{-1} g\partial_\sigma g^{-1} g\partial_\lambda g^{-1}\Big), \tag{4.123}$$

where the integration is performed over the three-dimensional sphere at infinity. In terms of it, the action bound reads

$$\frac{1}{2} \int d^4x \mathrm{Tr}\Big(F_{\mu\nu} F^{\mu\nu}\Big) \geqslant \frac{8\pi^2}{g^2_{\mathrm{YM}}} |Q|. \tag{4.124}$$

A look at (4.120) shows that the previous inequality is saturated if an only if the Euclidean gauge field is either selfdual or anti-selfdual, namely if its field strength tensor satisfies

$$F_{\mu\nu} = \pm \widetilde{F}_{\mu\nu}. \tag{4.125}$$

Euclidean solutions satisfying these conditions are called respectively *instantons* (+ sign) and *anti-instantons* (− sign). These are the configurations dominating the Euclidean amplitudes in the semiclassical limit within each topological sector. It is important to notice that any (anti-)selfdual gauge field is automatically a solution of the Euclidean field equations: the (anti)-selfduality condition reduces the equations of motion $D_\mu F^{\mu\nu} = 0$ to the Bianchi identities,

$$\varepsilon^{\mu\nu\sigma\lambda} D_\nu F_{\sigma\lambda} = 0, \tag{4.126}$$

that are identically satisfied by any field strength tensor (the reader is invited to prove it as an exercise). Finally, it is easy to see that instantons and anti-instantons have positive and negative topological charges respectively.

We study the solutions to the selfduality equation with instanton charge $Q = 1$. To keep things simple, we consider the case of a SU(2) gauge theory. In fact this does not mean a big loss of generality: the instanton solutions for other gauge groups can be constructed in terms of their SU(2) factors. The calculation of the instanton solution is rather long and its details can be found, for example, in [16]. The result for the gauge potential in a generic gauge is

$$A^a_\mu(x) = \frac{2}{g_{\mathrm{YM}}} \frac{\eta^a_{\mu\nu}\left(x^\nu - x^\nu_0\right)}{(x - x_0)^2 + \rho^2}, \tag{4.127}$$

where $a = 1, 2, 3$ is the SU(2) index and $\eta^a_{\mu\nu}$ are the 't Hooft symbols introduced in Chap. 3 (see page 35). The field strength

$$F^a_{\mu\nu}(x) = \frac{4}{g_{\mathrm{YM}}} \frac{\eta^a_{\mu\nu}\rho^2}{[(x - x_0)^2 + \rho^2]^2} \tag{4.128}$$

is selfdual and the Euclidean action saturates the bound (4.121) with unit instanton charge

$$S_E[A_\mu] = \frac{8\pi^2}{g_{\mathrm{YM}}^2}. \tag{4.129}$$

The solution (4.127) depends on a number of arbitrary parameters: the coordinates of its center x_0^μ and the size ρ. These are part of the so-called *collective coordinates* of the instanton. They are generated by applying to a given solution the invariances of the Euclidean action, in our case translations and dilatations

$$A_\mu^a(x) \longrightarrow A_\mu^a(x+\xi), \quad A_\mu^a(x) \longrightarrow \lambda A_\mu^a(\lambda x) \tag{4.130}$$

respectively. In addition to (x_0^μ, ρ) the general instanton solution have three additional collective coordinates associated with its orientation in SU(2) space, making a total of eight collective coordinates. This number might seem smaller than expected, since the Euclidean gauge action is invariant under the full conformal group that includes the Euclidean group (rotations and translations), dilatations and special conformal transformations. The reason why rotations and special conformal transformations do not generate collective coordinates is that the two can be combined with translations, dilatations and SU(2) rotations to leave the instanton solutions invariant up to a gauge transformation. As a result only 8 of the total 18 generators [15 of the Euclidean group plus 3 of SU(2)] give rise to collective coordinates.

Finite action classical solutions to the Euclidean field equations of motion represent tunneling between different vacua of the theory (see Sect. 2.5). Next we want to show how the instanton solutions (4.127) describe indeed the semiclassical tunneling between gauge field configurations with topological numbers differing by one unit (the topological charge of the instanton). In order to make the connection with the analysis of the gauge theory vacua presented in the previous section, we have to change from the generic gauge used in writing (4.127) to the gauge $A_0^a = 0$. This is accomplished by a gauge transformation $U(t, \mathbf{x})$ satisfying

$$U(t,\mathbf{x})^{-1}\partial_0 U(t,\mathbf{x}) = -ig_{\mathrm{YM}}A_0(t,\mathbf{x}), \tag{4.131}$$

such that in this new gauge

$$\begin{aligned} A_0'(t,\mathbf{x}) &= 0 \\ \mathbf{A}'(t,\mathbf{x}) &= -\frac{1}{ig_{\mathrm{YM}}}U\nabla U^{-1} + U\mathbf{A}(t,\mathbf{x})U^{-1}. \end{aligned} \tag{4.132}$$

The general solution to the differential equation (4.131) depends on an arbitrary function of $\mathbf{x}$. This is fixed by demanding that the spatial components of the instanton in the new gauge, $\mathbf{A}'(t,\mathbf{x})$, tend to zero at early Euclidean times, $t \to -\infty$. With this condition, the gauge transformation $U(t,\mathbf{x}) = \exp(i\chi^a T^a)$ is determined to be

$$\chi^a(t,\mathbf{x}) = \frac{2\left(x^a - x_0^a\right)}{\sqrt{(\mathbf{x}-\mathbf{x}_0)^2+\rho^2}}\left[\frac{\pi}{2} + \arctan\left(\frac{t-t_0}{\sqrt{(\mathbf{x}-\mathbf{x}_0)^2+\rho^2}}\right)\right]. \tag{4.133}$$

Since the spatial components of the instanton solution (4.127) vanish as $t \to \pm\infty$, the Euclidean gauge field (4.132) approaches a pure gauge configuration both at early and late Euclidean times. Therefore it can be interpreted as interpolating between two vacua of the SU(2) gauge theory. As $t \to -\infty$ the gauge field is identically zero, whereas when $t \to \infty$ the instanton solution approach the vacuum configuration

$$A_i'(t, \mathbf{x}) \longrightarrow -\frac{1}{ig_{\rm YM}} g(\mathbf{x})\partial_i g(\mathbf{x})^{-1} \tag{4.134}$$

with

$$g(\mathbf{x}) \equiv \lim_{t\to+\infty} U(t, \mathbf{x}) = \exp\left[\frac{2\pi i(x^a - x_0^a)}{\sqrt{(\mathbf{x}-\mathbf{x}_0)^2 + \rho^2}} T^a\right]. \tag{4.135}$$

This, unlike the $\mathbf{A}(t, \mathbf{x}) = \mathbf{0}$ vacuum in the asymptotic Euclidean "past", is a gauge configuration with nonvanishing topological number, namely [cf. equation (4.116)]

$$n[\mathbf{A}'] = \frac{1}{24\pi^2}\int d^3x \varepsilon_{ijk} \mathrm{Tr}\Big[\left(g\partial_i g^{-1}\right)\left(g\partial_i g^{-1}\right)\left(g\partial_i g^{-1}\right)\Big] = 1. \tag{4.136}$$

The final conclusion of our analysis is that the instanton solution (4.127) describes the tunneling from a gauge theory vacuum with vanishing winding number to a nontrivial vacuum with winding number equal to one, the difference being equal to the topological charge of the instanton. A similar analysis can be repeated for anti-instanton solutions, obtained from (4.127) by replacing the 't Hooft symbols by their duals $\overline{\eta}^a_{\mu\nu}$ [see Eq. (3.11)]. They have instanton charge $Q = -1$ and interpolate between gauge theory vacua with winding numbers that differ by this amount.

(Anti-)Instanton contributions to physical quantities are weighted by

$$\exp\left(-\frac{8\pi^2}{g_{\rm YM}^2}|Q|\right). \tag{4.137}$$

This factor is nonanalytic around $g_{\rm YM} = 0$, showing that the effect of tunneling between different gauge theory vacua is truly nonperturbative. We see that at weak coupling instanton effects are exponentially suppressed and therefore overshadowed by any perturbative contribution to the same process, that necessarily scales with a positive power of $g_{\rm YM}$. This is the reason why instantons are mostly relevant in physical situations where perturbative terms are known to be zero.

References

1. Aharonov, Y., Bohm, D.: Significance of the electromagnetic potentials in the quantum theory. Phys. Rev. **115**, 485 (1955)
2. Ehrenberg, W., Siday, R.E.: The refractive index in electron optics and the principles of dynamics. Proc. Phys. Soc. B **62**, 8 (1949)

3. Dirac, P.A.M.: Quantised singularities in the electromagnetic field. Proc. Roy. Soc. **133**, 60 (1931)
4. Schwinger, J.: Sources and magnetic charge. Phys. Rev. **173**, 1544 (1968)
5. Zwanziger, D.: Exactly soluble nonrelativistic model of particles with both electric and magnetic charges. Phys. Rev. **176**, 1480 (1968)
6. 't Hooft, G.: Magnetic monopoles in unified gauge theories. Nucl. Phys. B **79**, 276 (1974)
7. Polyakov, A.M.: Particle spectrum in the quantum field theory. JETP Lett **20**, 194 (1974)
8. Dirac, P.A.M.: Lectures on Quantum Mechanics, Dover, NY (2001)
9. Henneaux, M., Teitelboim, C.: Quantization of Gauge Systems, Princeton Press, Princeton,(1992)
10. Jackiw, R.: Quantum meaning of classical field theory. Rev. Mod. Phys. **49**, 681 (1977)
11. Jackiw, R.: Introduction to the Yang–Mills quantum theory. Rev. Mod. Phys. **52**, 661 (1980)
12. Jackiw, R.: Topological investigations of quantized gauge theories. In: DeWitt, B.S., Stora, R. (eds) Relativité, groupes et topologie II. Elsevier, London (1984)
13. Faddeev, L.D., Popov, V.N.: Feynman diagrams for the Yang–Mills field. Phys. Lett. B **25**, 29 (1967)
14. Nakahara, M.: Geometry, topology and physics. Institute of Physics, UK (1990)
15. 't Hooft, G.: The topological mechanism for permanent quark confinement in a Nonabelian gauge theory. Phys. Scripta **25**, 133 (1982)
16. Rubakov, V.: Theory of Gauge Fields, Princeton Press, Princeton (1999)

Chapter 5
Theories and Lagrangians III: The Standard Model

The previous two chapters were devoted to introducing the basic ingredients necessary in building up a physical description of elementary particles: the fermion matter fields and the gauge fields responsible for the interactions. The time has come to combine these elements into a description of the physics of elementary particles. The result will be the standard model.

In the next sections we are going to summarize the basic features of the standard model, also called the Glashow–Weinberg–Salam theory [1–3]. Our presentation here, however, will leave one important problem open: how particle masses in the standard model can be made compatible with gauge invariance. The missing ingredient to solve this problem, spontaneous symmetry breaking, will have to wait until Chap. 7. The presentation will remain mostly qualitative. The details of the construction of the standard model and a full study of its consequences for the phenomenology of elementary particles can be found in many textbooks (for example [4–8]).

5.1 Fundamental Interactions

Most of the phenomena we witness in our daily life can be explained in terms of two fundamental forces: gravity and electromagnetism. They are the only relevant interactions in a very wide range of phenomena that goes from the dynamics of galaxies to atomic and solid state physics.

These two interactions, however, do not suffice to give an account of all subnuclear physics. Gravity is indeed too weak to be of any relevance at the atomic level. The laws of electromagnetism, on the other hand, offer no explanation as to how a large number of positively charged protons can be confined in nuclei with a size of the order of 10^{-15} m. QED does not provide either any mechanism that could explain nuclear processes such as beta decay. These phenomena require invoking two nuclear interactions: a "strong" one responsible for binding protons and neutrons together in the atomic nuclei, and a "weak" one that, without producing bound states, accounts for nuclear disintegrations.

L. Álvarez-Gaumé and M. Á. Vázquez-Mozo, *An Invitation to Quantum Field Theory*,
Lecture Notes in Physics 839, DOI: 10.1007/978-3-642-23728-7_5,

To understand how the relevant interaction can be identified in subnuclear processes we need to recall some basic ideas from quantum mechanics. Take a system in a quantum state of energy E, $|\psi_E\rangle$. Let us assume that this state decays as a consequence of the interaction Hamiltonian H_{int}. Then, the lifetime τ of the state is equal to the inverse of its width Γ that, in turn, can be computed using Fermi's golden rule

$$\Gamma = 2\pi \sum_f \rho_f(E)|\langle f|H_{\text{int}}|\psi_E\rangle|^2. \tag{5.1}$$

Here the sum is over final states and $\rho_f(E)$ is the density of such states with energy E. The key point is that, generically, the overlap $\langle f|H_{\text{int}}|\psi_E\rangle$ is proportional to a power of the coupling constant (i.e., the charge) of the interaction involved in the process. Thus, the bottom line is that the lifetime of a quantum state is, roughly speaking, inversely proportional to the strength of the interaction responsible for its decay.

In high energy physics, this provides a good guiding principle to identify the interaction behind a decay process: the hierarchy in the strength of the three interactions should be reflected in a hierarchy of the characteristic times of the processes they mediate. This is indeed what happens. Strong interaction decays are characterized by very short lifetimes of the order

$$\tau_{\text{strong}} \sim 10^{-23}\ \text{s}. \tag{5.2}$$

Next in the hierarchy come electromagnetic processes, for which

$$\tau_{\text{em}} \sim 10^{-16}\ \text{s}. \tag{5.3}$$

Finally, the weak interaction is behind processes with typical times substantially longer than the ones above

$$\tau_{\text{weak}} \sim 10^{-8} - 10^{-6}\ \text{s}, \tag{5.4}$$

with some decays, such as the neutron β-decay, reaching characteristic times of the order of minutes.

Electromagnetic processes are described quantum mechanically using quantum electrodynamics (see Chap. 4). As for the strong and weak interactions, before entering into the details of their quantum field theory description we need to learn some basic facts about their phenomenology.

Strong Interaction

Let us begin with the strong interaction. The class of subatomic particles that feel the strong force, collectively denoted as *hadrons*, are classified in two types depending on their spin: *baryons* with half-integer spin (e.g. the proton and the neutron) and *mesons* with integer spin (e.g. the pions).

Approximate symmetries are a very useful tool in the study of physical processes mediated by the strong interaction. The best-known example is the isospin symmetry familiar from nuclear physics. Indeed, strong interactions do not seem to distinguish very much between protons and neutrons although both the weak and electromagnetic interactions do. This is shown by the similar energy levels of the so-called mirror nuclei, those related by replacing one or more protons by neutrons such as ^{11}B and ^{11}C. The slight differences in the spectrum of these nuclei can be explained by the small mass split between the proton and the neutron and by their different values for the electric charge.

That the strong interaction alone cannot tell apart neutrons from protons is codified in mathematical terms in a global $\mathrm{SU}(2)_I$ isospin symmetry that rotates these two particles into one another. Protons and neutrons form a doublet with isospin $I = \frac{1}{2}$ and third components $I_3(p) = \frac{1}{2}$ and $I_3(n) = -\frac{1}{2}$. The scheme is extended to other particles, such as the three pions π^0, $\pi^{\pm}$, that form an isospin triple ($I = 1$) where $I_3(\pi^{\pm}) = \pm 1$, $I_3(\pi^0) = 0$.

All this notwithstanding, isospin remains only an approximate symmetry of the strong force even after switching off both the electromagnetic and the weak interactions. This follows from the small but nonvanishing difference between the masses of the particles within an isospin multiplet. Isospin is nevertheless useful because the mass splitting is much smaller than the particle masses themselves and the symmetry breaking effects are small.

Besides isospin, the strong interaction preserves other quantum numbers, such as strangeness S. Adding this quantum number to isospin it is possible to extend $\mathrm{SU}(2)_I$ to the flavor $\mathrm{SU}(3)_f$ global symmetry. Strongly interacting particles are then classified in irreducible representations of this group: singlets, octets and decuplets but, interestingly, not triplets [the fundamental and antifundamental representations of $\mathrm{SU}(3)_f$]. To illustrate this we see that the isodoublet formed by the proton and the neutron is embedded into a $\mathrm{SU}(3)_f$ octet that also includes an isotriplet ($\Sigma^{\pm}$, Σ^0) with $S = -1$ and the isodoublet (Ξ^-, Ξ^0) with $S = -2$. We can get an idea of the accuracy of this approximate symmetry by noticing that

$$m(p,n) \approx 930\,\mathrm{MeV}, \quad m(\Sigma) \approx 1190\,\mathrm{MeV}, \quad m(\Xi) \approx 1320\,\mathrm{MeV}. \tag{5.5}$$

The mass split between the states with different strangeness in the octet is about 30% of the average mass, much larger than the 0.1–0.7% mass split within each isospin multiplet. Similarly, the addition of the lowest-lying strange mesons to the pion isotriplet results in the $\mathrm{SU}(3)_f$ octet and singlet shown in Fig. 5.1. Very soon we will learn how these approximate symmetries reflect the inner structure of the hadrons.

An important relation is the Gell-Mann–Nishijima formula giving the electric charge of a strong-interacting particle in terms of its third isospin component and strangeness

$$Q = I_3 + \frac{B+S}{2}, \tag{5.6}$$

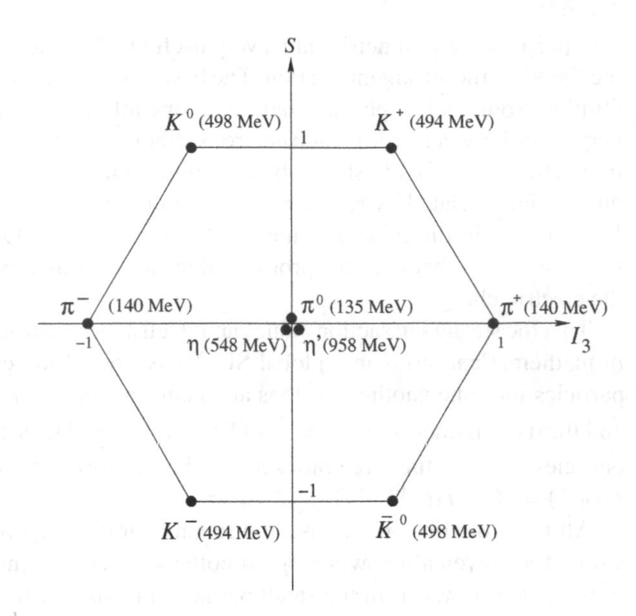

Fig. 5.1 The lowest-lying pseudoscalar mesons. The masses of the particles are indicated in parenthesis

where B is the baryon number, that takes the values $B = +1$ for baryons, $B = -1$ for antibaryons and $B = 0$ for mesons. The combination $Y = \frac{1}{2}(B + S)$ defines the strong hypercharge that is conserved in strong interaction processes.

Weak Interaction

After gravity, the weak interaction is the most universal force in Nature since every known matter particle takes part in it. This includes all hadrons as well as a number of nonhadronic particles called *leptons*. Although the weak interactions do not produce bound states, it is behind very important physical processes such as neutron beta decay

$$n \longrightarrow p + e^- + \overline{\nu}_e, \tag{5.7}$$

responsible for the radioactive disintegration of nuclei.

Neutron beta decay is an example of a process mediated by a so-called weak charged current: the hadronic (n, p) and leptonic $(e^-, \overline{\nu}_e)$ pairs contain particles whose electric charges differ in one unit. Another example of this kind of processes is provided by muon decay

$$\mu^- \longrightarrow e^- + \overline{\nu}_e + \nu_\mu. \tag{5.8}$$

Here the two pairs formed by the leptons of the same flavor, $(e^-, \overline{\nu}_e)$ and (μ^-, ν_μ), are composed of particles of different charge. Weak processes can also proceed through weak neutral currents in which the hadrons or the same-flavor leptons do not change their electric charge. One example is electron-neutrino scattering

$$e^- + \nu_\mu \longrightarrow e^- + \nu_\mu, \tag{5.9}$$

where the particles in each of the two same-flavor lepton pairs have the same electric charge.

One of the most distinctive features of the weak interaction is that it violates what once were cherished discrete symmetries. In the dominant decay channel of the negatively-charged pion into a muon and a muonic neutrino

$$\pi^- \longrightarrow \overline{\nu}_\mu + \mu^-, \tag{5.10}$$

it is experimentally observed that the muon is always emitted with positive helicity (i.e., it is right-handed). Since parity reverses the helicity of the particle, this result indicates that parity is violated by the weak interaction. Moreover, this violation is maximal because *all* muons emitted in the π^- decay are right-handed. This shows that any field-theoretical description of the weak interaction must necessarily be chiral, that is, the weak interaction coupling of the fermions should depend on their helicities. This feature singles out weak interaction among the fundamental forces in that it is the only one that distinguishes left from right. Why this is the case remains a mystery.

Charge conjugation, denoted by C, is a discrete operation that interchanges particles with their antiparticles. The properties of this discrete symmetry will be studied in detail in Chap. 11. Here we only need to know that the decay of the positively-charged pion is obtained by charge-conjugating (5.10)

$$\pi^+ \longrightarrow \nu_\mu + \mu^+. \tag{5.11}$$

An important property of the operation C is that it changes particles by antiparticles but does not modify the helicity of the fermions. This means that if charge conjugation is a symmetry of the weak interaction, the decay of the π^+ has to proceed by emission of a right-handed antimuon. Experimentally, however, it is observed that the antimuon emitted by the decaying pion is *always* left-handed! This shows that weak interactions not only violate parity but also charge conjugation and that this violation is also maximal.

This is not the end of the story. Not only P and C are violated by the weak interaction, but also their combination CP. How this happens is however more subtle (see Sect. 11.5).

5.2 Leptons and Quarks

One of the glaring features of the host of particles produced in high energy collisions is that there is only a small number of them that do not feel the strong nuclear force. The list is made of the following six leptons

e^-	electron	$q_e = -1$	ν_e	electron neutrino	$q_{\nu_e} = 0$
μ^-	muon	$q_\mu = -1$	ν_μ	muon neutrino	$q_{\nu_\mu} = 0$
τ^-	tau	$q_\tau = -1$	ν_τ	tau neutrino	$q_{\nu_\tau} = 0$

and their corresponding antiparticles. The rest of the over one hundred particles and resonances listed in the Review of Particle Physics [9], partake in physical processes mediated by the strong interaction.

Unlike the case of the leptons, the large number of hadronic particles strongly hints to them being composites of more fundamental objects. This idea is supported by the experimental evidence showing that hadrons are "extended" and have an internal structure. This is best seen in deep inelastic scattering where a hadron (typically a proton) is made to collide with a lepton (an electron, muon or neutrino). These processes are called inelastic because the hadron, as the result of the collision, is smashed into a bunch of hadrons. For example,

$$e^- + p \longrightarrow e^- + \text{hadrons}.$$

The incoming particles interact either electromagnetically or through the weak interaction. In either case the interchanged quanta probe the hadron with a resolution given by the inverse of the transferred momenta. The data obtained in these experiments is consistent with the interaction of the probe quanta with pointlike objects inside the hadron. In Sect. 5.3 we will see how the study of these processes provides plenty of useful information about the physical properties of the strong interaction. For the time being it suffices to know that they show that hadrons are made of pointlike objects.

In fact, the spectrum of hadrons can be reproduced by assuming that they are composed of particles with spin $\frac{1}{2}$ and fractional charge, called *quarks*. By simple addition of angular momentum we realize that the distinction between mesons and baryons comes out naturally. The first are bound states of a quark and an antiquark, whereas the second are composed of three quarks. All known hadrons can thus be explained as bound states of six different quarks. The quark types, called *flavors*, are conventionally denoted by the following names

u	up	$q_u = \frac{2}{3}$	d	down	$q_c = -\frac{1}{3}$
c	charm	$q_c = \frac{2}{3}$	s	strange	$q_s = -\frac{1}{3}$
t	top	$q_t = \frac{2}{3}$	b	bottom	$q_t = -\frac{1}{3}$

As a matter of fact, the top quark is too short-lived to give rise to bound states. Nevertheless it can be produced in the high energy collisions of protons, where its existence was verified in 1995 through the observation of its decay channels. One of the most remarkable properties of quarks is that, unlike leptons, they have fractional electric charge. Notice that, however, the charge of the bound state of a quark and an antiquark or of three quarks always results in a state with integer charge.

Many features of the hadronic spectrum can be predicted using the nonrelativistic quark model, where the quarks are taken to be nonrelativistic particles. In this model, the hadron wave function is constructed in terms of the wave functions of the constituent quarks. Thus, some quantum numbers of the hadrons can be obtained by doing "spectroscopy", in a similar fashion as it is done in atomic physics.

To see how this works we consider as an example the lightest hadrons composed only by the u and d quarks. We begin with the mesons for which we have four independent states in flavor space: $|u\overline{u}\rangle$, $|u\overline{d}\rangle$, $|d\overline{u}\rangle$ and $|d\overline{d}\rangle$. They have to be identified with the four lowest lying mesons, the pions $\pi^{\pm}$, π^{0} and the η meson. To identify who is who in this case, we begin by looking at the electric charge. This allow us to identify the flavor wave function of the charged pions as

$$|\pi^{+}\rangle = |u\overline{d}\rangle, \quad |\pi^{-}\rangle = |d\overline{u}\rangle. \tag{5.12}$$

The wave function of the two neutral mesons π^{0} and η, on the other hand, should be orthogonal combinations of the chargeless states $|u\overline{u}\rangle$ and $|d\overline{d}\rangle$. To identify them we need to invoke another quantum number that distinguishes betweeen the two particles. This is isospin. The neutral pion belongs, together with $\pi^{\pm}$, to a isospin triplet with $I_3(\pi^{0}) = 0$, whereas the η is an isospin singlet.

We have to assign then isospin quantum numbers of the u and d quarks. They are grouped together into an isodoublet transforming under isospin in the fundamental representation of SU(2), that is, $I_3(u) = \frac{1}{2}$ and $I_3(d) = -\frac{1}{2}$. With this choice we see that the flavor wave functions shown in (5.12) have the required isospin, $I_3(\pi^{\pm}) = \pm 1$. As for the third member of the $I = 1$ triplet, we have to decompose the product of two fundamental representations of SU(2) into irreducible representations. Using the rules familiar from the angular momentum algebra in quantum mechanics, we find the wave function of the neutral pion to be

$$|\pi^{0}\rangle = \frac{1}{\sqrt{2}}\left(|u\overline{u}\rangle - |d\overline{d}\rangle\right). \tag{5.13}$$

Since the pions have zero spin, the total (flavor + spin) wave function is the tensor product of (5.12) and (5.13) with the spin wave function

$$|s = 0\rangle = \frac{1}{\sqrt{2}}(|\uparrow\downarrow\rangle - |\downarrow\uparrow\rangle). \tag{5.14}$$

Having studied the mesons we proceed to the baryons, starting with the proton and the neutron. By just looking at the electric charge of these particles we see that their quark composition has to be uud and udd respectively. However, the obvious choice for the proton and neutron wave functions, $|uud\rangle$ and $|udd\rangle$, are not good candidates. The reason is that these states are eigenstate of the third component of the isospin I_3 but not of the total isospin I^2. Indeed, for the case of the proton the states with well defined total isospin are[1]

$$\begin{aligned}
|uud\rangle_S &= \frac{1}{\sqrt{6}}\left(|uud\rangle + |udu\rangle - 2|duu\rangle\right), \\
|uud\rangle_A &= \frac{1}{\sqrt{2}}\left(|uud\rangle - |udu\rangle\right).
\end{aligned} \tag{5.15}$$

[1] Here we have to remember that the isospin operators acting on the Hilbert space of three particles have the form $I_i = I_i^{(1)} \otimes \mathbf{1} \otimes \mathbf{1} + \mathbf{1} \otimes I_i^{(2)} \otimes \mathbf{1} + \mathbf{1} \otimes \mathbf{1} \otimes I_i^{(3)}$, where $I_i^{(a)}$ is the isospin operator acting on the Hilbert space of the a-th particle.

Both states have $I = \frac{1}{2}$, $I_3 = \frac{1}{2}$. The subscripts indicate that the states are symmetric and antisymmetric with respect to the interchange of the two last states. The proton is in fact a linear combination of these two states. To find the precise one we need to take into account that the total wave function, including the spin degrees of freedom, has to be antisymmetric under the interchange of any two quarks. Taking this into account we have

$$|p\uparrow\rangle = \frac{1}{\sqrt{2}}\left(|uud\rangle_S \otimes |\Uparrow\rangle_A + |uud\rangle_A \otimes |\Uparrow\rangle_S\right),$$
$$|p\downarrow\rangle = \frac{1}{\sqrt{2}}\left(|uud\rangle_S \otimes |\Downarrow\rangle_A + |uud\rangle_A \otimes |\Downarrow\rangle_S\right). \tag{5.16}$$

The spin states $|\Uparrow\rangle_{A,S}$, $|\Downarrow\rangle_{A,S}$ are eigenstates of the total spin (with $s = \frac{1}{2}$) and its third component ($s_z = \pm\frac{1}{2}$), the subscripts indicating again that the wave functions are symmetric and antisymmetric in the last two states. For example, for the spin-up states we have

$$|\Uparrow\rangle_S = \frac{1}{\sqrt{6}}\left(|\uparrow\uparrow\downarrow\rangle + |\uparrow\downarrow\uparrow\rangle - 2|\uparrow\uparrow\downarrow\rangle\right),$$
$$|\Uparrow\rangle_A = \frac{1}{\sqrt{2}}\left(|\uparrow\downarrow\uparrow\rangle - |\downarrow\uparrow\uparrow\rangle\right). \tag{5.17}$$

A similar analysis can be carried out for the neutron, whose flavor wave function is written in terms of the states $|ddu\rangle_{S,A}$ which have $I = \frac{1}{2}$, $I_3 = -\frac{1}{2}$.

Protons and neutrons are not the only hadrons made out of u and d quarks. By simple counting we see that there are $2^3 = 8$ possible baryon states. Keeping in mind that quarks transform in the fundamental representation of the $\mathrm{SU}(2)_I$ isospin group, these states are classified by the irreducible representations contained in the product representation

$$\mathbf{2}\otimes\mathbf{2}\otimes\mathbf{2} = \mathbf{4}\oplus\mathbf{2}_S\oplus\mathbf{2}_A. \tag{5.18}$$

The subscript in the last two terms indicates that these irreducible representations act on the spaces spanned by $\{|uud\rangle_S, |ddu\rangle_S\}$ and $\{|uud\rangle_A, |ddu\rangle_A\}$ respectively. The states transforming under the $\mathbf{4}$ are identified with the four Δ resonances: Δ^{++} (uuu), Δ^{+} (uud), Δ^{0} (udd) and Δ^{-} (ddd). They form an isoquadruplet with $I = \frac{3}{2}$. Notice that although Δ^+ and Δ^0 have the same quark composition as the proton and the neutron respectively, they differ in the spin, which is $S = \frac{3}{2}$ for the delta resonances. Their wave functions in flavor and spin spaces can be obtained along the lines showed above for the proton and the neutron.

The hadron spectroscopy described so far can be extended to include hadrons with nonvanishing strangeness. In the context of the quark model these are particles which contain a net number of s quarks. This quark has strangeness $S = -1$ and is an isospin singlet. The $\mathrm{SU}(2)_I$ isospin group is extended to flavor $\mathrm{SU}(3)_f$, where the three quarks u, d and s form a triplet that transforms in the fundamental representation **3**, with antiquarks transforming in the complex conjugate representation $\overline{\mathbf{3}}$. With this we can explain the hadron classification discussed in Sect. 5.1: the group theory identity

$$\mathbf{3} \otimes \overline{\mathbf{3}} = \mathbf{8} \oplus \mathbf{1} \tag{5.19}$$

means that the lightest mesons (including those with nonvahinish strangeness) come in octets and singlets, whereas baryons are classified in decuplets, octets and singlets according to

$$\mathbf{3} \otimes \mathbf{3} \otimes \mathbf{3} = \mathbf{10} \oplus \mathbf{8} \oplus \mathbf{8} \oplus \mathbf{1}. \tag{5.20}$$

The quark model gives a rationale for the existence of the approximate flavor symmetries of the strong interaction. The nine pseudoscalar mesons shown in Fig. 5.1 are the states on the right-hand side of the decomposition (5.19). The group theory analysis shows that the quark composition of the kaons is

$$|K^+\rangle = |u\overline{s}\rangle, \quad |K^0\rangle = |d\overline{s}\rangle. \tag{5.21}$$

In addition to the kaons, the multiplet also contains two more particles, η and η', with $I = 0$ and $S = 0$. The identification of the flavor wave function of these states requires a bit of extra work.

On purely group theoretical grounds, there are two possible ways to construct a state with vanishing isospin out of a quark and an antiquark triplet, namely

$$\begin{aligned} |\eta_1\rangle &= \frac{1}{\sqrt{3}}\left(|u\overline{u}\rangle + |d\overline{d}\rangle + |s\overline{s}\rangle\right), \\ |\eta_8\rangle &= \frac{1}{\sqrt{6}}\left(|u\overline{u}\rangle + |d\overline{d}\rangle - 2|s\overline{s}\rangle\right). \end{aligned} \tag{5.22}$$

With the subscript on the left-hand side we have indicated that the states come respectively from the singlet and the octet of $\mathrm{SU}(3)_f$. However, the identification of (5.22) with observed particles has to be done with care. Were $\mathrm{SU}(3)_f$ an exact symmetry of the strong interactions, $|\eta_1\rangle$ and $|\eta_8\rangle$ would be eigenstates of the Hamiltonian of the strong force. But we know that $\mathrm{SU}(3)_f$ is only an approximate symmetry and therefore time evolution mixes these two states. In fact, there are two particles, η and η', with the correct quantum numbers that are a mixture of the states (5.22)

$$\begin{aligned} |\eta\rangle &= \cos\theta_P|\eta_8\rangle - \sin\theta_P|\eta_1\rangle, \\ |\eta'\rangle &= \sin\theta_P|\eta_8\rangle + \cos\theta_P|\eta_1\rangle. \end{aligned} \tag{5.23}$$

The pseudoscalar mixing angle θ_P is experimentally found to be $\theta_P \simeq -17^\circ$.

We have encountered a general phenomenon called *mixing*. This happens whenever the propagation eigenstates (i.e., states with a well-defined mass) do not coincide with other quantum number eigenstates (in this case the flavor quantum number), and it is at the origin of many interesting phenomena in particle physics.

Looking back at the lowest lying mesons shown in Fig. 5.1, we immediately notice the rather small mass difference between the three pions. In the context of the quark model this experimental fact can be interpreted as indicating that the masses of the u and d quarks should be very similar. Using the same argument, the mass difference between pions and kaons hints to a larger mass for the strange quark, $m_s > m_u \simeq m_d$. This conclusion, however, has to be taken with a grain of salt: as quarks are confined inside the hadrons talking about their masses is a very delicate issue that we will elaborate upon in Chap. 10 (see Sect. 10.2).

5.3 Quantum Chromodynamics

The failure to detect isolated quarks indicates that some physical mechanism should be responsible for their confinement inside hadrons. This property of the quark interaction contrasts very much with the picture of the quark–quark interaction that emerges from the deep inelastic scattering experiments already discussed in the previous section. One of the surprising conclusions following from the study of these collisions is that the data extracted is compatible with the quarks inside the hadrons behaving as nearly free particles. More precisely, the results can be reproduced assuming that while the lepton interacts with the nucleon constituents, to a very good approximation these constituents can be considered as not interacting with each other.

This means that a successful theory of the strong interaction should account for these two curious features of the quark interaction force: it should grow at large distances in order to prevent quarks from being "ionized" out of the hadrons, while at the same time it should be negligible when the quarks are within a distance well below the nucleon radius, i.e., approximately 10^{-15} m.

The very implementation of the quark model leads to the realization that quarks have an extra quantum number beyond flavor and spin. This is most easily seen in the case of the Δ^{++}. As we discussed above, this resonance is made out of three u quarks and has total spin $s = \frac{3}{2}$. Then, its wave function with $s_z = \frac{3}{2}$ has to be

$$|\Delta^{++};\ s_z = \tfrac{3}{2}\rangle = |uuu\rangle \otimes |\uparrow\uparrow\uparrow\rangle \equiv |u\uparrow, u\uparrow, u\uparrow\rangle. \tag{5.24}$$

As it stands, the wave function is symmetric under the interchange of any of the identical three quarks. This is indeed a problem, the quarks are fermions and therefore their total wave function has to be completely antisymmetric. One way to avoid the problem is if each quark has an extra index taking three values, u_i with $i = 1, 2, 3$. Then, the wave function

$$|\Delta^{++};\ s_z = \tfrac{3}{2}\rangle = \frac{1}{\sqrt{3!}}\varepsilon_{ijk}|u_i\uparrow, u_j\uparrow, u_k\uparrow\rangle. \tag{5.25}$$

is antisymmetric under the interchange of any of the constituent quarks, as required by their fermionic statistics. This new quantum number is called *color*. The conclusion we have reached is that each quark flavor comes in tree different states labeled by this new index.

The color quantum number is the key to the formulation of a theory of strong interaction able to account for the phenomenology. This theory is called Quantum Chromodynamics (QCD) and is a nonabelian gauge theory based on the gauge group SU(3). This group acts on the color index of the quark spinor field as

$$Q_i^f \longrightarrow U(g)_{ij} Q_j^f, \quad \text{with} \quad g \in \text{SU}(3), \tag{5.26}$$

where $f = 1, \ldots, 6$ runs over the six quark flavors and $U(g)$ is an element of the fundamental representation of the gauge group. The Lagrangian of the theory can be constructed using what we learned in Sect. 4.4

$$\mathscr{L}_{\text{QCD}} = -\frac{1}{4} F^a_{\mu\nu} F^{a\mu\nu} + \sum_{f=1}^{6} \overline{Q}^f \left(i \not{D} - m_f\right) Q^f. \tag{5.27}$$

To keep the notation simple we have omitted the color indices. The nonabelian gauge field strength $F^a_{\mu\nu}$ (with $a = 1, \ldots, 8$) and the covariant derivative D_μ are given in terms of the SU(3) gauge field A^a_μ by (4.52) and (4.46) respectively. In the latter case the generators T^a_R are the Gell–Mann matrices listed in Eq. (B.16).

The QCD Lagrangian (5.27) leads to a theory where the interactions between quarks have the features required to explain both quark confinement and the deep inelastic scattering experiments. Unfortunately, at this point we cannot be more explicit. We still have to learn how to quantize an interacting field theory such as QCD. The most we can say now is that quantum effects result in an effective force between quarks that grows at large distances, whereas it tends to zero at short distances. The clarification of this statement will have to wait until Chap. 8.

From the point of view of the quark model it seems rather arbitrary that hadrons result form bound states of either a quark and an antiquark or of three quarks. Why not, say, having hadrons made of two quarks? QCD offers an explanation of this fact. What happens is that hadrons are colorless objects, i.e., they transform as singlets under SU(3). Then, since quarks (resp. antiquarks) transform under the fundamental $\mathbf{3}_c$ (resp. antifundamental $\overline{\mathbf{3}}_c$) of SU(3), it is impossible to produce a colorless object out of two quarks

$$\mathbf{3}_c \otimes \mathbf{3}_c = \mathbf{6}_c \oplus \overline{\mathbf{3}}_c. \tag{5.28}$$

Here, to avoid confusion with the notation of previous sections, we have introduced a subscript to indicate that we are referring to irreducible representations of color SU(3). On the other hand, using the identities

$$\begin{aligned} \mathbf{3}_c \otimes \overline{\mathbf{3}}_c &= \mathbf{8}_c \oplus \mathbf{1}_c, \\ \mathbf{3}_c \otimes \mathbf{3}_c \otimes \mathbf{3}_c &= \mathbf{10}_c \oplus \mathbf{8}_c \oplus \mathbf{8}_c \oplus \mathbf{1}_c, \end{aligned} \tag{5.29}$$

we find that there is no problem in constructing colorless mesons and baryons. One example is the Δ^{++} wave function shown in equation (5.25). Notice that on purely group theoretical grounds there are ways other than (5.29) of producing color singlets. For example, the product of four fundamental and one antifundamental representations of SU(3) contains several singlets. These exotic baryons, however, have not been observed experimentally to date.

QCD includes, besides the six quarks, eight gauge fields mediating the strong interaction, one for each generator of SU(3). These intermediate vector bosons are the *gluons*. It is rather counterintuitive that a short-ranged force such as the strong interaction is mediated by massless particles. However, we have to recall that the strong nuclear force that we referred to in Sect. 5.1 is a force between colorless hadrons. The nuclear force between nucleons emerges as a residual interaction very much in the same fashion as the van der Waals force does in molecular physics between electrically neutral atoms, where the Coulomb force produces a residual potential falling off as r^{-6}. The problem is that in the case of QCD the complexity of the theory makes it very difficult to give a concrete form to this general idea. In spite of recent progresses [10], there is still no precise understanding of how nuclear effective potentials emerge from the gluon-mediated QCD interaction between quarks.

The approximate symmetries of the strong interaction are (approximate) global symmetries of the QCD Lagrangian. Focusing on the two lightest quark flavors, u and d, the fermionic part of this Lagrangian can be written as

$$\mathscr{L} = (\overline{u}, \overline{d}) \begin{pmatrix} i\not{D} - \frac{m_u+m_d}{2} & 0 \\ 0 & i\not{D} - \frac{m_u+m_d}{2} \end{pmatrix} \begin{pmatrix} u \\ d \end{pmatrix} - \frac{m_u - m_d}{2} (\overline{u}, \overline{d}) \begin{pmatrix} 1 & 0 \\ 0 & -1 \end{pmatrix} \begin{pmatrix} u \\ d \end{pmatrix}. \tag{5.30}$$

In the limit when $m_u \simeq m_d$, the second term can be ignored and the Lagrangian is approximately invariant under the global $SU(2)_I$ isospin transformations

$$\begin{pmatrix} u \\ d \end{pmatrix} \longrightarrow M \begin{pmatrix} u \\ d \end{pmatrix}, \tag{5.31}$$

where M is a SU(2) matrix. Acting on the flavor wave function of the nucleons and the pions this gives the usual isospin transformation. In a similar fashion, $SU(3)_f$ can be seen to emerge from the approximate global symmetry of the QCD Lagrangian in the limit in which the mass differences between the masses of the u, d and s are neglected. As in (5.31), these transformation acts linearly on the quark triplet.

5.4 The Electroweak Theory

At low energies weak processes such as those described in Sect. 5.1 can be phenomenologically described by interaction terms of the form

$$\mathscr{L}_{\text{int}} = \frac{G_F}{\sqrt{2}} J^\mu(x) J_\mu(x)^\dagger. \tag{5.32}$$

The dimensionful coupling constant G_F, called the Fermi constant, has the value

$$G_F = 1.166 \times 10^{-5}\,\text{GeV}^{-2}. \tag{5.33}$$

The current in the Lagrangian (5.32) is split into hadronic and leptonic contributions, $J_\mu(x) = J_\mu^{(h)}(x) + J_\mu^{(\ell)}(x)$. As hadrons are composite objects, the hadronic current has to be expressed in terms of form factors. The leptonic current, on the other hand, is written in terms of the lepton fields as

$$J_\mu^{(\ell)} = \overline{\nu}_e(x)\gamma_\mu(1-\gamma_5)e(x) + \cdots \tag{5.34}$$

The dots stand for other fields. Currents like (5.34) are known as charged currents because the two fields forming it have electric charges that differ by one unit. We also have contributions to the Lagrangian coming from neutral currents made of like-charge leptons. It is important that all the terms appearing in the lepton current have the so-called V–A form including the chirality-sensitive factor $1-\gamma_5$. This is imposed by the fact that weak interactions maximally violate parity.

The interaction Lagrangian (5.32) describes weak interaction processes very successfully at low energies. However, for various reasons the theory runs into trouble when the energy gets close to the characteristic energy scale $1/\sqrt{G_F}$.

A way to deal with these problems is to give up the "contact" interaction (5.32) in favor of an intermediate boson, in analogy with QED or QCD. The only problem is that the intermediate boson now has to be massive if we want to recover the effective current–current interaction at low energies. This we can illustrate with a simple toy model of a massive abelian gauge field coupled to a real current J_μ

$$\mathscr{L}_{\text{int}} = -\frac{1}{4}F_{\mu\nu}F^{\mu\nu} + \frac{m^2}{2}A_\mu A^\mu + gJ_\mu A^\mu, \tag{5.35}$$

with g a dimensionless coupling constant. At energies below the mass m, the kinetic term of the gauge field is subleading with respect to the mass term. Solving the equations of motion for A_μ in this limit, and substituting the result in (5.35), we arrive at the low energy "contact" interaction

$$\mathscr{L}_{\text{int}} = \frac{g^2}{2m^2}J_\mu J^\mu. \tag{5.36}$$

Extrapolation of this result to the weak interaction leads to the conclusion that both charged and neutral weak currents are mediated by *massive* gauge bosons.[2]

The construction of a theory of weak interactions based on the interchange of vector bosons leads in fact to the unification of the weak and electromagnetic interaction based on a gauge theory with gauge group $\text{SU}(2)\times\text{U}(1)_Y$. There are four generators: two charged and one neutral bosons, responsible respectively for charged

---

[2] At the end of this chapter we will see that this is itself not free of problems. How these are overcome will be explained in Chap. 10.

and neutral weak currents, and the photon. As group generators we use $\{T^{\pm}, T^3, Y\}$. The first three are the ladder generators (B.10) of the SU(2) factor, called the weak isospin. In addition, the so-called weak hypercharge is the generator of the $\mathrm{U}(1)_Y$ factor where the subscript is intended to avoid confusion with the electromagnetic U(1) gauge group. It is important to keep in mind that, despite their similar names, the weak isospin and hypercharge are radically different from the strong interaction namesakes introduced in Sect. 5.1. This notwithstanding, the value of the weak hypercharge of the different fields will be fixed in such a way that the analog of the Gell-Mann–Nishijima relation is satisfied

$$Q = T^3 + Y. \tag{5.37}$$

Once the gauge group is chosen, we exhibit the vector bosons of the theory. For this we introduce the Lie algebra valued gauge fields

$$\mathbf{W}_\mu = W_\mu^+ T^- + W_\mu^- T^+ + W_\mu^3 T^3, \quad \mathbf{B}_\mu = B_\mu Y. \tag{5.38}$$

Using the Gell-Mann–Nishijima formula (5.37) and the commutation relations of the SU(2) algebra shown in Eq. (B.10), we have

$$[Q, T^{\pm}] = \pm T^{\pm}, \quad [Q, T^3] = [Q, Y] = 0. \tag{5.39}$$

This means that the gauge fields $W_\mu^{\pm}$ are electrically charged, while W_μ^3 and B_μ are neutral fields.

We still have to identify the electromagnetic U(1) factor in the gauge group. Since the photon has no electric charge, the Maxwell gauge field A_μ must be a combination of the two neutral gauge bosons, W_μ^3 and B_μ. We define a new pair of neutral gauge fields (A_μ, Z_μ) by

$$\begin{aligned} A_\mu &= B_\mu \cos\theta_w + W_\mu^3 \sin\theta_w, \\ Z_\mu &= -B_\mu \sin\theta_w + W_\mu^3 \cos\theta_w, \end{aligned} \tag{5.40}$$

where the transformation is parametrized by an angle θ_w called the weak mixing angle. The form of the linear combination is not arbitrary: it is the most general one guaranteeing that the new gauge fields A_μ and Z_μ have canonical kinetic terms in the action. The field A_μ is now identified with the electromagnetic potential. In short, what we have done is to parametrize our ignorance of how QED is embedded in the electroweak gauge theory by introducing the weak mixing angle. Its value will have to be determined experimentally. The fact that it is nonzero indicates that the weak and electromagnetic interactions are mixed.

This concludes our study of gauge bosons. Next we fix the representation of the matter fields, i.e. how matter fields transform under the gauge group. Here the experiment is our guiding principle. For example, we know that charged weak currents couple left-handed leptons to their corresponding left-handed neutrinos. Since these interactions are mediated by the charged gauge fields $\mathbf{W}_\mu^{\pm} = W_\mu^{\pm} T^{\mp}$, we are led to include both fields in a SU(2) doublet

Table 5.1 Transformation properties of the lepton fields under the electroweak gauge group $SU(2) \times U(1)_Y$

Leptons i (generation)	1	2	3	T^3	Y
$\mathbf{L}^i$	$\begin{pmatrix} \nu_e \\ e^- \end{pmatrix}_L$	$\begin{pmatrix} \nu_\mu \\ \mu^- \end{pmatrix}_L$	$\begin{pmatrix} \nu_\tau \\ \tau^- \end{pmatrix}_L$	$\begin{pmatrix} \frac{1}{2} \\ -\frac{1}{2} \end{pmatrix}$	$-\frac{1}{2}$
ℓ_R^i	e_R^-	μ_R^-	τ_R^-	0	-1

In the last two columns on the right the values of the weak isospin and the hypercharge are shown for the different fields

$$\begin{pmatrix} \nu_e \\ e^- \end{pmatrix}_L, \quad \begin{pmatrix} \nu_\mu \\ \mu^- \end{pmatrix}_L, \quad \begin{pmatrix} \nu_\tau \\ \tau^- \end{pmatrix}_L. \tag{5.41}$$

In addition, we also know that the right-handed component of the electron does not take part in interactions mediated by weak charged currents. This indicates that they should be taken to be singlets under the SU(2) factor.

The Gell-Mann–Nishijima formula can be used now to fix the weak hypercharge of the leptons, i.e. their transformations under the $U(1)_Y$ factor of the gauge group. Using that the left-handed isodoublets (5.41) transform in the fundamental ($s = \frac{1}{2}$) representation of SU(2) where $T^3 = \frac{1}{2}\sigma_3$, we have

$$Y(\nu_\ell) = -\frac{1}{2}, \quad Y(\ell) = -\frac{1}{2}, \tag{5.42}$$

where ℓ denotes e^-, μ^- or τ^-. For right-handed leptons, being singlets under SU(2), we have $T^3 = 0$ and therefore

$$Y(\ell_R) = -1. \tag{5.43}$$

We summarize the results in Table 5.1. We have introduced the compact notation $\mathbf{L}^i$ and ℓ^i to denote respectively the left-handed isodoublets and right-handed singlets.[3]

In all this discussion we have ignored the possibility of having a right-handed component for the neutrino. Being a SU(2) singlet and having zero charge, this particle would have also vanishing hypercharge. Thus, such a particle would be a singlet under all gauge groups of the standard model. This is called a sterile neutrino. It would only interact gravitationally or via some yet unknown interaction making their detection extremely difficult.

In the case of quarks we proceed along similar lines. We look first at the charged weak current that couples protons with neutrons. Taking into account the quark content of these particles, we see how this current in fact couples the u and d quark,

[3] It should be stressed that the quantum numbers appearing in Tables 5.1 and 5.2 summarize a great deal of experimental data resulting from decades of work.

Table 5.2 Transformation properties of the quarks in the electroweak sector of the standard model

Quarks i (generation)	1	2	3	T^3	Y
$\mathbf{Q}^i$	$\begin{pmatrix} u \\ d \end{pmatrix}_L$	$\begin{pmatrix} c \\ s \end{pmatrix}_L$	$\begin{pmatrix} t \\ b \end{pmatrix}_L$	$\begin{pmatrix} \frac{1}{2} \\ -\frac{1}{2} \end{pmatrix}$	$\frac{1}{6}$
U_R^i	u_R	c_R	t_R	0	$\frac{2}{3}$
D_R^i	d_R	s_R	b_R	0	$-\frac{1}{3}$

suggesting that they form an isodoublet. This structure is repeated for the three quark generations[4]

$$\begin{pmatrix} u \\ d \end{pmatrix}_L, \begin{pmatrix} c \\ s \end{pmatrix}_L, \begin{pmatrix} t \\ b \end{pmatrix}_L. \tag{5.44}$$

As with leptons, the right-handed quark components are singlet under SU(2). The hypercharges of the different quarks are shown in Table 5.2, where the notation $\mathbf{Q}^i$, U_L^i and D_L^i is respectively introduced to denote the left-handed doublets and right-handed SU(2) singlets.

The next task is to determine the couplings of the different matter fields (leptons and quarks) to the intermediate vector bosons. From Eq. (4.46), the covariant derivative acting on the matter fields in a representation R of the gauge field is of the form

$$\begin{aligned} D_\mu &= \partial_\mu - ig\mathbf{W}_\mu - ig'\mathbf{B}_\mu \\ &= \partial_\mu - igW_\mu^+ T_R^- - igW_\mu^- T_R^+ - igW_\mu^3 T_R^3 - ig' B_\mu Y_R, \end{aligned} \tag{5.45}$$

It is important to notice that we have introduced two distinct coupling constants g and g' associated with the two factors of the gauge group, SU(2) and U(1)$_Y$. The reason is that gauge transformations do not mix the gauge field $\mathbf{W}_\mu$ with $\mathbf{B}_\mu$ and therefore gauge invariance does not require the coupling constants to be related. Applying Eq. (5.40), we express D_μ in terms of the gauge fields A_μ and Z_μ

$$\begin{aligned} D_\mu = \partial_\mu - igW_\mu^+ T_R^- - igW_\mu^- T_R^+ - iA_\mu(g \sin\theta_w T_R^3 + g' \cos\theta_w Y_R) \\ - iZ_\mu(gT_R^3 \cos\theta_w - g' Y_R \sin\theta_w). \end{aligned} \tag{5.46}$$

We have identified A_μ with the electromagnetic gauge field. Thus, the third term in the covariant derivative gives the coupling of the matter field to electromagnetism

[4] A word of warning is in order here. Although denoted by the same letter, the fields in the quark doublets are not necessarily the same ones that appear as the hadron constituents in the quark model. The two are related by a linear combination. This for the time being cryptic remark will find clarification in Chap. 10 (see page 197).

and, as a consequence, it should be of the form $-ieQA_\mu$, with Q the charge operator. With this in mind and using once more the Gell-Mann–Nishijima relation (5.37) we conclude that the electric charge e is related to the coupling constants g and g' by

$$e = g\sin\theta_w = g'\cos\theta_w. \tag{5.47}$$

This equation gives the physical interpretation of the weak mixing angle. It measures the ratio between the two independent coupling constants in the electroweak sector of the standard model

$$\tan\theta_w = \frac{g'}{g}. \tag{5.48}$$

Precise calculations with the standard model require writing a Lagrangian from where to start a quantization of the theory. A first part contains the dynamics of gauge fields and can be constructed using what we learned in Chap. 4 about nonabelian gauge fields

$$\begin{aligned}\mathscr{L}_{\text{gauge}} =& -\frac{1}{2}W^+_{\mu\nu}W^{-\mu\nu} - \frac{1}{4}Z_{\mu\nu}Z^{\mu\nu} - \frac{1}{4}F_{\mu\nu}F^{\mu\nu} + \frac{ig}{2}\cos\theta_w W^+_\mu W^-_\nu Z^{\mu\nu} \\ &+ \frac{ie}{2}W^+_\mu W^-_\nu F^{\mu\nu} - \frac{g^2}{2}\Big[(W^+_\mu W^{+\mu})(W^-_\mu W^{-\mu}) - (W^+_\mu W^{-\mu})^2\Big]\end{aligned} \tag{5.49}$$

where we have introduced the notation

$$\begin{aligned}W^\pm_{\mu\nu} &= \partial_\mu W^\pm_\nu - \partial_\nu W^\pm_\mu \mp ie\left(W^\pm_\mu A_\nu - W^\pm_\nu A_\mu\right) \mp ig\cos\theta_w\left(W^\pm_\mu Z_\nu - W^\pm_\nu Z_\mu\right) \\ Z_{\mu\nu} &= \partial_\mu Z_\nu - \partial_\nu Z_\mu,\end{aligned} \tag{5.50}$$

while $F_{\mu\nu}$ is the familiar field strength of the Maxwell field A_μ. The gauge part of the Lagrangian is a bit cumbersome because we have chosen to write it in terms of the fields A_μ and Z_μ. The SU(2) $\times$ U(1)$_Y$ gauge symmetry is not obvious in this expression, but it has the advantage of making the invariance under the gauge transformations of electromagnetism manifest. We have also eliminated the coupling constant g' in favor of g and the weak mixing angle θ_w. Moreover, whenever the combination $g\sin\theta_w$ appeared we further used (5.47) and wrote the electric charge e.

For the matter fields we can write the following gauge invariant Lagrangian

$$\begin{aligned}\mathscr{L}_{\text{matter}} = \sum_{i=1}^{3}\Big(&i\overline{\mathbf{L}}^j \not{D}\mathbf{L}^j + i\overline{\ell}^j_R\not{D}\ell^j_R \\ &+ i\overline{\mathbf{Q}}^j\not{D}\mathbf{Q}^j + i\overline{U}^j_R\not{D}U^j_R + i\overline{D}^j_R\not{D}D^j_R\Big).\end{aligned} \tag{5.51}$$

The covariant derivatives appearing in this Lagrangian can be written explicitly from (5.46) taking into account the representation for the different matter fields.

A glimpse to the Lagrangians $\mathscr{L}_{\text{gauge}}$ and $\mathscr{L}_{\text{matter}}$ and to the covariant derivative (5.46) shows the coupling between the standard model particles. As right-handed fields are singlets under SU(2), the $W^{\pm}$ boson only couples to the left-handed doublets. Using the expression of the $T^{\pm}$ generators in the fundamental representation of SU(2), we find that the terms in the standard model Lagrangian coupling the $W^{\pm}$ boson to the leptons take the form

$$gW_{\mu}^{+}\overline{\nu}_{\ell}\gamma^{\mu}\ell_{L}, \quad gW_{\mu}^{-}\overline{\ell}_{L}\gamma^{\mu}\nu_{\ell}. \tag{5.52}$$

Notice that the strength of these couplings is given by g.

From the covariant derivative (5.46), we see that the Z^0 couples to a combination of the two generators of the Cartan subalgebra of $\mathrm{SU}(2)\times\mathrm{U}(1)_Y$, namely T^3 and Y. Since they can be simultaneously diagonalized, this gauge boson couples to fermions of the same kind. In the case of the leptons the couplings are

$$\frac{g}{2\cos\theta_w}Z_{\mu}\overline{\nu}_{\ell}\gamma^{\mu}\nu_{\ell}, \quad \frac{g}{\cos\theta_w}\left(-\frac{1}{2}+\sin^2\theta_w\right)Z_{\mu}\overline{\ell}_{L}\gamma^{\mu}\ell_{L}, \tag{5.53}$$

and

$$\frac{g\sin^2\theta_w}{\cos\theta_w}Z_{\mu}\overline{\ell}_{R}\gamma^{\mu}\ell_{R}. \tag{5.54}$$

Unlike the $W^{\pm}$, the Z^0 boson couples to the right-handed components through the hypercharge.

The analysis can be repeated for quarks. The result is that once again right-handed quarks only couple to the Z^0, while the $W^{\pm}$ couple the upper and lower components of the left-handed doublets. Both left- and right-handed quarks, being charged, couple also to the electromagnetic field A_{μ}. The derivation of the form of these terms as well as the corresponding couplings is left as an exercise. Finally, the couplings between the gauge bosons can be read from the gauge Lagrangian (5.49).

5.5 Closing Remarks: Particle Masses in the Standard Model

The alert reader surely has noticed that in our discussion of the electroweak theory we have been conspicuously silent about particle masses. That particles such as the electron or the muon have nonvanishing masses is a well known experimental fact. Moreover, we have seen that phenomena such as beta decay cannot be explained by the model unless a mass is assumed for the intermediate vector bosons.

At the time when the standard model was developed in the 1960s, QED was the archetype of a successful quantum field theory: physical processes could be accurately computed at arbitrary high energies in terms of a small number of experimentally fixed parameters. From this point of view, there were fundamental obstacles to giving masses to the fields in the electroweak theory described in the previous section.

Adding explicit mass terms for fermions and gauge bosons to the Lagrangian breaks gauge invariance: in the case of the fermions a Dirac mass term mixes fields transforming in different representations of $SU(2) \times U(1)_Y$, whereas a term $\text{Tr}(A_\mu A^\mu)$ is obviously not invariant under the gauge transformation of the vector field (4.42).

Giving up gauge invariance means destroying the possibility of building a theory of electroweak interactions valid to all energies. In more precise terms, gauge invariance is crucial for the *renormalizability* of the theory, a property whose physical relevance will be discussed in Chap. 8. Moreover, gauge invariance restricts the ways the standard model fields couple among themselves.

Note that the conflict between fermion masses and gauge invariance does not appear in the pure QCD sector. The reason is that the action of SU(3) is vector-like, i.e., the same for left- and right-handed quarks. Therefore a Dirac mass term is gauge invariant, and it can be included in the Lagrangian (5.27) without endangering desirable properties of the theory.

There is nevertheless a way of constructing massive intermediate gauge bosons and fermions in a manner compatible with $SU(2) \times U(1)_Y$ gauge invariance. It consists of generating the mass terms at low energies rather than putting them by hand in the Lagrangian, so gauge invariance is not broken, just hidden. This is achieved through the implementation of the Brout-Englert-Higgs mechanism to be presented in Chap. 7.

To conclude we must say that, as a matter of fact, there is nothing fundamentally wrong with adding explicit mass terms to the standard model Lagrangian, so long as we are only interested in describing the physics at energies *below* the mass scales appearing in the Lagrangian. We will elaborate on this statement in Chaps. 10 and 12, once we learn more about the quantum properties of interacting field theories.

References

1. Glashow, S.L.: Partial symmetries of weak interactions. Nucl. Phys. **22**, 579 (1961)
2. Weinberg, S.: A model of leptons. Phys. Rev. Lett. **19**, 1264 (1967)
3. Salam, A.: Weak and electromagnetic interactions. In: Svartholm, N. (ed.) Elementary Particle Theory: Relativistic Groups and Analiticity, Almqvist and Wiksell, Stockholm (1968)
4. Perkins, D.H.: Introduction to High Energy Physics, 4th edn. Cambridge University Press, Cambridge (2000)
5. Burgess, C., Moore, G.: The Standard Model. A Primer, Cambridge (2007)
6. Paschos, E.A.: Electroweak Theory. Cambridge University Press, Cambridge (2007)
7. Cottingham, W.N., Greenwood, D.A.: An Introduction to the Standard Model of Particle Physics. Cambridge University Press, Cambridge (2007)
8. Bettini, A.: Introduction to Elementary Particle Physics. Cambridge University Press, Cambridge (2008)
9. Nakamura, K. et al.: Review of particle physics. J. Phys. G **37**, 075021 (2010)
10. Epelbaum, E., Hammer, H.W., Meißner, U.-G.: Modern theory of nuclear forces. Rev. Mod. Phys. **81**, 1773 (2009)

Chapter 6
Towards Computational Rules: Feynman Diagrams

As the basic tool to describe the physics of elementary particles, the final aim of quantum field theory is the calculation of observables. Most of the information we have about the physics of subatomic particles comes from scattering experiments. Typically, these experiments consist of arranging two or more particles to collide with a certain energy and to setup an array of detectors, sufficiently far away from the region where the collision takes place, that register the outgoing products of the collision and their momenta (together with other relevant quantum numbers).

Next we discuss how these cross sections can be computed from quantum mechanical amplitudes and how these amplitudes themselves can be evaluated in perturbative quantum field theory. We keep our discussion rather heuristic and avoid technical details that can be found in standard texts (see Ref. [1–15] of Chap. 1). The techniques described will be illustrated with the calculation of the cross section for Compton scattering at low energies and its application to the study of the polarization of the cosmic microwave background radiation. Exceptionally, and in order to better show the computational power of the diagrammatic tools in quantum field theory, these calculations will be presented in some detail.

6.1 Cross Sections and S-Matrix Amplitudes

In order to fix ideas, we consider the simplest case of a collision experiment where two particles collide to produce again two particles in the final state. The aim of such an experiments is a direct measurement of the number of particles per unit time $\frac{dN}{dt}(\theta, \varphi)$ registered by the detector within a solid angle $d\Omega$ in the direction specified by the polar angles θ, φ (see Fig. 6.1). On general grounds, we know that this quantity has to be proportional to the flux of incoming particles[1] f_{in}. The proportionality constant defines the differential cross section

[1] This is defined as the number of particles that enter the interaction region per unit time and per unit area perpendicular to the direction of the beam.

L. Álvarez-Gaumé and M. Á. Vázquez-Mozo, *An Invitation to Quantum Field Theory*, Lecture Notes in Physics 839, DOI: 10.1007/978-3-642-23728-7_6,

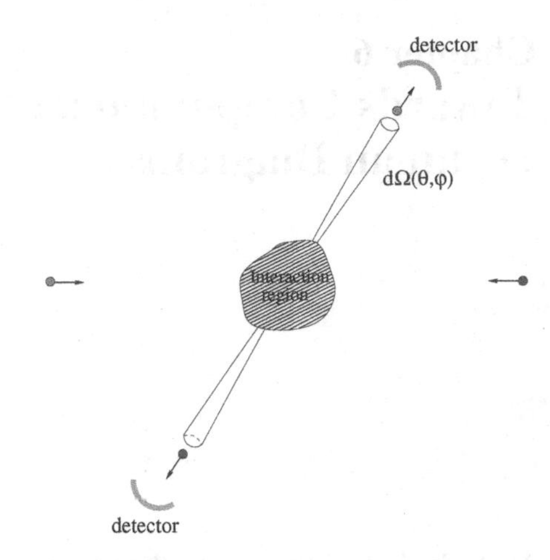

Fig. 6.1 Schematic setup of a two-to-two-particles scattering event in the center of mass reference frame

$$\frac{dN}{dt}(\theta,\varphi) = f_{\text{in}}\,\frac{d\sigma}{d\Omega}(\theta,\varphi). \tag{6.1}$$

In natural units f_{in} has dimensions of (length)$^{-3}$, so the differential cross section has dimensions of (length)^2. It depends, apart from the direction (θ,φ), on the parameters of the collision (energy, impact parameter, etc.) as well as on the masses and spins of the incoming and outgoing particles.

The differential cross section measures the angular distribution of the products of the collision. It is also physically interesting to quantify how effective the interaction between the particles is in order to produce a nontrivial dispersion. This is measured by the total cross section, which is obtained by integrating the differential cross section over all directions

$$\sigma = \int_{-1}^{1} d(\cos\theta) \int_{0}^{2\pi} d\varphi\,\frac{d\sigma}{d\Omega}(\theta,\varphi). \tag{6.2}$$

To gain some physical intuition on the meaning of the total cross section, we can think of the classical scattering of a point particle off a sphere of radius R. The particle undergoes a collision only when the impact parameter is smaller than the radius of the sphere and a calculation of the total cross section yields $\sigma = \pi R^2$. This is precisely the cross area that the sphere presents to incoming particles.

The starting point for the calculation of cross sections is the probability amplitude for the corresponding process. In a scattering experiment, one prepares a system with a given number of particles with definite momenta $\mathbf{p}_1,\ldots,\mathbf{p}_n$. In the Heisenberg picture this is described by a time independent state labelled by the incoming momenta of the particles (to keep things simple we consider spinless particles) that we denote by

$$|p_1,\ldots,p_n;\text{in}\rangle. \tag{6.3}$$

As a result of the scattering, a number k of particles with momenta $\mathbf{p}'_1, \ldots, \mathbf{p}'_k$ are detected. Thus, the system is now in the "out" Heisenberg picture state

$$|p'_1, \ldots, p'_k; \text{out}\rangle \tag{6.4}$$

labelled by the momenta of the particles detected at late times. The probability amplitude of detecting k particles in the final state with momenta $\mathbf{p}'_1, \ldots, \mathbf{p}'_k$ in the collision of n particles with initial momenta $\mathbf{p}_1, \ldots, \mathbf{p}_n$ defines the S-matrix amplitude

$$S(\text{in} \to \text{out}) = \langle p'_1, \ldots, p'_k; \text{out}|p_1, \ldots, p_n; \text{in}\rangle. \tag{6.5}$$

It is very important to keep in mind that both (6.3) and (6.4) are time-independent states in the Hilbert space of a very complicated interacting theory. However, since both at early and late times the incoming and outgoing particles are far apart from each other, the "in" and "out" states can be thought as two states $|p_1, \ldots, p_n\rangle$ and $|p'_1, \ldots, p'_k\rangle$ in the Fock space of the corresponding free theory. Then, the overlaps (6.5) can be written in terms of the matrix elements of an S-matrix operator $\hat{S}$ acting on the free Fock space

$$\langle p'_1, \ldots, p'_k; \text{ out}|p_1, \ldots, p_n; \text{in}\rangle = \langle p'_1, \ldots, p'_k|\hat{S}|p_1, \ldots, p_n\rangle. \tag{6.6}$$

The operator $\hat{S}$ is unitary, $\hat{S}^\dagger = \hat{S}^{-1}$, Lorentz invariant and its matrix elements are analytic in the external momenta.

In a scattering experiment there is the possibility that the particles do not interact at all and the system is left in the same initial state. It is useful to factor out this possibility from the S-matrix elements between initial and final states by writing

$$\hat{S} = \mathbf{1} + i\hat{T}, \tag{6.7}$$

where $\mathbf{1}$ represents the identity operator. In this way, all nontrivial interactions are encoded in the matrix elements of the T-operator, $\langle p'_i, \ldots, p'_k|i\hat{T}|p_1, \ldots, p_n\rangle$. Furthermore, in these matrix elements it is convenient to factor out a delta function implementing momentum conservation to define the invariant scattering amplitude, $i\mathscr{M}_{i\to f}$

$$\langle f|\hat{S}|i\rangle = \langle f|i\rangle + (2\pi)^4\delta^{(4)}\Bigg(\sum_{\text{final}} p'_i - \sum_{\text{initial}} p_j\Bigg) i\mathscr{M}_{i\to f}. \tag{6.8}$$

Using the Lorentz invariance of the S-matrix it is not difficult to show that $i\mathscr{M}_{i\to f}$ is a relativistic invariant as well (hence its name). Our next task is to show how observable quantities such as decay rates or cross sections can be obtained from the knowledge of this invariant amplitude. Then we will turn to the problem of computing the amplitude itself in quantum field theory.

In studying a scattering problem in the infinite volume limit we would have to consider localized wave packets for the asymptotic in and out states. Although this can be done, it is rather cumbersome. This is the reason why we are going to employ a common trick consisting in working with plane waves for the in and out states, while putting the system at the same time in a space-time box of finite but large volume VT. We will see how at the end of the calculation all dependence on the size of the box drops out and the limit $V \to \infty$, $T \to \infty$ can be taken safely.

The probability amplitude for the process is $|\langle f|\hat{S}|i\rangle|^2$ which, for nontrivial transitions, is given by the modulus squared of the second term on the right-hand side of Eq. (6.8). The presence of the momentum conservation delta function makes the computation problematic. This is precisely where working at finite volume comes handy. The idea is to write one of the delta functions in terms of its Fourier transform

$$\left|(2\pi)^4\delta^{(4)}\left(\sum_{\text{final}} p_i' - \sum_{\text{initial}} p_j\right)\right|^2 = (2\pi)^4\delta^{(4)}\left(\sum_{\text{final}} p_i' - \sum_{\text{initial}} p_j\right) \times \int d^4x \exp\left[i\left(\sum_{\text{final}} p_i' - \sum_{\text{initial}} p_j\right)\cdot x\right]. \tag{6.9}$$

The remaining delta function then sets the argument of the exponential to zero and we have

$$\left|(2\pi)^4\delta^{(4)}\left(\sum_{\text{final}} p_i' - \sum_{\text{initial}} p_j\right)\right|^2 = VT(2\pi)^4\delta^{(4)}\left(\sum_{\text{final}} p_i' - \sum_{\text{initial}} p_j\right). \tag{6.10}$$

With this result we can compute the non-diagonal probability amplitude $|\langle f|\hat{S}|i\rangle|^2$. Dividing by T, the transition probability per unit time is given by

$$w_{i\to f} = V(2\pi)^4\delta^{(4)}\left(\sum_{\text{final}} p_i' - \sum_{\text{initial}} p_j\right)|i\mathscr{M}_{i\to f}|^2. \tag{6.11}$$

In both scattering and decay processes, the final states have a continuous energy spectrum. To compute the probability for the particles in the final state to have momenta in a volume element $d^3p_1' \cdots d^3p_k'$ around $(\mathbf{p}_1', \ldots, \mathbf{p}_k')$, we should multiply (6.11) by the number of states contained in it. Let us assume we have one particle in the volume V (or in other words, that the state is normalized to one). The number of available states within a momentum space volume element d^3p can be directly written as

$$\frac{V d^3p}{(2\pi\hbar)^3}, \tag{6.12}$$

where we have momentarily restored the powers of $\hbar$ so the reader can clearly identify the density of states in phase space.

In the calculation of the S-matrix elements at infinite volume, the one-particle states $|p\rangle$ satisfy the relativistic normalization (2.20). By putting the system in a box of large volume V, the states become normalizable with $\langle p|p\rangle = 2E_{\mathbf{p}}V$. Thus, if we insist in using (6.12) for the final density of states of each outgoing particle we should not compute the S-matrix element between the relativistically normalized states $|p\rangle$ but rather between the properly normalized ones $(2E_{\mathbf{p}}V)^{-1/2}|p\rangle$. Thus, since the probability amplitude involves the modulus *squared* of the amplitude, to find the correct expression for the number of states we should correct Eq. (6.12) by the normalization factor of the final states

$$\frac{Vd^3p}{(2\pi\hbar)^3}\frac{1}{2E_{\mathbf{p}}V} = \frac{d^3p}{(2\pi\hbar)^3}\frac{1}{2E_{\mathbf{p}}}. \tag{6.13}$$

We see how the volume cancels out and the infinite volume limit can be taken safely. Moreover, the resulting expression is relativistically invariant. Doing this for every particle in the final state of the scattering/decay process leads to the so-called *phase space factor*

$$d\Phi_k = \prod_{i=1}^{k}\frac{d^3p'_i}{(2\pi)^3}\frac{1}{2E'_i}, \tag{6.14}$$

where $E'_i = \sqrt{m_i^2 + \mathbf{p}_i'^2}$ and we have restored natural units.

After this preliminary discussion we can compute the particle decay rate where we have a single particle in the initial state with momentum $\mathbf{p}$. As explained above, the proper normalization of the initial state introduces an extra factor of $(2E_{\mathbf{p}}V)^{-1}$ in the square of the S-matrix element leading to (6.11). This has the effect of removing the remaining volume dependence, and we obtain the decay width

$$d\Gamma = \frac{1}{2E_{\mathbf{p}}}(2\pi)^4\delta^{(4)}\left(p - \sum_{j=1}^{k}p'_j\right)|\mathcal{M}_{i\to f}|^2 d\Phi_k. \tag{6.15}$$

To calculate the total rate for this particular decay channel we should integrate over all final momenta. In doing that it is important to bear in mind that one has to divide the expression by a factor $\prod_a n_a!$, where n_a is the number of identical particles of type a. This is crucial to avoid overcounting the number of final states.

Here we notice that the factor of $E_{\mathbf{p}}^{-1}$ in front of (6.15) has a simple physical meaning. Its suppression effect for large $|\mathbf{p}|$ accounts for the relativistic effect of time dilation due to the motion of the decaying particle. In the rest frame this is equal to the rest mass m. The calculation of the total decay width for a particle requires not only to integrate over final momenta but also to sum over all possible decay channels, namely

$$\Gamma_{\text{total}} = \sum_{\text{channels}}\Gamma_i, \tag{6.16}$$

where Γ_i is the width of the ith decay channel. The lifetime of the particle is given by the inverse of the total width Γ_{total}.

We study now the calculation of the differential and total cross sections in the scattering of two particles with an arbitrary number of particles in the final state. The differential cross section for this problem is given by the number of particles scattered within an infinitesimal solid angle along the final momenta $\mathbf{p}'_1, \ldots, \mathbf{p}'_k$ divided by the flux of incoming particles f_{in}, thus generalizing Eq. (6.1). In terms of the probability density per unit time computed in Eq. (6.11), this gives

$$d\sigma = \frac{1}{4E_1E_2V^2}\frac{w_{i\to f}}{f_{\text{in}}}d\Phi_k \tag{6.17}$$

To get this expression we have multiplied by $(2E_1V)^{-1}(2E_2V)^{-1}$ to take care of the normalization of the incoming states as discussed previously.

We need to compute the incoming flux f_{in}. The number of particles approaching the target (say particle 2) in a time dt across a surface dS orthogonal to the beam is given by $n|\mathbf{v}_1 - \mathbf{v}_2|dtdS$, with n the number density of projectiles (in this case the particle 1). Since in the calculation of the S-matrix amplitude we have normalized our states such that there is one particle per unit volume, we have that $n = V^{-1}$ and the incoming flux is $f_{\text{in}} = |\mathbf{v}_1 - \mathbf{v}_2|/V$. Plugging this result into (6.17), we see how the powers of the volume cancel out and the differential cross section in the infinite volume limit reads

$$d\sigma = \frac{|\mathscr{M}_{i\to f}|^2}{4E_1E_2|\mathbf{v}_1 - \mathbf{v}_2|}(2\pi)^4\delta^{(4)}\left(p_1 + p_2 - \sum_{j=1}^{n} p'_j\right)d\Phi_k, \tag{6.18}$$

where $d\Phi_k$ is the phase space factor for the k particles in the final state. To calculate the total cross section we have to integrate over all final momenta and include the necessary symmetry factors if identical particles are produced as the result of the collision.

An inspection of Eq. (6.18) shows that the only piece depending on the observer's frame is the denominator $F \equiv 4E_1E_2|\mathbf{v}_1 - \mathbf{v}_2|$. The presence of this term implies that the measurement of the differential and total cross sections of the same collision in various reference frames takes different values. This is an important point that we discuss now in some detail.

We consider first a collinear reference frame in which the momenta of the two colliding particles lie along the same direction, $\mathbf{p}_1 \parallel \mathbf{p}_2$. This class of frames include two cases of particular interest: the laboratory frame, where one of the particles is at rest (for example $\mathbf{p}_2 = 0$), and the center of mass frame where the center of mass is at rest, $\mathbf{p}_2 = -\mathbf{p}_1$.

It is not difficult to check that in the collinear case the combination $E_1E_2|\mathbf{v}_1 - \mathbf{v}_2|$ is invariant under boosts along the direction of the two incoming momenta. This means that the value of the differential and total cross section is the same in *all* collinear frames. Moreover, we can write

$$\begin{aligned} F_{\text{coll}} &= 4E_1E_2|\mathbf{v}_1 - \mathbf{v}_2| = 4E_1E_2\left|\frac{\mathbf{p}_1}{E_1} - \frac{\mathbf{p}_2}{E_2}\right| \\ &= 4|E_2\mathbf{p}_1 - E_1\mathbf{p}_2| = 4\left(E_2|\mathbf{p}_1| + E_1|\mathbf{p}_2|\right), \end{aligned} \tag{6.19}$$

where in writing the last identity we have used that in the collinear frames the two particles are approaching each other from opposite directions. It can be written in a Lorentz invariant form

$$F_{\text{coll}} = 4\sqrt{(p_1 \cdot p_2)^2 - m_1^2 m_2^2}. \tag{6.20}$$

Then, the differential cross section measured in a collinear frame is given by

$$d\sigma_{\text{coll}} = \frac{|\mathscr{M}_{i\to f}|^2}{4\sqrt{(p_1 \cdot p_2)^2 - m_1^2 m_2^2}}(2\pi)^4\delta^{(4)}\left(p_1 + p_2 - \sum_{j=1}^{n} p'_j\right) d\Phi_k. \tag{6.21}$$

The corresponding total cross section is obtained by integrating over all momenta in the final state, namely

$$\sigma_{\text{coll}} = \frac{1}{4\sqrt{(p_1 \cdot p_2)^2 - m_1^2 m_2^2}} \int \left[\prod_{\substack{\text{final}\\\text{states}}} \frac{d^3 p'_i}{(2\pi)^3}\frac{1}{2E'_i}\right] \times |\mathscr{M}_{i\to f}|^2 (2\pi)^4 \delta^{(4)}\left(p_1 + p_2 - \sum_{\substack{\text{final}\\\text{states}}} p'_i\right). \tag{6.22}$$

We will make use of this expression in Sect. 6.4 when studying Compton scattering.

Due to their invariance under Lorentz transformations, Eqs. (6.21) and (6.22) allow the computation in an arbitrary frame of the cross section measured by the collinear observer. For example, in a general frame where the two particles collide with velocities $\mathbf{v}_1$ and $\mathbf{v}_2$ the collinear cross section is obtained by using the following expression for F_{coll}

$$F_{\text{coll}} = 4E_1E_2\sqrt{(\mathbf{v}_1 - \mathbf{v}_2)^2 - (\mathbf{v}_1 \times \mathbf{v}_2)^2}. \tag{6.23}$$

In various physical setups, most notably in astrophysics, one needs to compute the cross sections *measured by a generic observer* with respect to whom the momenta of the colliding particles form an arbitrary angle. This requires the evaluation of the denominator in Eq. (6.18) in a generic "oblique" frame,

$$F_{\text{obl}} = 4E_1E_2|\mathbf{v}_1 - \mathbf{v}_2| = 4|E_2\mathbf{p}_1 - E_1\mathbf{p}_2|. \tag{6.24}$$

To relate F_{obl} to the corresponding factor for the collinear observer, F_{coll}, we split the incoming momenta into their components parallel and perpendicular to the center of mass momentum $\mathbf{P}_{\text{cm}} = \mathbf{p}_1 + \mathbf{p}_2$,

$$\mathbf{p}_{i\parallel} = \left(\frac{\mathbf{p}_i \cdot \mathbf{P}_{\text{cm}}}{\mathbf{P}_{\text{cm}}^2}\right)\mathbf{P}_{\text{cm}}, \quad \mathbf{p}_{i\perp} = \mathbf{p}_i - \mathbf{p}_{i\parallel}, \tag{6.25}$$

with $i = 1, 2$. It is not difficult to show that $\mathbf{p}_{1\perp} = -\mathbf{p}_{2\perp} \equiv \mathbf{p}_{\perp}$. Applying now this decomposition to Eq. (6.24) we arrive at

$$F_{\text{obl}} = 4\sqrt{(E_2\mathbf{p}_{1\parallel} - E_1\mathbf{p}_{2\parallel})^2 + (E_1 + E_2)^2\mathbf{p}_{\perp}^2}. \tag{6.26}$$

To go from the oblique frame to the center of mass frame, we only have to perform a boost with velocity

$$\mathbf{V} = \frac{1}{E_1 + E_2}\mathbf{P}_{\text{cm}}. \tag{6.27}$$

This boost only transforms the parallel components of the momenta, $\mathbf{p}_{i\parallel}$. It is possible to show that the combination $E_2\mathbf{p}_{1\parallel} - E_1\mathbf{p}_{2\parallel}$ appearing under the square root in Eq. (6.26) is left invariant by the boost. Hence, it can be computed either in the oblique or the center of mass frame and consequently we can write

$$F_{\text{obl}}^2 = F_{\text{coll}}^2 + 16\left[(E_1 + E_2)^2 - (E_1^{\text{cm}} + E_2^{\text{cm}})^2\right]\mathbf{p}_{\perp}^2, \tag{6.28}$$

where we have used superscripts to indicate the quantities that are referred to the center of mass frame. Finally, we notice that the second term inside the square brackets is just the Lorentz invariant quantity $(p_1 + p_2)^2$. Evaluating it in the oblique frame we arrive at the final expression

$$F_{\text{obl}}^2 = F_{\text{coll}}^2 + 16\mathbf{p}_{\perp}^2\mathbf{P}_{\text{cm}}^2. \tag{6.29}$$

We have seen that in a collision experiment all collinear observers measure the same value of the cross section.[2] This is not the case, however, for the cross section measured by another observer boosted with respect to the collinear ones along a direction forming a non-zero angle with the beams. In this oblique frame, both $\mathbf{P}_{\text{cm}}$ and $\mathbf{p}_{\perp}$ are different from zero and from Eq. (6.29) the cross section is suppressed by a larger value in the denominator. This can be understood heuristically by thinking that, as the result of this transverse boost, the area of the sections normal to the beams are Lorentz contracted.

We have learned how particle cross sections are given in terms of the invariant amplitude for the corresponding processes, which in turn are related to the S-matrix amplitudes. Generically, an *exact* computation of these amplitudes in quantum field theory is not feasible. Nevertheless, in many physical situations it can be argued that interactions are weak enough to allow for a perturbative evaluation. In the remainder of this chapter we will describe how S-matrix elements can be computed in perturbation theory using Feynman diagrams and rules. These are very convenient bookkeeping techniques allowing both to track all contributions to a process at a given order in perturbation theory and to compute them.

[2] This is a particular case of Eq. (6.29) where $\mathbf{p}_{\perp} = 0$ for all collinear observers.

6.2 From Green's Functions to Scattering Amplitudes

The basic quantities to be computed in quantum field theory are the vacuum expectation values of products of the operators of the theory. Particularly useful are time-ordered Green's functions of a number of local operators $\mathcal{O}_i(x)$

$$\langle\Omega|T\left[\mathcal{O}_1(x_1)\ldots\mathcal{O}_n(x_n)\right]|\Omega\rangle, \tag{6.30}$$

where $|\Omega\rangle$ is the ground state of the theory and the time ordered product has been defined in Eq. (2.63).

The interest of these correlation functions lies in the fact that they can be related to S-matrix amplitudes through the so-called reduction formula. The idea consists of replacing a particle of momentum $\mathbf{p}$ in the in- or out-state by the insertion of a certain quantum field $\phi(x)$ interpolating between the vacuum and the one-particle states with the normalization

$$\langle\Omega|\phi(t,\mathbf{x})|p\rangle = \varphi(\mathbf{p})e^{-iE_{\mathbf{p}}t+i\mathbf{p}\cdot\mathbf{x}}, \tag{6.31}$$

where $\varphi(\mathbf{p})$ is the one-particle wave function, carrying the corresponding indices and quantum numbers: for example, $\varphi = 1$ for a scalar field while $\varphi = \varepsilon_\mu(\mathbf{p},\lambda)$ for the electromagnetic field.[3] This expression fixes the global normalization of the field, while the coordinate dependence is completely determined by the translational invariance of the vacuum state $|\Omega\rangle$ [see Eq. (2.36)].

To keep our discussion as simple as possible, we will not derive the reduction formula, or even write it down in full detail. Suffice it to say that the reduction formula states that any S-matrix amplitude

$$\langle p'_1,\ldots,p'_k;\text{out}|p_1,\ldots,p_n;\text{in}\rangle \tag{6.32}$$

can be written in terms of the Fourier transform of a time-ordered correlation function

$$\begin{aligned}\int d^4x_1\ldots d^4x_k\int d^4y_1\ldots d^4y_n\langle\Omega|T\left[\phi(x_1)^\dagger\ldots\phi(x_k)^\dagger\phi(y_1)\ldots\phi(y_n)\right]|\Omega\rangle \\ \times\, e^{ip'_1\cdot x_1+\cdots+ip'_k\cdot x_k}e^{-ip_1\cdot y_1-\cdots-ip_n\cdot y_n},\end{aligned} \tag{6.33}$$

where $\phi(x)$ is a field that "creates" the particles out of the vacuum. Since the momenta of the particles in the asymptotic states are on-shell, the expression (6.33) has to be evaluated in the limit $p_i^2, p_j'^2 \to m^2$, where it diverges. In the reduction formula connecting (6.32) with (6.33) these poles are cancelled by factors of

[3] The field $\phi(x)$ appearing in Eq. (6.31) is the so-called "renormalized" field, and it is not canonically normalized. It is related to the canonically normalized "bare" field $\phi_0(x)$ by an overall numerical factor, $\phi_0(x) = \sqrt{Z_\phi}\phi(x)$, where $Z_\phi = 1$ in the case of a free field. The difference between bare and renormalized fields will become clear in Chap. 8 (see Sect. 8.3).

$p_i^2 - m^2$ and $p_j'^2 - m^2$. The technical details and the form taken by the reduction formula for various quantum field theories can be found in the textbooks listed in Ref. [1–15] of Chap. 1.

The "interpolating field" used to write the scattering amplitude in terms of Green's functions is not uniquely determined. Any local field satisfying the normalization Eq. (6.31) can be used for this purpose. The scattering amplitudes calculated from a quantum field theory are invariant under local field redefinitions. For example, for a massive field $\phi(x)$ we could use in Eq. (6.33) instead of $\phi(x)$ the local field

$$\phi'(x) = -\frac{1}{m^2}\Box\phi(x) \tag{6.34}$$

that also interpolates between the one-particle states and the vacuum with the correct normalization (6.31).

6.3 Feynman Rules

The reduction formula transforms the problem of computing S-matrix elements to the evaluation of time-ordered correlation functions. These quantities are easy to compute exactly for free fields. For an interacting theory, generically we can only evaluate them perturbatively. Using path integrals, the vacuum expectation value of the time-ordered product of a number of operators can be written as

$$\langle\Omega|T\Big[\mathscr{O}_1(x_1)\ldots\mathscr{O}_n(x_n)\Big]|\Omega\rangle = \frac{\int \mathscr{D}\phi\mathscr{D}\phi^\dagger\,\mathscr{O}_1(x_1)\ldots\mathscr{O}_n(x_n)e^{iS[\phi,\phi^\dagger]}}{\int \mathscr{D}\phi\mathscr{D}\phi^\dagger\,e^{iS[\phi,\phi^\dagger]}}. \tag{6.35}$$

For a theory with interactions, neither the path integral in the numerator or in the denominator are Gaussian and cannot be computed exactly. In spite of this, Eq. (6.35) is still very useful to implement a perturbative calculation. The action $S[\phi,\phi^\dagger]$ can be split into the free (quadratic) and the interaction parts

$$S[\phi,\phi^\dagger] = S_0[\phi,\phi^\dagger] + S_{\text{int}}[\phi,\phi^\dagger]. \tag{6.36}$$

All dependence on the coupling constants of the theory comes from the second piece. Expanding $\exp(iS_{\text{int}})$ in power series of the coupling, we find that each term in the series expansion of the integrals in Eq. (6.35) has the following structure

$$\int \mathscr{D}\phi\mathscr{D}\phi^\dagger[\ldots]e^{iS_0[\phi,\phi^\dagger]}, \tag{6.37}$$

where "[...]" denotes certain monomial of fields.

The crucial point is that the integration measure $\mathscr{D}\phi\mathscr{D}\phi^\dagger\exp(iS_0)$ only involves the free action, so the path integrals (6.37) are Gaussian and therefore can be computed exactly. The same conclusion can be reached using the operator formalism.

In this case the correlation function (6.30) can be expressed in terms of correlation functions of operators in the interaction picture $\phi_I(x)$. The advantage of using this picture is that the field operators satisfy the free equations of motion

$$i\dot{\phi}_I = [\phi_I, H_0] \tag{6.38}$$

and therefore can be expanded in creation–annihilation operators. Time-ordered correlation functions are then computed using Wick's theorem.

The previous discussion outlines the strategy to calculate S-matrix amplitudes in perturbation theory: using the reduction formula they are expressed in terms of time-ordered correlation functions that in turn are calculated in terms of a series expansion in the coupling constants. The most convenient way to carry out this program is by using Feynman diagrams and rules. They provide a very economical way not only to keep track of each term in the expansion but also to compute their contributions. In what follows we will refrain from giving a detailed derivation of the Feynman rules. Instead we will present them using heuristic arguments.

For the sake of concreteness we focus on the case of QED first. We use the action (4.87) with the gauge fixing term included (see Sect. 4.6). Expanding the covariant derivative and setting $\xi = 1$ (called the Feynman gauge) we have

$$S = \int d^4x \left[-\frac{1}{4} F_{\mu\nu} F^{\mu\nu} + \overline{\psi}(i\partial\!\!\!/ - m)\psi - \frac{1}{2}(\partial_\mu A^\mu)^2 - e\overline{\psi}\gamma^\mu \psi A_\mu \right]. \tag{6.39}$$

We begin with the quadratic part. Integrating by parts we have

$$S_0 = \int d^4x \left[\frac{1}{2} A_\mu (\eta^{\mu\nu} \partial_\sigma \partial^\sigma) A_\nu + \overline{\psi}_\beta (i\partial\!\!\!/ - m)_{\beta\alpha} \psi_\alpha \right]. \tag{6.40}$$

The action contains two types of propagating particles, photons and fermions, represented by wavy and straight lines respectively:

A_μ ∿∿∿∿∿∿ A_ν $\qquad$ ψ_α ——▶—— $\overline{\psi}_\beta$

The arrow in the fermion line does not represent the direction of the momentum but the flux of (negative) charge. This distinguishes particles form antiparticles: if the fermion propagates from left to right (i.e. in the direction of the charge flux) it represents a particle, whereas when it does from right to left it corresponds to an antiparticle. Photons are not charged and therefore wavy lines have no orientation.

Next we turn to the cubic part of the action containing a photon field, a spinor and its conjugate

$$S_{\text{int}} = -e \int d^4x \overline{\psi}_\beta \gamma^\mu_{\beta\alpha} \psi_\alpha A_\mu. \tag{6.41}$$

In a Feynman diagram this interaction is represented by the vertex:

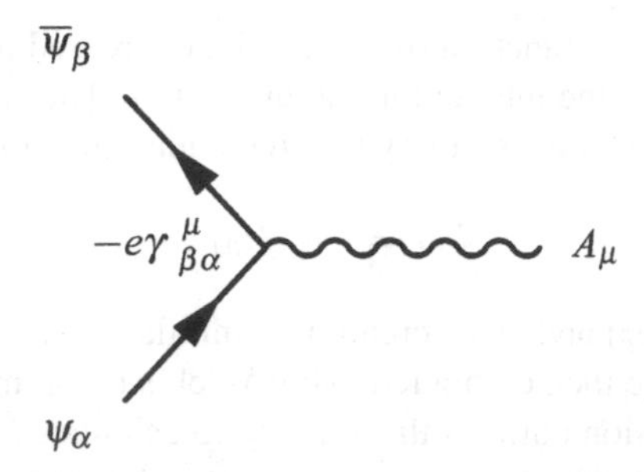

To compute an S-matrix amplitude to a given order in e, one should draw all possible diagrams with as many vertices as the order in perturbation theory, and the number and type of external legs dictated by the in and out states of the amplitude. It is very important to keep in mind that in joining the fermion lines among the different building blocks of the diagram one has to respect their orientation. This reflects the conservation of the electric charge. In addition, one should only consider diagrams that are topologically non-equivalent, i.e. that cannot be smoothly deformed into one another while keeping the external legs fixed.[4]

To show practically how Feynman diagrams are drawn, we consider Bhabha scattering: elastic electron–positron scattering

$$e^+ + e^- \longrightarrow e^+ + e^-.$$

Our problem is to compute the S-matrix amplitude to leading order in the electric charge. Since the QED vertex contains a photon line and our process does not have photons in the initial or the final states, drawing a Feynman diagram requires at least two vertices. In fact, the leading contribution is of order e^2 and comes from the following two diagrams

e^- e^- e^- e^-

$+(-1)\times$

e^+ e^+ e^+ e^+

Incoming and outgoing particles appear respectively on the left and the right of these diagrams. The identification of electrons and positrons is done by comparing the direction of the charge flux with the direction of propagation. For electrons the flux of charge goes in the direction of propagation, whereas for positrons they go in

4 From the point of view of the operator formalism, the requirement of considering only diagrams that are topologically nonequivalent comes from the fact that each diagram represents a certain Wick contraction in the correlation function of interaction–picture operators.

opposite directions. These are the only two diagrams that can be drawn to this order in perturbation theory.

It should be noticed that the two diagrams contribute with opposite signs. The reason is that the second diagram can be obtained from the first one by interchanging the incoming positron external line attached to the vertex on the left with that of the outgoing electron coming from the vertex on the right. This permutation of two fermions introduces the minus sign.

We have learned how to draw Feynman diagrams in QED. Now it is time to compute the contribution of each one to the amplitude using the Feynman rules. The idea is simple: each of the diagram's building blocks (vertices as well as external and internal lines) comes associated with a term. Putting all of them together according to certain rules results in the contribution of the corresponding diagram to the amplitude. In the case of QED in the Feynman gauge ($\xi = 1$), we have the following correspondence for vertices and internal propagators:

$$\alpha \longrightarrow \beta \quad \Longrightarrow \quad \left(\frac{i}{\not{p} - m + i\varepsilon}\right)_{\beta\alpha}$$

$$\mu \sim\sim\sim \nu \quad \Longrightarrow \quad \frac{-i\eta_{\mu\nu}}{p^2 + i\varepsilon}$$

$$\beta,\ \alpha,\ \mu \text{ (vertex)} \quad \Longrightarrow \quad -ie\gamma^{\mu}_{\beta\alpha}.$$

In addition, each vertex carries a factor $(2\pi)^4\delta^{(4)}(p_1 + p_2 + p_3)$ implementing momentum conservation, where we take the convention that all momenta are entering the vertex. The Feynman rules for other values of the gauge fixing parameter ξ only differ from the ones above by an extra term in the photon propagator. In addition, one has to perform an integration over the momenta running in internal lines with the measure

$$\int \frac{d^4 p}{(2\pi)^4}, \tag{6.42}$$

and introduce a factor of -1 for each fermion loop in the diagram.[5]

[5] The contribution of each diagram comes also multiplied by a symmetry factor that takes into account in how many ways a given Wick contraction can be done. In QED, however, these factors are equal to one for many diagrams.

A number of integrations over the internal momenta can be eliminated using the delta functions from the vertices. The result is a global delta function implementing the total momentum conservation in the diagram [cf. Eq. (6.8)]. In fact, there is a whole class of diagrams for which *all* integrations can be eliminated in this way. These are the so-called tree level diagrams containing no closed loops. As a general rule, there will be as many remaining integrations as the number of independent loops in the diagram.

Generically, finding the contribution of a Feynman diagram with ℓ independent loops involves the calculation of integrals of the form

$$I(p_1, \ldots, p_n) = \int \frac{d^4 q_1}{(2\pi)^4} \cdots \frac{d^4 q_\ell}{(2\pi)^4} f(q_1, \ldots, q_\ell; p_1, \ldots, p_n), \tag{6.43}$$

where $f(q_1, \ldots, q_\ell; p_1, \ldots, p_n)$ is a rational function of its arguments and $p_1, \ldots, p_n$ are the external momenta. In many cases these integrals are divergent. When the divergence is associated with the limit of small loop momenta it is called an *infrared divergence*. They usually cancel once all diagrams contributing to a given order in perturbation theory are added together. The second type of divergences that one expects in the integrals (6.43) comes from the region of large loop momenta. These are called *ultraviolet divergences*. They cannot be cancelled by adding the contribution of different diagram and have to be dealt with using the procedure of renormalization. We will discuss this problem in some detail in Chaps. 8 and 12.

This is not the end of the story. In the calculation of S-matrix amplitudes the contribution of the Feynman diagram contains factors associated with the external legs. These are the wave functions and/or polarization tensor of the corresponding asymptotic states containing all the information about the spin and polarization of the incoming and outgoing particles. In the case of QED these factors are:

Incoming fermion: α  $\Longrightarrow$ $u_\alpha(\mathbf{p}, s)$

Incoming antifermion: α  $\Longrightarrow$ $\overline{v}_\alpha(\mathbf{p}, s)$

Outgoing fermion:  α $\Longrightarrow$ $\overline{u}_\alpha(\mathbf{p}, s)$

Outgoing antifermion:  α $\Longrightarrow$ $v_\alpha(\mathbf{p}, s)$

Incoming photon: μ $\Longrightarrow$ $\varepsilon_\mu(\mathbf{p})$

Outgoing photon: μ $\Longrightarrow$ $\varepsilon_\mu(\mathbf{p})^*$

Here $u_\alpha(\mathbf{p}, s)$, $v_\alpha(\mathbf{p}, s)$ are the positive and negative energy solutions of the Dirac equation introduced in Chap. 3, whereas $\varepsilon_\mu(\mathbf{p}, \lambda)$ is the polarization tensor of the photon with polarization λ. Here we have assumed that the momenta for incoming (resp. outgoing) particles are entering (resp. leaving) the diagram, and all external momenta are on-shell, $p_i^2 = m_i^2$.

The use of Feynman diagrams is not restricted to quantum field theory, they can also be found in condensed matter physics and statistical mechanics. Their calculation is not an easy task. The number of diagrams contributing to a process grows very fast with the order of perturbation theory and the integrals arising in calculating loop diagrams soon get very complicated.

Feynman rules can be constructed for any interacting quantum field theory with scalar, vector or spinor fields. For the nonabelian gauge theories introduced in Chap. 4 these are:

α, i —— β, j $\Longrightarrow$ $\left(\dfrac{i}{\not{p} - m + i\varepsilon}\right)_{\beta\alpha} \delta_{ij}$

μ, A ν, B $\Longrightarrow$ $\dfrac{-i\eta_{\mu\nu}}{p^2 + i\varepsilon}\delta^{AB}$

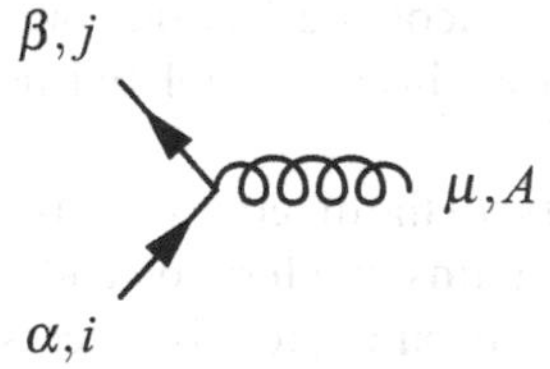

β, j; α, i; μ, A $\Longrightarrow$ $-ig\gamma^\mu_{\beta\alpha} t^A_{ij}$

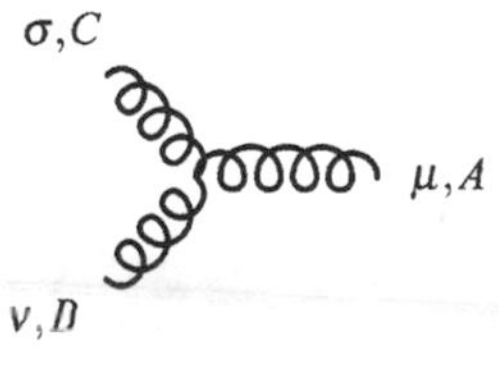

σ, C; ν, D; μ, A $\Longrightarrow$ $g f^{ABC}\left[\eta^{\mu\nu}(p_1^\sigma - p_2^\sigma) + \text{permutations}\right]$

σ, C λ, D

μ, A ν, B

$$\Longrightarrow \quad -ig^2\Big[f^{ABE}f^{CDE}\Big(\eta^{\mu\sigma}\eta^{\nu\lambda} - \eta^{\mu\lambda}\eta^{\nu\sigma}\Big) + \text{permutations}\Big]$$

As in the case of QED, each vertex includes a delta function implementing momentum conservation.

It is not our aim here to give a full and detailed description of the Feynman rules for nonabelian gauge theories. We only point out that, unlike the case of QED, here the gauge fields interact among themselves. These three and four gauge field vertices are a consequence of the cubic and quartic terms in the Lagrangian (4.54). The self-interactions of the nonabelian gauge field theories have crucial dynamical consequences and its at the very heart of their physical successes.

6.4 An Example: Compton Scattering at Low Energies

We illustrate now the use of Feynman diagrams in the calculation of observables in physical processes by studying an example with important physical applications. This is the calculation of the cross section for the dispersion of photons by free electrons: Compton scattering

$$\gamma(k, \varepsilon) + e^-(p, s) \longrightarrow \gamma(k', \varepsilon') + e^-(p', s'). \tag{6.44}$$

Inside the parenthesis we have indicated the momenta for the different particles, as well as the polarizations and spins of the incoming and outgoing photons and electrons respectively. We study this scattering in the nonrelativistic limit for the electrons.

The first step in our calculation is to identify all the diagrams contributing to (6.44) at leading order. Since the vertex of QED contains two fermion and one photon leg it is immediate to realize that any diagram contributing to this process must contain at least two vertices, so the leading contribution is of order e^2. A first diagram that can be drawn is:

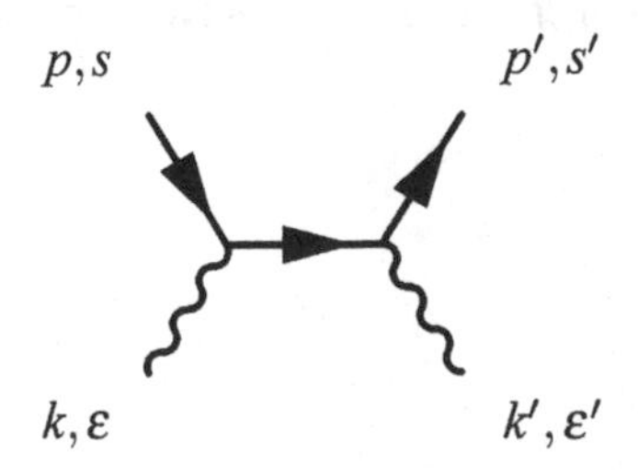

There is however a second possibility given by the following diagram:

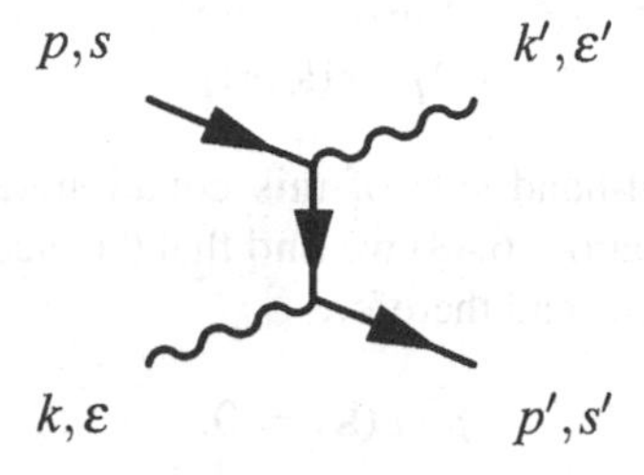

These two diagrams are topologically nonequivalent, since deforming one into the other requires changing the label of the external legs. In addition, unlike the example of the Bhabha scattering studied in the previous section, both diagrams contribute with the same sign. This is because they are related by interchanging the incoming with the outgoing photon. Since photons are bosons, no minus sign comes from this permutation.

Using the Feynman rules of QED we find the contribution of the two diagrams to be

$$
\begin{aligned}
&= (ie)^2 \overline{u}(\mathbf{p}',s') \not{\varepsilon}'(\mathbf{k}')^* \frac{\not{p}+\not{k}+m_e}{(p+k)^2-m_e^2} \not{\varepsilon}(\mathbf{k}) u(\mathbf{p},s) \\
&+ (ie)^2 \overline{u}(\mathbf{p}',s') \not{\varepsilon}(\mathbf{k}) \frac{\not{p}-\not{k}'+m_e}{(p-k')^2-m_e^2} \not{\varepsilon}'(\mathbf{k}')^* u(\mathbf{p},s).
\end{aligned}
\tag{6.45}
$$

where m_e is the electron mass and we have factored out $(2\pi)^4$ times the delta function implementing momentum conservation. As explained in Sect. 6.3, all incoming and outgoing particles are on-shell,

$$
p^2 = m_e^2 = p'^2 \quad \text{and} \quad k^2 = 0 = k'^2. \tag{6.46}
$$

Our calculation involves only tree-level diagrams, so there is no integration left over internal momenta. To get an explicit result we begin by simplifying the numerators. The following simple identity turns out to be very useful

$$
\not{a}\not{b} = -\not{b}\not{a} + 2(a \cdot b)\mathbf{1}. \tag{6.47}
$$

In addition, we are interested in Compton scattering at low energy when electrons are nonrelativistic particles. This is known in the literature as Thomson scattering. To be more precise, we take all spatial momenta much smaller than the electron mass

$$
|\mathbf{p}|,\ |\mathbf{k}|,\ |\mathbf{p}'|,\ |\mathbf{k}'| \ll m_e. \tag{6.48}
$$

In this approximation, the amplitude for Compton scattering simplifies substantially. Let us begin with the first term in Eq. (6.45). Applying the identity (6.47) we obtain

$$
\begin{aligned}
(\not{p} + \not{k} + m_e)\not{\varepsilon}(\mathbf{k})u(\mathbf{p}, s) = & -\not{\varepsilon}(\mathbf{k})(\not{p} - m_e)u(\mathbf{p}, s) + \not{k}\not{\varepsilon}(\mathbf{k})u(\mathbf{p}, s) \\
& + 2p \cdot \varepsilon(\mathbf{k})u(\mathbf{p}, s).
\end{aligned} \tag{6.49}
$$

The first term on the right-hand side of this equation vanishes using Eq. (3.45). Moreover, in the approximation (6.48) we find that the electrons' four-momenta can be written $p^\mu, p'^\mu \approx (m_e, 0)$ and therefore

$$
p \cdot \varepsilon(\mathbf{k}) = 0. \tag{6.50}
$$

This follows from the absence of the temporal photon polarization, $\varepsilon^0(\mathbf{k}) = 0$. Thus, we conclude that at low energies

$$
(\not{p} + \not{k} + m_e)\not{\varepsilon}(\mathbf{k})u(\mathbf{p}, s) = \not{k}\not{\varepsilon}(\mathbf{k})u(\mathbf{p}, s) \tag{6.51}
$$

and similarly for the second term in Eq. (6.45)

$$
(\not{p} - \not{k}' + m_e)\not{\varepsilon}'(\mathbf{k}')^* u(\mathbf{p}, s) = -\not{k}'\not{\varepsilon}'(\mathbf{k}')^* u(\mathbf{p}, s). \tag{6.52}
$$

Next, we turn to the denominators in (6.45). Using the mass-shell condition we find

$$
\begin{aligned}
(p + k)^2 - m_e^2 &= p^2 + k^2 + 2p \cdot k - m_e^2 = 2p \cdot k \\
&= 2\omega_p|\mathbf{k}| - 2\mathbf{p} \cdot \mathbf{k}
\end{aligned} \tag{6.53}
$$

and

$$
\begin{aligned}
(p - k')^2 - m_e^2 &= p^2 + k'^2 + 2p \cdot k' - m_e^2 = -2p \cdot k' \\
&= -2\omega_p|\mathbf{k}'| + 2\mathbf{p} \cdot \mathbf{k}'.
\end{aligned} \tag{6.54}
$$

Working again in the low energy approximation (6.48), these two expressions simplify to

$$
(p + k)^2 - m_e^2 \approx 2m_e|\mathbf{k}|, \quad (p - k')^2 - m_e^2 \approx -2m_e|\mathbf{k}'|. \tag{6.55}
$$

Collecting all results we obtain

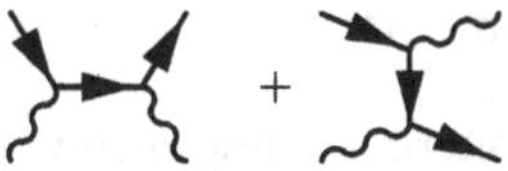

$$
\begin{aligned}
\approx \; & \frac{(ie)^2}{2m_e}\overline{u}(\mathbf{p}', s')\left[\not{\varepsilon}'(\mathbf{k}')^* \frac{\not{k}}{|\mathbf{k}|}\not{\varepsilon}(\mathbf{k}) \right. \\
& \left. + \not{\varepsilon}(\mathbf{k})\frac{\not{k}'}{|\mathbf{k}'|}\not{\varepsilon}'(\mathbf{k}')^*\right] u(\mathbf{p}, s).
\end{aligned} \tag{6.56}
$$

Using the identity (6.47) a number of times, as well as the transversality condition of the polarization vectors (4.32), we end up with a simpler expression

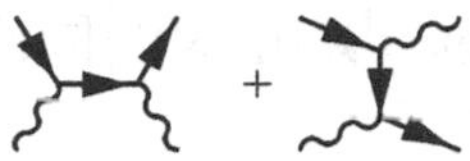

$$\approx \frac{e^2}{m_e}\left[\varepsilon(\mathbf{k})\cdot\varepsilon'(\mathbf{k}')^*\right]\overline{u}(\mathbf{p}',s')\frac{\not k}{|\mathbf{k}|}u(\mathbf{p},s) + \frac{e^2}{2m_e}\overline{u}(\mathbf{p}',s')\not\varepsilon(\mathbf{k})\not\varepsilon'(\mathbf{k}')^*\left(\frac{\not k}{|\mathbf{k}|}-\frac{\not k'}{|\mathbf{k}'|}\right)u(\mathbf{p},s). \tag{6.57}$$

With a little extra effort one can show that the second term on the right-hand side of this equation vanishes. First we notice that in the low energy limit $|\mathbf{k}| \approx |\mathbf{k}'|$. If, in addition, we use the conservation of momentum $k - k' = p' - p$ and the identity (3.45) we can write

$$\overline{u}(\mathbf{p}',s')\not\varepsilon(\mathbf{k})\not\varepsilon'(\mathbf{k}')^*\left(\frac{\not k}{|\mathbf{k}|}-\frac{\not k'}{|\mathbf{k}'|}\right)u(\mathbf{p},s) \approx \frac{1}{|\mathbf{k}|}\overline{u}(\mathbf{p}',s')\not\varepsilon(\mathbf{k})\not\varepsilon'(\mathbf{k}')^*(\not p'-m_e)u(\mathbf{p},s). \tag{6.58}$$

Next we use the identity (6.47) to take the term $(\not p' - m_e)$ to the right. Finally, keeping in mind that in the low energy limit the electron four-momenta are orthogonal to the photon polarization vectors [see Eq. (6.50)], we conclude that

$$\overline{u}(\mathbf{p}',s')\not\varepsilon(\mathbf{k})\not\varepsilon'(\mathbf{k}')^*(\not p' - m_e)u(\mathbf{p},s) = \overline{u}(\mathbf{p}',s')(\not p' - m_e)\not\varepsilon(\mathbf{k})\not\varepsilon'(\mathbf{k}')^*u(\mathbf{p},s) = 0 \tag{6.59}$$

where the last identity follows from the equation satisfied by the conjugate positive–energy spinor, $\overline{u}(\mathbf{p}',s')(\not p' - m_e) = 0$.

After all these lengthy manipulations we have finally arrived at the expression of the invariant amplitude for the Compton scattering at low energies

$$i\mathscr{M}_{i\to f} = \frac{e^2}{m_e}\left[\varepsilon(\mathbf{k})\cdot\varepsilon'(\mathbf{k}')^*\right]\overline{u}(\mathbf{p}',s')\frac{\not k}{|\mathbf{k}|}u(\mathbf{p},s). \tag{6.60}$$

To calculate the cross section we need to compute $|\mathscr{M}_{i\to f}|^2$, as shown in Eq. (6.18). For many physical applications, however, one is interested in the dispersion of photons with a given polarization by electrons that are not polarized, i.e. whose spins are randomly distributed. To describe this physical setup we have to average over initial electron polarization (since we do not know them) and sum over all possible final electron polarization (because our detector is blind to this quantum number),

$$\overline{|i\mathscr{M}_{i\to f}|^2} = \frac{1}{2}\left(\frac{e^2}{m_e|\mathbf{k}|}\right)^2\left|\varepsilon(\mathbf{k})\cdot\varepsilon'(\mathbf{k}')^*\right|^2\sum_{s=\pm\frac{1}{2}}\sum_{s'=\pm\frac{1}{2}}\left|\overline{u}(\mathbf{p}',s')\not k u(\mathbf{p},s)\right|^2. \tag{6.61}$$

The factor of $\frac{1}{2}$ comes from averaging over the two possible polarizations of the incoming electrons. The sums in this expression can be calculated without much difficulty. Expanding the absolute value

$$\sum_{s=\pm\frac{1}{2}}\sum_{s'=\pm\frac{1}{2}}\left|\overline{u}(\mathbf{p}',s')\not{k}u(\mathbf{p},s)\right|^2=\sum_{s=\pm\frac{1}{2}}\sum_{s'=\pm\frac{1}{2}}\left[u(\mathbf{p},s)^{\dagger}\not{k}^{\dagger}\overline{u}(\mathbf{p}',s')^{\dagger}\right]\left[\overline{u}(\mathbf{p}',s')\not{k}u(\mathbf{p},s)\right], \tag{6.62}$$

and using that $\gamma^{\mu\dagger}=\gamma^0\gamma^\mu\gamma^0$ one finds, after some manipulations,

$$\begin{aligned}\sum_{s=\pm\frac{1}{2}}\sum_{s'=\pm\frac{1}{2}}\left|\overline{u}(\mathbf{p}',s')\not{k}u(\mathbf{p},s)\right|^2&=\left[\sum_{s=\pm\frac{1}{2}}u_\alpha(\mathbf{p},s)\overline{u}_\beta(\mathbf{p},s)\right](\not{k})_{\beta\sigma}\left[\sum_{s'=\pm\frac{1}{2}}u_\sigma(\mathbf{p}',s')\overline{u}_\rho(\mathbf{p}',s')\right](\not{k})_{\rho\alpha}\\&=\mathrm{Tr}\left[(\not{p}+m_e)\not{k}(\not{p}'+m_e)\not{k}\right],\end{aligned} \tag{6.63}$$

where the final result has been obtained using the completeness relations (3.51). The final evaluation of the trace can be done using the relation (6.47) to commute $\not{p}'$ and $\not{k}$. Using $k^2=0$ and that we are working in the low energy limit, we have[6]

$$\mathrm{Tr}\left[(\not{p}+m_e)\not{k}(\not{p}'+m_e)\not{k}\right]=2(p\cdot k)(p'\cdot k)\mathrm{Tr}\mathbf{1}\approx 8m_e^2|\mathbf{k}|^2. \tag{6.64}$$

With this we arrive at the following value for the invariant amplitude for the Compton scattering at low energies

$$\overline{|i\mathscr{M}_{i\to f}|^2}=4e^4\left|\varepsilon(\mathbf{k})\cdot\varepsilon'(\mathbf{k}')^*\right|^2. \tag{6.65}$$

We have reached the end of our calculation. Plugging $\overline{|i\mathscr{M}_{i\to f}|^2}$ into (6.22) and dropping the integration over the direction of the outgoing particles we find the differential cross section for the scattering of a photon by an electron at rest

$$\frac{d\sigma}{d\Omega}=\frac{1}{64\pi^2m_e^2}\overline{|i\mathscr{M}_{i\to f}|^2}=\left(\frac{e^2}{4\pi m_e}\right)^2\left|\varepsilon(\mathbf{k})\cdot\varepsilon'(\mathbf{k}')^*\right|^2. \tag{6.66}$$

The prefactor of the last expression is precisely the square of the classical electron radius r_{cl}. In fact, the result can be rewritten as

$$\frac{d\sigma}{d\Omega}=\frac{3}{8\pi}\sigma_T\left|\varepsilon(\mathbf{k})\cdot\varepsilon'(\mathbf{k}')^*\right|^2, \tag{6.67}$$

where σ_T is the total Thomson cross section

$$\sigma_T=\frac{e^4}{6\pi m_e^2}=\frac{8\pi}{3}r_{\mathrm{cl}}^2, \tag{6.68}$$

obtained from integrating (6.66) over angles.

One of the most important physical consequences of Eq. (6.67) is that a net polarization is produced in the scattering of unpolarized radiation off nonrelativistic charges. To see this, we take the Thomson differential cross section and average over

---

6 We use also the fact that the trace of the product of an odd number of Dirac matrices is always zero.

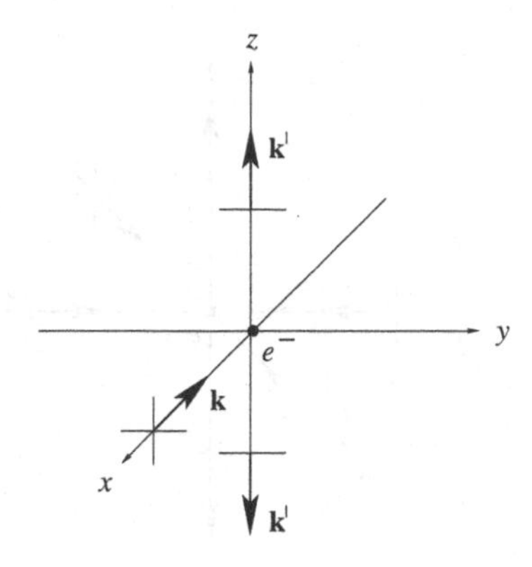

Fig. 6.2 This figure illustrate Eq. (6.70). The "vertical" component of the unpolarized radiation arriving from the x direction is suppressed in the photons scattered along the z axis. This results in a linear polarization of the scattered radiation

the polarization of the incoming photon. Denoting by $\varepsilon(\mathbf{k}, a)$, with $a = 1, 2$, a basis for the photon polarizations, this average gives

$$\frac{1}{2}\sum_{a=1,2}\left|\varepsilon(\mathbf{k},a)\cdot\varepsilon'(\mathbf{k}')^*\right|^2 = \left[\frac{1}{2}\sum_{a=1,2}\varepsilon_i(\mathbf{k},a)\varepsilon_j(\mathbf{k},a)^*\right]\varepsilon_j(\mathbf{k}')\varepsilon_i(\mathbf{k}')^*. \quad (6.69)$$

The sum inside the brackets can be computed using the normalization of the polarization vectors, $|\varepsilon(\mathbf{k}, n)|^2 = 1$, and the transversality condition $\mathbf{k}\cdot\varepsilon(\mathbf{k}, n) = 0$

$$\begin{aligned}\frac{1}{2}\sum_{a=1,2}\left|\varepsilon(\mathbf{k},a)\cdot\varepsilon'(\mathbf{k}')^*\right|^2 &= \frac{1}{2}\left(\delta_{ij}-\frac{k_ik_j}{|\mathbf{k}|^2}\right)\varepsilon'_j(\mathbf{k}')\varepsilon'_i(\mathbf{k}')^* \\ &= \frac{1}{2}\left[1-|\hat{\mathbf{k}}\cdot\varepsilon'(\mathbf{k}')|^2\right], \end{aligned} \quad (6.70)$$

where $\hat{\mathbf{k}} = \frac{\mathbf{k}}{|\mathbf{k}|}$ is the unit vector in the direction of the incoming photon.

From the last equation we conclude that Thomson scattering suppresses all polarizations parallel to the direction of the incoming photon. At the same time, the differential cross section reaches its maximum values when the polarization of the scattered photon lies in the plane normal to $\hat{\mathbf{k}}$. This is represented in Fig. 6.2, where nonpolarized radiation coming from the x direction is scattered by a nonrelativistic electron. According to Eq. (6.70) the vertical polarization is fully suppressed in the radiation scattered along the z direction, thus producing linear polarization.

6.5 Polarization of the Cosmic Microwave Background Radiation

The differential cross section of Thomson scattering we have derived is relevant in many areas of physics, but its importance is paramount in the study of the cosmological microwave background radiation (CMB). Here we are going to review briefly

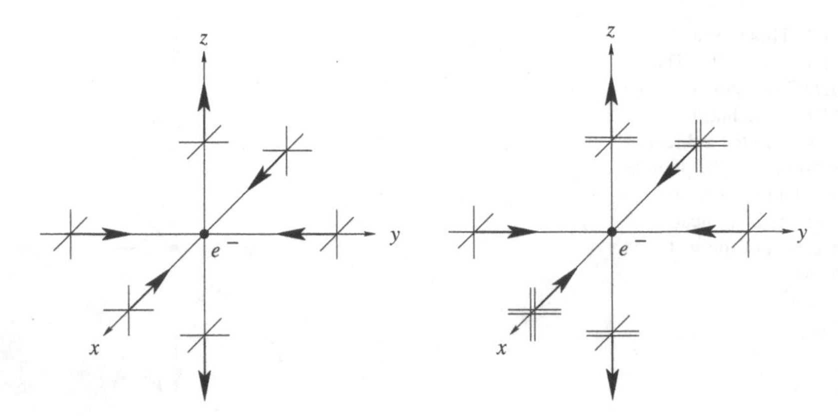

Fig. 6.3 In these figures the larger density of unpolarized photons arriving from different directions is represented through two parallel lines indicating the polarization. The *left panel* shows the scattering of isotropic radiation by a free electron and how this does not produce any net polarization in the scattered photons. On the *right panel*, on the other hand, the anisotropy in the intensity of the radiation has a quadrupole component, being larger along the x direction. The result is a net polarization in the photons scattered along the z axis

how polarization emerges in the cosmic radiation and discuss why its detection could serve as a window to the physics of the very early universe. Our presentation will be rather sketchy. A thorough analysis of this problem can be found in many places, such as [1, 2].

Just before recombination the universe is filled with a plasma of electrons interacting with photons via Compton scattering. This plasma has a temperature of the order of $T \approx 1\,\text{keV}$ and therefore electrons are nonrelativistic ($T \ll m_e \sim 0.5\,\text{MeV}$), so the approximations leading to the Thomson differential cross section apply. At the last scattering surface there is no way to know the polarization state of the photons in the plasma before they are scattered by electrons to produce the CMB radiation that we detect today. Therefore we have to average over incoming polarizations as shown at the end of the previous section.

The relation between the polarization of the CMB and the anisotropies in the density of photons at last scattering can be understood with the help of Fig. 6.2. We consider the polarization of photons traveling along the z direction resulting from the scattering of photons traveling along the x and y axis. Since Thomson scattering suppresses all polarizations in the direction of the incoming photons we find that the two polarizations in the scattered radiation come from the "horizontal" polarizations of the incoming photons. If the number of photons coming from the x and y directions are the same no net polarization is produced. This is shown in the left panel of Fig. 6.3. It is an instructive exercise to check that no polarization is produced either in the presence of a dipolar anisotropy.

When the anisotropy has a quadrupole component, on the other hand, the situation changes. Then the intensity of the unpolarized radiation approaching from the x and

y directions is different and so is the relative intensity of the two polarizations in the scattered radiation along the z axis. The outgoing radiation is then polarized.

The previous heuristic arguments show that the presence of a net polarization in the CMB is the smoking gun of quadrupole anisotropies in the photon distribution at the last scattering surface. There are several possible physical causes for such an anisotropy. One of them, however, is specially glaring. Gravitational waves propagating through the plasma induce changes in its density with precisely the quadrupole component necessary to produce the polarization in the CMB radiation.

Now we make this discussion more precise. The polarization of radiation can be described using three Stokes parameters: Q measures the excess of horizontal versus vertical, U of diagonal versus antidiagonal and V of left versus right polarization. CMB experiments allow the measurements of these parameters for the background radiation arriving from a direction in the sky specified by a unit vector $\hat{\mathbf{n}}$.

To compute the parameter $Q(\hat{\mathbf{n}})$ we consider the polarizations along the directions defined by the unit vectors $\hat{\mathbf{e}}_{\leftrightarrow} = -\hat{\mathbf{e}}_{\varphi}$ and $\hat{\mathbf{e}}_{\updownarrow} = -\hat{\mathbf{e}}_{\theta}$, normal to the plane defined by $\hat{\mathbf{n}}$ (see left panel in Fig. 6.4). We denote by $f(\hat{\mathbf{k}}, \hat{n})$ the distribution function of photons in the plasma with momentum along the unit vector $\hat{\mathbf{k}}$ at the last scattering surface in the sky direction $\hat{\mathbf{n}}$. This distribution function does not depend on the polarization of the photons because the incoming radiation is taken to be unpolarized. Using the expression of the Thomson cross section (6.67), the Stokes parameter $Q(\hat{\mathbf{n}})$ can be written as

$$Q(\hat{\mathbf{n}}) \sim \sum_{a=1,2} \int d\Omega(\hat{\mathbf{k}}) f(\hat{\mathbf{k}}, \hat{\mathbf{n}}) \left[|\varepsilon(\mathbf{k}, a) \cdot \hat{\mathbf{e}}_{\leftrightarrow}|^2 - |\varepsilon(\mathbf{k}, a) \cdot \hat{\mathbf{e}}_{\updownarrow}|^2 \right], \qquad (6.71)$$

where we integrate over the directions of the incoming photons and have omitted a global normalization constant. To write this expression we have taken the intensity of scattered radiation to be proportional to the Thomson differential cross section averaged over polarizations. The result is integrated over the direction of the incoming photons weighted by the distribution function. The sum over polarizations can be explicitly done using the result derived in Eq. (6.69) to give

$$Q(\hat{\mathbf{n}}) \sim -\frac{1}{2} \int d\Omega(\hat{\mathbf{k}}) f(\hat{\mathbf{k}}, \hat{\mathbf{n}}) \left[(\hat{\mathbf{k}} \cdot \hat{\mathbf{e}}_{\leftrightarrow})^2 - (\hat{\mathbf{k}} \cdot \hat{\mathbf{e}}_{\updownarrow})^2 \right]. \qquad (6.72)$$

In order to evaluate the parameter $U(\hat{\mathbf{n}})$ we need to consider the polarizations along the unit vectors defined by (see right panel in Fig. 6.4)

$$\hat{\mathbf{e}}_{\nearrow} = -\frac{1}{\sqrt{2}}(\hat{\mathbf{e}}_{\varphi} + \hat{\mathbf{e}}_{\theta}), \quad \hat{\mathbf{e}}_{\searrow} = -\frac{1}{\sqrt{2}}(\hat{\mathbf{e}}_{\varphi} - \hat{\mathbf{e}}_{\theta}). \qquad (6.73)$$

This parameter is then given by the difference in intensity of the scattered radiation with these polarizations, namely

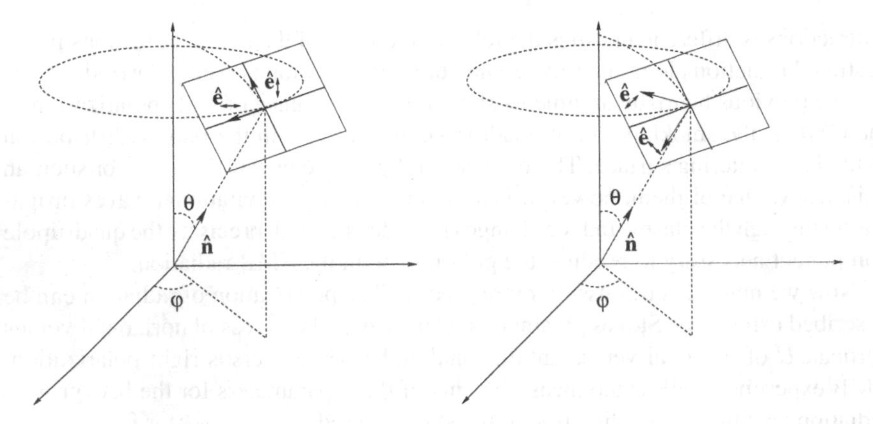

Fig. 6.4 Polarization states used to define the Stokes parameter $Q(\hat{\mathbf{n}})$ and $U(\hat{\mathbf{n}})$ for a photon scattered by a nonrelativistic electron and arriving from the direction $\hat{\mathbf{n}}$. The notation used in the unit vectors $\hat{\mathbf{e}}_{\swarrow\!\nearrow}$ and $\hat{\mathbf{e}}_{\nwarrow\!\searrow}$ reflects the point of view of an observer located at the origin looking in the direction defined by $\hat{\mathbf{n}}$

$$\begin{aligned} U(\mathbf{n}) &\sim \sum_{a=1,2}\int d\Omega(\hat{\mathbf{k}})f(\hat{\mathbf{k}},\mathbf{n})\Big[|\varepsilon(\mathbf{k},a)\cdot\hat{\mathbf{e}}_{\swarrow\!\nearrow}|^2-|\varepsilon(\mathbf{k},a)\cdot\hat{\mathbf{e}}_{\nwarrow\!\searrow}|^2\Big] \\ &= -\frac{1}{2}\int d\Omega(\hat{\mathbf{k}})f(\hat{\mathbf{k}},\hat{\mathbf{n}})\Big[(\hat{\mathbf{k}}\cdot\hat{\mathbf{e}}_{\swarrow\!\nearrow})^2-(\hat{\mathbf{k}}\cdot\hat{\mathbf{e}}_{\nwarrow\!\searrow})^2\Big], \end{aligned} \tag{6.74}$$

where in the second line we have carried out the sum over incoming polarizations. A look at Fig. 6.4 shows that $Q(\hat{\mathbf{n}})$ and $U(\hat{\mathbf{n}})$ can be transformed into one another, up to a sign, by a rotation of $\frac{\pi}{4}$ along the line of sight $\hat{\mathbf{n}}$.

Finally, the Stokes parameter $V(\hat{\mathbf{u}})$ measures the net circular polarization of the CMB photons arriving from the last scattering surface

$$\begin{aligned} V(\hat{\mathbf{n}}) &\sim \sum_{a=1,2}\int d\Omega(\hat{\mathbf{k}})f(\hat{\mathbf{k}},\mathbf{n})\Big[|\varepsilon(\mathbf{k},a)\cdot\hat{\mathbf{e}}_+|^2-|\varepsilon(\mathbf{k},a)\cdot\hat{\mathbf{e}}_-|^2\Big] \\ &= \int d\Omega(\hat{\mathbf{k}})f(\hat{\mathbf{k}},\mathbf{n})\Big[|\hat{\mathbf{k}}\cdot\hat{\mathbf{e}}_+|^2-|\hat{\mathbf{k}}\cdot\hat{\mathbf{e}}_-|^2\Big]=0, \end{aligned} \tag{6.75}$$

where $\hat{\mathbf{e}}_\pm=-\frac{1}{\sqrt{2}}(\hat{\mathbf{e}}_\varphi\pm i\hat{\mathbf{e}}_\theta)$ and the last identity follows immediately from $\hat{\mathbf{e}}_\pm^*=\hat{\mathbf{e}}_\mp$. This result reflects the fact that Thomson scattering does not distinguish between left and right polarizations.

The measurement of $Q(\hat{\mathbf{n}})$ and $U(\hat{\mathbf{n}})$ provides important information about the distribution function of photons at decoupling $f(\hat{\mathbf{k}},\hat{\mathbf{n}})$, as we will see shortly. In order to carry out the integration over $\hat{\mathbf{k}}$ in Eqs. (6.72) and (6.74) we use the system of coordinates defined by the three unit vectors $\hat{\mathbf{e}}_\updownarrow$, $\hat{\mathbf{e}}_\leftrightarrow$ and $\hat{\mathbf{n}}$, as shown in Fig. 6.5. After a bit of algebra we arrive at

$$Q(\hat{\mathbf{n}})\pm iU(\hat{\mathbf{n}})\sim-\int d\Omega(\theta',\varphi')f(\theta',\varphi';\hat{\mathbf{n}})\sin^2\theta' e^{\pm 2i\varphi'}, \tag{6.76}$$

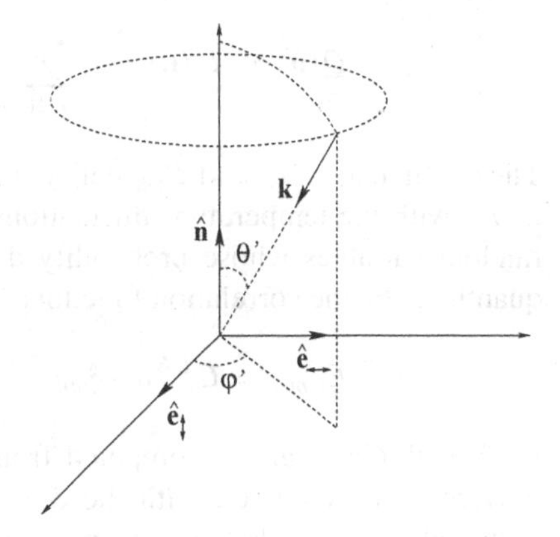

Fig. 6.5 A photon with momentum $\mathbf{k}$ is scattered by a nonrelativistic electron located at the origin. The frame vectors $\hat{\mathbf{n}}$, $\hat{\mathbf{e}}_{\leftrightarrow}$ and $\hat{\mathbf{e}}_{\updownarrow}$ are the ones shown in the *left panel* of Fig. 6.4

where the dependence on the unit vector $\hat{\mathbf{k}}$ is indicated by its polar coordinates (φ', θ').

There is something very interesting about this expression. The functional dependence on $\hat{\mathbf{k}}$ of the term multiplying $f(\hat{\mathbf{k}}, \hat{\mathbf{n}})$ is that of the spherical harmonics

$$Y_2^{\pm 2}(\theta', \varphi') = 3\sqrt{\frac{5}{96\pi}}\sin^2\theta' e^{\pm 2i\varphi'}. \tag{6.77}$$

Thus, the only way to make the integral (6.76) nonzero is that the distribution function $f(\hat{\mathbf{k}}, \hat{\mathbf{n}})$ contains a quadrupole anisotropy. In other words, what we have concluded is that the measurement of the polarization of the CMB gives direct information about the quadrupole component of the distribution function of photons at decoupling!

The distinction between $Q(\hat{\mathbf{n}})$ and $U(\hat{\mathbf{n}})$ is rather arbitrary, since one parameter can be transformed into the other by an appropriate rotation along $\hat{\mathbf{n}}$. In fact, under such a rotation of angle ϕ the complex combinations of the two Stokes parameters in Eq. (6.76) transform as

$$Q(\hat{\mathbf{n}}) \pm iU(\hat{\mathbf{n}}) \longrightarrow e^{\mp 2i\phi}\left[Q(\hat{\mathbf{n}}) \pm iU(\hat{\mathbf{n}})\right]. \tag{6.78}$$

Now, $Q(\hat{\mathbf{n}}) \pm iU(\hat{\mathbf{n}})$ defines two complex functions on the two–dimensional sphere whose points are labelled by the unit vector $\hat{\mathbf{n}}$. Eq. (6.78) defines a local SO(2) rotations in the sphere under which $Q(\hat{\mathbf{n}}) \pm iU(\hat{\mathbf{n}})$ transform as quantities with spin ± 2. Were they scalars, we could expand them using the ordinary spherical harmonics $Y_\ell^m(\hat{\mathbf{n}})$. Due however to their nontrivial transformation properties, the expansion has to be made in terms of a basis of eigenfunctions of the Laplace operators on the sphere S^2 with the appropriate transformations under SO(2) local rotations. The sought for basis of functions are generalizations of the standard spherical harmonics called the spin–weighted spherical harmonics of spin ± 2, denoted by ${}_{\pm 2}Y_\ell^m(\hat{\mathbf{n}})$. Here we will not elaborate on their properties (see [2] for details). For us it is enough to know that they can be used to write the expansion

$$Q(\hat{\mathbf{n}}) \pm iU(\hat{\mathbf{n}}) = -\sum_{\ell=0}^{\infty} \sum_{m=-\ell}^{\ell} (E_{\ell m} \pm i B_{\ell m})_{\pm 2} Y_{\ell}^{m}(\hat{\mathbf{n}}). \tag{6.79}$$

The coefficients $E_{\ell m}$ and $B_{\ell m}$ define the E- and B-mode of the CMB polarization .

As with the temperature fluctuations, the CMB polarization can be handled as random variables whose probability distributions are characterized, among other quantities, by the correlation functions

$$\langle E_{\ell m}^{*} E_{\ell' m'} \rangle = C_{\ell}^{EE} \delta_{\ell\ell'} \delta_{mm'}, \quad \langle B_{\ell m}^{*} B_{\ell' m'} \rangle = C_{\ell}^{BB} \delta_{\ell\ell'} \delta_{mm'}. \tag{6.80}$$

C_{ℓ}^{EE} and C_{ℓ}^{BB} can be computed from different theoretical models of the early universe and compared with the direct measurements of the CMB polarization. Although the E-mode has been measured by WMAP, the detection of the B-mode remains one of the big observational challenges in CMB physics. This is an important issue: among other possible sources, a nonvanishing B-mode would be produced by primordial gravitational waves, a generic prediction of inflation.

References

1. Dodelson, S.: Modern Cosmology. Academic Press, London (2003)
2. Durrer, R.: The Cosmic Microwave Background. Cambridge University Press, Cambridge (2008)

Chapter 7
Symmetries I: Continuous Symmetries

The concept of symmetry is paramount in modern Physics. In this chapter we are going to deal with the implementation of symmetries in quantum field theory. After reviewing the relation between continuous symmetries and conservations laws, we study how symmetries are realized quantum mechanically and in which way different realizations reflect in the spectrum of the theory. Our aim is to describe the concept of spontaneous symmetry breaking, which is crucial to our current understanding of how particle masses emerge in the standard model. A number of subtleties in how and when spontaneous symmetry breaking can occur are described towards the end of the chapter. The focus of the present chapter centers on continuous symmetries. The physics of discrete symmetries will be taken up in Chap. 11.

7.1 Noether's Theorem

In classical mechanics and classical field theory there is a basic result relating symmetries and conserved charges. This is called Noether's theorem and states that for each continuous symmetry of the system there is a conserved current. In its simplest version in classical mechanics it is easy to prove. Let us consider a system whose action $S[q_i]$ is invariant under a transformation $q_i(t) \rightarrow q_i'(t,\varepsilon)$ labelled by a continuous parameter ε. This means that, without using the equations of motion, the Lagrangian changes at most by a total time derivative

$$L(q',\dot{q}') = L(q,\dot{q}) + \frac{d}{dt}f(q,\varepsilon), \tag{7.1}$$

where $f(q,\varepsilon)$ is a function of the coordinates. If $\varepsilon \ll 1$ we can consider an infinitesimal variation of the coordinates $\delta_\varepsilon q_i(t)$ and the transformation (7.1) of the Lagrangian implies

L. Álvarez-Gaumé and M. Á. Vázquez-Mozo, *An Invitation to Quantum Field Theory*, Lecture Notes in Physics 839, DOI: 10.1007/978-3-642-23728-7_7,

$$
\begin{aligned}
\frac{d}{dt} f(q, \delta\varepsilon) = \delta_\varepsilon L(q_i, \dot{q}_i) &= \frac{\partial L}{\partial q_i}\delta_\varepsilon q_i + \frac{\partial L}{\partial \dot{q}_i}\delta_\varepsilon \dot{q}_i \\
&= \left[\frac{\partial L}{\partial q_i} - \frac{d}{dt}\frac{\partial L}{\partial \dot{q}_i}\right]\delta_\varepsilon q_i + \frac{d}{dt}\left(\frac{\partial L}{\partial \dot{q}_i}\delta_\varepsilon q_i\right),
\end{aligned} \tag{7.2}
$$

When $\delta_\varepsilon q_i$ is applied on a solution to the equations of motion, the term inside the square brackets vanishes and we conclude that there is a conserved quantity

$$
\dot{Q} = 0 \quad \text{with} \quad Q \equiv \frac{\partial L}{\partial \dot{q}_i}\delta_\varepsilon q_i - f(q, \delta\varepsilon). \tag{7.3}
$$

Notice that in this derivation it is crucial that the symmetry depends on a continuous parameter since otherwise the infinitesimal variation of the Lagrangian in Eq. (7.2) does not make sense.

In classical field theory a similar result holds. Let us consider for simplicity a theory of a single field $\phi(x)$. We say that the variation $\delta_\varepsilon \phi$ depending on a continuous parameter ε is a symmetry of the theory if, again without using the equations of motion, the Lagrangian density changes by

$$
\delta_\varepsilon \mathscr{L} = \partial_\mu K^\mu. \tag{7.4}
$$

If this happens, the action remains invariant and so do the equations of motion. Working out now the variation of $\mathscr{L}$ under $\delta_\varepsilon \phi$ we find

$$
\begin{aligned}
\delta_\varepsilon \mathscr{L} &= \frac{\partial \mathscr{L}}{\partial(\partial_\mu \phi)}\partial_\mu \delta_\varepsilon \phi + \frac{\partial \mathscr{L}}{\partial \phi}\delta_\varepsilon \phi \\
&= \partial_\mu \left(\frac{\partial \mathscr{L}}{\partial(\partial_\mu \phi)}\delta_\varepsilon \phi\right) + \left[\frac{\partial \mathscr{L}}{\partial \phi} - \partial_\mu \left(\frac{\partial \mathscr{L}}{\partial(\partial_\mu \phi)}\right)\right]\delta_\varepsilon \phi \\
&= \partial_\mu K^\mu.
\end{aligned} \tag{7.5}
$$

If $\phi(x)$ is a solution to the equations of motion, the last term in the second line disappears, and we find a conserved current

$$
\partial_\mu J^\mu = 0 \quad \text{with} \quad J^\mu = \frac{\partial \mathscr{L}}{\partial(\partial_\mu \phi)}\delta_\varepsilon \phi - K^\mu. \tag{7.6}
$$

A conserved current implies the existence of a charge

$$
Q \equiv \int d^3x\, J^0(t, \mathbf{x}) \tag{7.7}
$$

which is conserved

$$
\frac{dQ}{dt} = \int d^3x\, \partial_0 J^0(t, \mathbf{x}) = -\int d^3x\, \partial_i J^i(t, \mathbf{x}) = 0, \tag{7.8}
$$

provided the fields vanish at infinity fast enough. Moreover, the conserved charge Q is a Lorentz scalar. After canonical quantization Q is promoted to an operator generating the symmetry on the fields

$$\delta\phi = i[\phi, Q]. \tag{7.9}$$

As an example of how Noether's theorem works, we consider a scalar field $\phi(x)$ with Lagrangian density $\mathscr{L}$. Being $\phi(x)$ a scalar, its transformation under the Poincaré group $x \to x'$ is $\phi'(x') = \phi(x)$. Performing in particular a space-time translation $x^{\mu'} = x^\mu + a^\mu$ we have

$$\phi'(x) - \phi(x) = -a^\mu \partial_\mu \phi + \mathscr{O}(a^2) \Longrightarrow \delta\phi = -a^\mu \partial_\mu \phi. \tag{7.10}$$

That the theory is invariant under the Poincaré group means that the Lagrangian density is also a scalar quantity. Thus, it should also transform under translations as

$$\delta\mathscr{L} = -a^\mu \partial_\mu \mathscr{L}. \tag{7.11}$$

Noether's theorem implies then the existence of a conserved current. Applying the previous results we conclude that this is given by

$$J^\mu = -\frac{\partial\mathscr{L}}{\partial(\partial_\mu\phi)} a^\nu \partial_\nu \phi + a^\mu \mathscr{L} \equiv -a_\nu T^{\mu\nu}, \tag{7.12}$$

where we introduced the energy-momentum tensor

$$T^{\mu\nu} = \frac{\partial\mathscr{L}}{\partial(\partial_\mu\phi)} \partial^\nu \phi - \eta^{\mu\nu} \mathscr{L}. \tag{7.13}$$

We have found that associated with the invariance of the theory with respect to space-time translations there are four conserved currents defined by $T^{\mu\nu}$ with $\nu = 0, \ldots, 3$, each one associated with the translation along a space-time direction. These four currents form a rank-two tensor under Lorentz transformations satisfying

$$\partial_\mu T^{\mu\nu} = 0. \tag{7.14}$$

The associated conserved charges are given by

$$P^\nu = \int d^3x T^{0\nu} \tag{7.15}$$

and correspond to the total energy-momentum content of the field configuration. Therefore the energy density of the field is given by T^{00} while T^{0i} is the momentum density. In the quantum theory P^μ are the generators of space-time translations.

Another example of a symmetry related with a physically relevant conserved charge is the global phase invariance of the Dirac Lagrangian (3.36), $\psi \to e^{i\theta}\psi$.

For small θ this corresponds to the variations $\delta_\theta \psi = i\theta\psi$, $\delta_\theta \overline{\psi} = -i\theta\overline{\psi}$ and using Noether's theorem we obtain the conserved current

$$j^\mu = \overline{\psi}\gamma^\mu\psi, \quad \partial_\mu j^\mu = 0. \tag{7.16}$$

The associated charge is

$$Q = \int d^3x \overline{\psi}\gamma^0\psi = \int d^3x \psi^\dagger\psi. \tag{7.17}$$

In physics there are several instances of global U(1) symmetries acting as phase shifts on spinors. This is the case, for example, of the baryon and lepton number symmetries in the standard model. A more familiar case is the U(1) local symmetry associated with electromagnetism. Although this is a local symmetry, $\theta \to q\varepsilon(x)$, the Lagrangian is invariant also under global transformations with $\varepsilon(x)$ constant and there is a conserved current $j^\mu = q\overline{\psi}\gamma^\mu\psi$. In Eq. (4.40) we learned how spinors in QED are coupled to the photon field precisely through this current. Its time component is the electric charge density ρ, while the spatial components make the current density vector $\mathbf{j}$.

The previous analysis can be extended also to nonabelian unitary global symmetries acting on N species of fermions as

$$\psi_i \longrightarrow U_{ij}\psi_j, \tag{7.18}$$

where U_{ij} is a $N \times N$ unitary matrix, $U^\dagger U = UU^\dagger = \mathbf{1}$. This transformation leaves invariant the Lagrangian

$$\mathscr{L} = i\overline{\psi}_j \not{\partial}\psi_j - m\overline{\psi}_j\psi_j, \tag{7.19}$$

where we sum over repeated indices. If we write the matrix U in terms of the N^2 hermitian group generators T^A of U(N) as

$$U = \exp\left(i\alpha^A T^A\right), \quad (T^A)^\dagger = T^A, \tag{7.20}$$

the conserved currents are found to be

$$j^{\mu A} = \overline{\psi}_i T^A_{ij}\gamma^\mu\psi_j, \quad \partial_\mu j^\mu = 0. \tag{7.21}$$

with N^2 conserved charges

$$Q^A = \int d^3x \psi_i^\dagger T^A_{ij}\psi_j \tag{7.22}$$

The group U(N) of $N \times N$ unitary matrices admits the decomposition U(N) = $U(1) \times \mathrm{SU}(N)$. The U(1) factor corresponds to the element $U = e^{i\alpha^0}\mathbf{1}$ multiplying all spinor fields by the same phase. The corresponding charge

$$Q^0 = \int d^3x \psi_i^\dagger \psi_i \tag{7.23}$$

measures, in the quantum theory, the number of fermions minus the number of antifermions. It commutes with the other N^2-1 charges associated with the nontrivial SU(N) part of the global symmetry group.

As an example of these internal unitary symmetries, we mention the approximate flavor symmetries in hadron physics. Ignoring charge and mass differences, the QCD Lagrangian is invariant under the following unitary symmetry acting on the quarks u and d

$$\begin{pmatrix} u \\ d \end{pmatrix} \longrightarrow M \begin{pmatrix} u \\ d \end{pmatrix}, \tag{7.24}$$

where $M \in \mathrm{U}(2) = \mathrm{U}(1)_B \times \mathrm{SU}(2)$. The $\mathrm{U}(1)_B$ factor corresponds to the baryon number, whose conserved charge assigns $\pm\frac{1}{3}$ to quarks and antiquarks respectively. On the other hand, the SU(2) part mixes the u and d quarks. Since the proton is a bound state of two quarks u and one quark d, while the neutron is made out of one quark u and two quarks d, this symmetry reduces at low energies to the well-known isospin transformations of nuclear physics mixing protons and neutrons.

7.2 Quantum Mechanical Realizations of Symmetries

In a quantum theory physical symmetries are maps in the Hilbert space of the theory preserving the probability amplitudes. In more precise terms, a symmetry is a one-to-one transformation that, acting on two arbitrary states $|\alpha\rangle, |\beta\rangle \in \mathscr{H}$

$$|\alpha\rangle \longrightarrow |\alpha'\rangle, \quad |\beta\rangle \longrightarrow |\beta'\rangle, \tag{7.25}$$

satisfies

$$|\langle\alpha|\beta\rangle| = |\langle\alpha'|\beta'\rangle|. \tag{7.26}$$

Wigner's theorem states that these transformations are implemented by operators that are either unitary or antiunitary. Unitary operators are well-known objects from any quantum mechanics course. They are *linear* operators $\mathscr{U}$ satisfying[1]

$$\langle\mathscr{U}\alpha|\mathscr{U}\beta\rangle = \langle\alpha|\beta\rangle, \tag{7.27}$$

for any two states in the Hilbert space. In addition, the transformation of an operator $\mathscr{O}$ under $\mathscr{U}$ is

$$\mathscr{O} \longrightarrow \mathscr{O}' = \mathscr{U}\mathscr{O}\mathscr{U}^{-1}, \tag{7.28}$$

from where it follows that $\langle\alpha|\mathscr{O}|\beta\rangle = \langle\alpha'|\mathscr{O}'|\beta'\rangle$.

---

[1] Here we use the notation $|\mathscr{U}\alpha\rangle \equiv \mathscr{U}|\alpha\rangle$ and $\langle\mathscr{U}\alpha| \equiv \langle\alpha|\mathscr{U}^\dagger$.

Antiunitary operators, on the other hand, have the property

$$\langle \mathscr{U}\alpha|\mathscr{U}\beta\rangle = \langle\alpha|\beta\rangle^* \tag{7.29}$$

and are *antilinear*, i.e.

$$\mathscr{U}\left(a|\alpha\rangle + b|\beta\rangle\right) = a^*|\mathscr{U}\alpha\rangle + b^*|\mathscr{U}\beta\rangle, \quad a, b \in \mathbb{C}. \tag{7.30}$$

To find the transformation of operator matrix elements under an antiunitary transformation we compute

$$\langle\alpha|\mathscr{O}|\beta\rangle = \langle\mathscr{O}^\dagger\alpha|\beta\rangle = \langle\mathscr{U}\beta|\mathscr{U}\mathscr{O}^\dagger\alpha\rangle. \tag{7.31}$$

Writing now $|\mathscr{U}\mathscr{O}^\dagger\alpha\rangle = \mathscr{U}\mathscr{O}^\dagger|\alpha\rangle$ and inserting the identity we arrive at the final result

$$\langle\alpha|\mathscr{O}|\beta\rangle = \langle\beta'|\mathscr{U}\mathscr{O}^\dagger\mathscr{U}^{-1}|\alpha'\rangle. \tag{7.32}$$

Continuous symmetries are implemented only by unitary operators. This is because they are continuously connected with the identity, which is a unitary operator. Discrete symmetries, on the other hand, can be implemented by either unitary or antiunitary operators. An example of the latter is time reversal, that we will study in detail in Chap. 11.

In the previous section we have seen that in canonical quantization the conserved charges Q^a associated with a continuous symmetry by Noether's theorem are operators generating the infinitesimal transformations of the quantum fields. The conservation of the classical charges $\{Q^a, H\}_{\text{PB}} = 0$ implies that the operators Q^a commute with the Hamiltonian

$$[Q^a, H] = 0. \tag{7.33}$$

The symmetry group generated by the operators Q^a is implemented in the Hilbert space of the theory by a set of unitary operators $\mathscr{U}(\alpha)$, where α^a (with $a = 1, \ldots, \dim\mathfrak{g}$) labels the transformation.[2] That the group is generated by the conserved charges means that in a neighborhood of the identity, the operators $\mathscr{U}(\alpha)$ can be written as

$$\mathscr{U}(\alpha) = e^{i\alpha^a Q^a}. \tag{7.34}$$

A symmetry group can be realized in the quantum theory in two different ways, depending on how its elements act on the ground state of the theory. Implementing it in one way or the other has important consequences for the spectrum of the theory, as we now learn.

---

2 A quick survey of group theory can be found in Appendix B.

Wigner–Weyl Realization

In this case the ground state of the theory $|0\rangle$ is invariant under all the elements of the symmetry group $\mathscr{U}(\alpha)|0\rangle = |0\rangle$. Equation (7.34) implies that the vacuum is annihilated by them

$$Q^a|0\rangle = 0. \tag{7.35}$$

The field operators of the quantum theory have to transform according to some irreducible representation of the symmetry group. It is easy to see that the finite form of the infinitesimal transformation (7.9) is given by

$$\mathscr{U}(\alpha)\phi_i\mathscr{U}(\alpha)^{-1} = U_{ij}(\alpha)\phi_j, \tag{7.36}$$

where the matrices $U_{ij}(\alpha)$ form the group representation in which the field ϕ_i transforms. If we consider now the quantum state associated with the operator ϕ_i

$$|i\rangle = \phi_i|0\rangle \tag{7.37}$$

we find that, due to the invariance of the vacuum (7.35), the states $|i\rangle$ have to transform in the same representation as ϕ_i

$$\mathscr{U}(\alpha)|i\rangle = \mathscr{U}(\alpha)\phi_i\mathscr{U}(\alpha)^{-1}\mathscr{U}(\alpha)|0\rangle = U_{ij}(\alpha)\phi_j|0\rangle = U_{ij}(\alpha)|j\rangle. \tag{7.38}$$

Therefore the spectrum of the theory is classified in multiplets of the symmetry group.

Any two states within a multiplet can be "rotated" into one another by a symmetry transformation. Now, since $[H, \mathscr{U}(\alpha)] = 0$ the conclusion is that all states in the same multiplet have the same energy. If we consider one-particle states, then going to the rest frame we see how all states in the same multiplet have exactly the same mass.

Nambu–Goldstone Realization

In our previous discussion we have seen how the invariance of the ground state of a theory under a symmetry group has as a consequence that the spectrum splits into multiplets transforming under irreducible representations of the symmetry group. This shows in degeneracies in the mass spectrum.

The condition (7.35) is not mandatory and can be relaxed by considering theories where the vacuum state is not preserved by the symmetry

$$e^{i\alpha^a Q^a}|0\rangle \neq |0\rangle \Longrightarrow Q^a|0\rangle \neq 0. \tag{7.39}$$

The symmetry is said to be spontaneously broken by the vacuum.

To illustrate the consequences of (7.39) we consider the example of a number of scalar fields φ^i $(i = 1, \ldots, N)$ whose dynamics is governed by the Lagrangian

$$\mathscr{L} = \frac{1}{2}\partial_\mu\varphi^i\partial^\mu\varphi^i - V(\varphi^i), \tag{7.40}$$

where we assume that $V(x)$ is bounded from below and depends on the fields through the combination $\varphi^i\varphi^i$. The theory is invariant under the transformations

$$\delta\varphi^i = \varepsilon^a (T^a)^i_{\ j}\varphi^j, \tag{7.41}$$

with T^a, $a = 1, \ldots, \frac{1}{2}N(N-1)$ the generators of the group SO(N).

To analyze the structure of vacua in this theory we construct its Hamiltonian

$$H[\pi^i, \varphi^i] = \int d^3x \left[\frac{1}{2}\pi^i\pi^i + \frac{1}{2}\nabla\varphi^i\cdot\nabla\varphi^i + V(\varphi^i)\right] \tag{7.42}$$

and look for the minimum of the potential energy functional, given by

$$\mathscr{V}[\varphi^i] = \int d^3x \left[\frac{1}{2}\nabla\varphi^i\cdot\nabla\varphi^i + V(\varphi^i)\right]. \tag{7.43}$$

We want the vacuum to preserve translational invariance, so we will be looking for field configurations satisfying $\nabla\varphi = \mathbf{0}$. This means that the vacua of the potential $\mathscr{V}[\varphi^i]$ coincides with those of $V(\varphi^i)$. The corresponding values of the scalar fields φ^i we denote by[3]

$$\langle\varphi^i\rangle : V(\langle\varphi^i\rangle) = 0, \quad \left.\frac{\partial V}{\partial\varphi^i}\right|_{\varphi^i = \langle\varphi^i\rangle} = 0. \tag{7.44}$$

Let us divide now the generators T^a of SO(N) into two groups: the first set consists of H^α ($\alpha = 1, \ldots, h$) satisfying

$$(H^\alpha)^i_{\ j}\langle\varphi^j\rangle = 0. \tag{7.45}$$

Thus, the vacuum configuration $\langle\varphi^i\rangle$ is left invariant by the group transformations generated by H^α. For this reason they are called *unbroken generators*. Notice that the commutator of two unbroken generators also annihilates the vacuum expectation value, $[H^\alpha, H^\beta]_{ij}\langle\varphi^j\rangle = 0$. They form a subalgebra of the algebra of the generators of SO(N). The subgroup they generate preserves the vacuum and hence it is realized à la Wigner–Weyl. This means in particular that the spectrum is classified in multiplets with respect to this unbroken subgroup.

The remaining generators we denote by K^A, with $A = 1, \ldots, \frac{1}{2}N(N-1) - h$, and by definition they satisfy

$$(K^A)^i_{\ j}\langle\varphi^j\rangle \neq 0. \tag{7.46}$$

These are called the *broken generators*. They generate group transformations that do not preserve the vacuum expectation value of the field. Next we prove a very

---

[3] For simplicity we consider that the minima of $V(x)$ occur at $V = 0$.

important result concerning these broken generators known as Goldstone's theorem: for each generator broken by the vacuum there is a massless excitation in the theory.

The mass matrix of the field excitations around the vacuum $\langle\varphi^i\rangle$ is determined by the quadratic part of the potential. Since we have assumed that $V(\langle\varphi^i\rangle) = 0$ and we are expanding around a minimum, the leading term in the expansion of the potential around the vacuum expectation values is given by

$$V(\varphi^i\varphi^i) = \left.\frac{\partial^2 V}{\partial\varphi^i\partial\varphi^j}\right|_{\varphi=\langle\varphi\rangle} (\varphi^i - \langle\varphi^i\rangle)(\varphi^j - \langle\varphi^j\rangle) + \mathscr{O}\left[(\varphi - \langle\varphi\rangle)^3\right] \quad (7.47)$$

and the mass matrix is

$$M^2_{ij} \equiv \left.\frac{\partial^2 V}{\partial\varphi^i\partial\varphi^j}\right|_{\varphi=\langle\varphi\rangle}. \quad (7.48)$$

In order to avoid a cumbersome notation, we do not indicate explicitly the dependence of the mass matrix on $\langle\varphi^i\rangle$.

To extract information about the possible zero modes of M^2_{ij}, we write down the conditions that follow from the invariance of the potential $V(\varphi^i)$ under the field transformations $\delta\varphi^i = \varepsilon^a (T^a)^i{}_j\varphi^j$. At first order in ε^a

$$\delta V(\varphi) = \varepsilon^a \frac{\partial V}{\partial\varphi^i}(T^a)^i{}_j\varphi^j = 0. \quad (7.49)$$

Differentiating this expression with respect to φ^k we arrive at

$$\frac{\partial^2 V}{\partial\varphi^i\partial\varphi^k}(T^a)^i{}_j\varphi^j + \frac{\partial V}{\partial\varphi^i}(T^a)^i{}_k = 0. \quad (7.50)$$

We evaluate this expression in the vacuum $\varphi^i = \langle\varphi^i\rangle$. The derivative in the second term cancels while the second derivative in the first one gives the mass matrix. Hence we have found

$$M^2_{ik}(T^a)^i{}_j\langle\varphi^j\rangle = 0. \quad (7.51)$$

Now we can write this expression for both broken and unbroken generators. For the unbroken ones, since $(H^\alpha)^i{}_j\langle\varphi^j\rangle = 0$, we find a trivial identity 0=0. Things are more interesting for the broken generators, for which we have

$$M^2_{ik}(K^A)^i{}_j\langle\varphi^j\rangle = 0. \quad (7.52)$$

Since $(K^A)^i{}_j\langle\varphi^j\rangle \neq 0$ this equation implies that the mass matrix has as many zero modes as broken generators. Therefore we have proven Goldstone's theorem: associated with each broken symmetry there is a massless mode in the theory. These modes are known in the literature as Nambu–Goldstone modes. Here we have presented a classical proof of the theorem. In the quantum theory the proof follows the same lines as the one presented here but one has to consider the effective action containing the effects of the quantum corrections to the classical Lagrangian.

7.3 Some Applications of Goldstone's Theorem

To illustrate Goldstone's theorem we consider a three-component real scalar field $\Phi = (\varphi^1, \varphi^2, \varphi^3)$ with the SO(3)-invariant "mexican hat" potential

$$V(\Phi) = \frac{\lambda}{4}\left(\Phi^2 - a^2\right)^2. \tag{7.53}$$

The vacua of the theory are the field configurations satisfying the condition $\langle\Phi\rangle^2 = a^2$. In field space this equation describes a two-dimensional sphere and each vacuum is represented by a point in that sphere. It is easy to visualize geometrically how choosing one of these vacua results in symmetry breaking: while the whole sphere is invariant under the global SO(3) symmetry, each vacuum (i.e. each point) is preserved only by the SO(2) rotations around the axis of the sphere that passes through that point. Hence, the vacuum expectation value of the scalar field breaks the symmetry according to

$$\langle\Phi\rangle : \mathrm{SO}(3) \longrightarrow \mathrm{SO}(2). \tag{7.54}$$

The symmetry group SO(3) has three generators while the symmetry of the vacuum SO(2) has only one. This means that the vacuum breaks two generators and using the Goldstone theorem we conclude that the system should have two massless Nambu–Goldstone bosons. These are easy to identify heuristically because they correspond to excitations along the surface of the sphere $\langle\Phi\rangle^2 = a^2$. That they are indeed massless follows from the fact that the potential (7.53) is flat along the directions of these excitations.

Once a minimum of the potential has been chosen, we can proceed to quantize the excitations around it. Since the vacuum only leaves invariant a SO(2) subgroup of the original SO(3) global symmetry group, it seems that in expanding around a particular vacuum expectation value of the scalar field we have lost part of the symmetry of the Lagrangian. This is however not the case. The full quantum theory is indeed symmetric under the whole SO(3). This is reflected in the fact that the physical properties of the theory do not depend on the particular point of the sphere $\langle\Phi\rangle^2 = a^2$ that we have chosen for our vacuum. In fact, different vacua are related by the full SO(3) symmetry and therefore should give the same physics.

A very important point to keep in mind is that once the system described by the theory chooses a vacuum determined by a value of $\langle\Phi\rangle$, all other possible vacua of the theory are inaccessible in the infinite volume limit. This means that any two vacuum states $|0_1\rangle$, $|0_2\rangle$ corresponding to different vacuum expectation values of the scalar field are orthogonal $\langle 0_1|0_2\rangle = 0$ and, moreover, cannot be connected by any local observable $\mathscr{O}(x)$, $\langle 0_1|\mathscr{O}(x)|0_2\rangle = 0$. Heuristically, this can be understood by thinking that in the infinite volume limit switching from one vacuum into another requires changing the vacuum expectation value of the field everywhere in space at the same time, something that cannot be done by any local operator of the theory. Notice that this is radically different from our expectations based on the quantum mechanics of a system with a finite number of degrees of freedom where symmetries do not break spontaneously, i.e. the ground state is always symmetrical.

Let us make these arguments a bit more explicit since they are very important in understanding how symmetry breaking works. Consider a relatively simple system: a set of spin-$\frac{1}{2}$ magnets, the Heisenberg ferromagnet model, with nearest neighbors interactions. Space is replaced by a lattice with spacing a and lattice vectors $\mathbf{x} = (n_1 a, n_2 a, n_3 a)$. At each lattice site $\mathbf{x}$ there is a spin-$\frac{1}{2}$ degree of freedom

$$\mathbf{s} = \left(\frac{1}{2}\sigma_1, \frac{1}{2}\sigma_2, \frac{1}{2}\sigma_3\right), \tag{7.55}$$

with σ_i the Pauli matrices. The Heisenberg Hamiltonian is defined by

$$H = -J \sum_{\langle \mathbf{x},\mathbf{x}' \rangle} \mathbf{s}(\mathbf{x}) \cdot \mathbf{s}(\mathbf{x}') \quad \text{with} \quad J > 0, \tag{7.56}$$

where the symbol $\langle \mathbf{x}, \mathbf{x}' \rangle$ indicates that we are summing over nearest neighbors on the lattice.

At each lattice site we have a two-dimensional Hilbert space whose basis we can take to be the two $s_3(\mathbf{x})$ eigenstates $\{|\mathbf{x};\uparrow\rangle, |\mathbf{x};\downarrow\rangle\}$. The state corresponding to the spin at the site $\mathbf{x}$ being aligned along the direction $\hat{\mathbf{r}}$

$$\hat{\mathbf{r}} \cdot \mathbf{s}(\mathbf{x})|\mathbf{x}; \hat{\mathbf{r}}\rangle = \frac{1}{2}|\mathbf{x}; \hat{\mathbf{r}}\rangle, \tag{7.57}$$

can be written in this basis as

$$|\mathbf{x}; \hat{\mathbf{r}}\rangle = \cos\left(\frac{\theta}{2}\right)|\mathbf{x};\uparrow\rangle + e^{i\phi}\sin\left(\frac{\theta}{2}\right)|\mathbf{x};\downarrow\rangle, \tag{7.58}$$

where θ and ϕ are the polar an azimuthal angle associated with the unit vector $\hat{\mathbf{r}}$. Using rotational invariance it is an easy exercise to show that

$$\langle \mathbf{x}; \hat{\mathbf{r}}|\mathbf{x}; \hat{\mathbf{r}}'\rangle = \cos\left(\frac{\alpha}{2}\right), \tag{7.59}$$

where α is the angle between the unit vectors $\hat{\mathbf{r}}$ and $\hat{\mathbf{r}}'$, i.e. $\hat{\mathbf{r}} \cdot \hat{\mathbf{r}}' = \cos\alpha$.

We can construct now the ground states of the Hamiltonian (7.56). They correspond to states where all spins in the ferromagnet are aligned along the same direction, that we indicate by the unit vector $\hat{\mathbf{r}}$. Thus we write

$$|\hat{\mathbf{r}}\rangle = \bigotimes_{\mathbf{x}} |\mathbf{x}; \hat{\mathbf{r}}\rangle, \tag{7.60}$$

From this result we conclude that the overlap between two ground states characterized by unit vectors $\hat{\mathbf{r}}$ and $\hat{\mathbf{r}}'$ is given by

$$\langle \hat{\mathbf{r}}|\hat{\mathbf{r}}\rangle = \left[\cos\left(\frac{\alpha}{2}\right)\right]^N \tag{7.61}$$

where $N = Va^{-3}$ is the number of lattice sites, with V the spatial volume.

It is clear that unless the directions $\hat{\mathbf{r}}, \hat{\mathbf{r}}'$ are parallel, the scalar product vanishes in the infinite volume limit. Hence we can build disconnected Hilbert spaces for each different direction $\hat{\mathbf{r}}$. If the spatial volume is finite, the scalar product is non-vanishing and the ground states associated to different directions will mix, so that the lowest ground state will preserve the symmetry. It is only in the limit $V \to \infty$, when the states are orthogonal, that we obtain spontaneous symmetry breaking. It is clear that if the volume is finite but large, the mixing of the different ground states is very highly suppressed, so that for many practical purposes we can approximate this finite volume theory by the theory with Goldstone bosons.

A similar argument can be carried out in field theory. The simplest theory with a Nambu–Goldstone boson is a free real massless scalar with Lagrangian

$$\mathscr{L} = \frac{1}{2}\partial_\mu \varphi \partial^\mu \varphi. \tag{7.62}$$

It is invariant under shifts of the field by a real constant, $\varphi(x) \to \varphi(x) + \alpha$, with $\alpha \in \mathbb{R}$. This symmetry has an associated Noether current given by $j_\mu = \partial_\mu \varphi$, whose conservation can be checked by applying the equations of motion. The vacuum of the theory is not invariant under the symmetry. Indeed, when we quantize the theory by expanding about some particular constant value of the field $\varphi(x) = \varphi_0$ the symmetry is broken, and the massless particle created by $\varphi(x)$ is the associated Goldstone boson.

In fact, at low energies all Nambu–Goldstone bosons are well represented by this approximation. Consider for instance a scalar doublet $\Phi = (\varphi_1, \varphi_2)$ with a "Mexican hat" potential of the type (7.53). The vacuum breaks the global SO(2) symmetry completely. We parametrize the fields $\varphi_1(x)$ and $\varphi_2(x)$ in terms of a single complex scalar field

$$\zeta(x) = \frac{1}{\sqrt{2}}[\varphi_1(x) + i\varphi_2(x)] \equiv \frac{1}{\sqrt{2}}[a + h(x)]\, e^{i\theta(x)}. \tag{7.63}$$

The field $\theta(x)$ is the Nambu–Goldstone boson. Using this parametrization, the action can be written as

$$\begin{aligned} \mathscr{L} &= \frac{1}{2}\partial_\mu \Phi \cdot \partial^\mu \Phi - \frac{\lambda}{4}\left(\Phi^2 - a^2\right)^2 \\ &= \partial_\mu \zeta^* \partial^\mu \zeta - \lambda\left(|\zeta|^2 - \frac{a^2}{2}\right)^2 = \frac{a^2}{2}\partial_\mu \theta \partial^\mu \theta + \cdots, \end{aligned} \tag{7.64}$$

where the dots stand for the other terms involving the field $h(x)$ and its couplings to the Nambu–Goldstone boson $\theta(x)$. If we do not excite the $h(x)$ field, the Lagrangian for $\theta(x)$ is at leading order of the form (7.62). This shows that, although we consider a simplified example, the analysis can be extended to more general situations.

We wish to study the quantization of (7.62) in a spatial box of side L and periodic boundary conditions. A complete set of properly normalized plane waves solutions to the field equation $\partial_\mu \partial^\mu \varphi(x) = 0$ is provided by

$$\varphi_{\mathbf{k}}(t, \mathbf{x}) = \frac{1}{\sqrt{V}} e^{-i|\mathbf{k}|t + i\mathbf{k}\cdot\mathbf{x}}, \quad \mathbf{k} = \frac{2\pi}{L}\mathbf{n} \tag{7.65}$$

where $\mathbf{n} = (n_1, n_2, n_3)$, with $n_i \in \mathbb{Z}$, and $V = L^3$. Their completeness relation reads

$$\delta(\mathbf{x}_1 - \mathbf{x}_2) = \sum_{\mathbf{k}} \varphi_{\mathbf{k}}(t, \mathbf{x}_1)\varphi_{\mathbf{k}}(t, \mathbf{x}_2)^* = \frac{1}{V} + \frac{1}{V}\sum_{\mathbf{k}\neq\mathbf{0}} e^{i\mathbf{k}\cdot(\mathbf{x}_1 - \mathbf{x}_2)}, \tag{7.66}$$

where we have extracted explicitly the zero mode $\mathbf{k} = \mathbf{0}$.

There is a very important difference between the quantization of this massless scalar field in a spatial box and the one we carried out in Chap. 2 for a free scalar field in $\mathbb{R}^3$. This difference lies in the treatment of the zero mode that in finite volume is a normalizable state that has to be quantized independently. In our case the most general position-independent solution to the equation of motion compatible with the periodic boundary conditions is linear in time. Taking this into account, we write the following mode expansion

$$\varphi(t, \mathbf{x}) = \varphi_0 + \pi_0 t + \sum_{\mathbf{k}\neq\mathbf{0}} \frac{1}{\sqrt{2V|\mathbf{k}|}} \left[\alpha(\mathbf{k}) e^{-i|\mathbf{k}|t + i\mathbf{k}\cdot\mathbf{x}} + \alpha^\dagger(\mathbf{k}) e^{i|\mathbf{k}|t - i\mathbf{k}\cdot\mathbf{x}}\right]. \tag{7.67}$$

Imposing the canonical equal-time commutation relations

$$[\varphi(t, \mathbf{x}_1), \dot{\varphi}(t, \mathbf{x}_2)] = i\delta(\mathbf{x}_1 - \mathbf{x}_2) = \frac{i}{V} + \frac{i}{V}\sum_{\mathbf{k}\neq\mathbf{0}} e^{i\mathbf{k}\cdot(\mathbf{x}_1 - \mathbf{x}_2)} \tag{7.68}$$

yields the standard canonical commutation relations for creation-annihilation operators $\alpha(\mathbf{k})$, $\alpha^\dagger(\mathbf{k})$, as well as the commutation relations for φ_0 and π_0

$$[\varphi_0, \pi_0] = \frac{i}{V}. \tag{7.69}$$

After a little work, we find a simple form for the normal-ordered Hamiltonian

$$:H:= \frac{V}{2}\pi_0^2 + \sum_{\mathbf{k}\neq\mathbf{0}} |\mathbf{k}|\alpha^\dagger(\mathbf{k})\alpha(\mathbf{k}). \tag{7.70}$$

Since we are interested in states where $\varphi(x)$ acquires an expectation value, we follow by analogy the treatment of coherent states in elementary quantum mechanics. Let us introduce the operators a and $a^\dagger$ associated to the zero modes

$$a = \frac{1}{\sqrt{2}}\left(\varphi_0 + iV^{\frac{1}{3}}\pi_0\right), \quad a^\dagger = \frac{1}{\sqrt{2}}\left(\varphi_0 - iV^{\frac{1}{3}}\pi_0\right), \tag{7.71}$$

that satisfy

$$[a, a^\dagger] = V^{-\frac{2}{3}}. \tag{7.72}$$

The conserved charge associated with the Noether current $j_\mu = \partial_\mu \varphi$ has the simple form

$$Q = \int d^3x \partial_0 \varphi = V \pi_0 = \frac{V^{\frac{2}{3}}}{i\sqrt{2}} \left(a - a^\dagger \right). \tag{7.73}$$

This charge generates constants shifts in the value of the field, namely

$$e^{-i\xi Q} \varphi(x) e^{i\xi Q} = \varphi(x) + \xi, \tag{7.74}$$

for any real ξ.

We consider a ground state $|0\rangle$ defined by $a|0\rangle = 0$, $\alpha(\mathbf{k})|0\rangle = 0$, for all $\mathbf{k} \neq \mathbf{0}$. It can be immediately shown that the field (7.67) has zero expectation value in this vacuum, $\langle 0|\varphi(x)|0\rangle = 0$. For every real ξ we define the state

$$|\xi\rangle \sim e^{i\xi Q}|0\rangle = e^{-\frac{1}{\sqrt{2}}\xi V^{\frac{2}{3}}(a^\dagger - a)}|0\rangle. \tag{7.75}$$

Using the properties of the creation-annihilation operators a, $a^\dagger$ we can compute the overlap

$$\langle 0|\xi\rangle = e^{-\frac{1}{4}\xi^2 V^{\frac{2}{3}}} \langle 0|0\rangle. \tag{7.76}$$

This vanishes exponentially as $V \to \infty$, as we found for the Heisenberg Hamiltonian. This shows once again that, strictly speaking, Goldstone bosons only appear in the infinite volume limit.

To be fair we must say that we have been a bit sloppy with the argument. In the cases considered so far, the field $\varphi(x)$ is itself an angle. Hence we should also impose the condition that in field space $\varphi(x) \sim \varphi(x) + 2\pi$, that means that the π_0 is not actually a momentum but an angular momentum variable. This complicates the argument technically, but does not change the conclusion: the overlap $\langle 0|\xi\rangle$ still vanishes exponentially in the infinite volume limit.

We close this discussion with a further comment. The argument presented works in space-times of dimension higher than two. In two-dimensions, space is a line and a number of specific subtleties appear. The conclusion is that in two-dimensions there are no Goldstone bosons. The quantum fluctuations always restore the original symmetry. This theorem appeared first in statistical mechanics, where it is known as the Mermin–Wagner theorem [1, 2]. Its field-theoretical version (Coleman theorem) was proved in [3].

A typical example of a Goldstone boson in high energy physics are the pions, associated with the spontaneous breaking of the global chiral isospin $SU(2)_L \times SU(2)_R$ symmetry and that we will study in some detail in Chap. 9. This symmetry acts independently in the left- and right-handed u and d quark spinors as

$$\begin{pmatrix} u_{L,R} \\ d_{L,R} \end{pmatrix} \longrightarrow M_{L,R} \begin{pmatrix} u_{L,R} \\ d_{L,R} \end{pmatrix}, \quad M_{L,R} \in SU(2)_{L,R} \tag{7.77}$$

Since quarks are confined at low energies, this symmetry is expected to be spontaneously broken by a nonvanishing vacuum expectation value of quark bilinears of the type $\langle \overline{u}_R u_L \rangle \neq 0$.

This breaking of the global $\mathrm{SU}(2)_L \times \mathrm{SU}(2)_R$ symmetry to the diagonal SU(2) acting in the same way on the two chiralities has three Nambu–Goldstone modes which are identified with the pions (see Sect. 9.3). This identification, however, might seem a bit puzzling at first sight, because pions are massive contrary to what is expected of a Goldstone boson. The solution to this apparent riddle is that the $\mathrm{SU}(2)_L \times \mathrm{SU}(2)_R$ would be an exact global symmetry of the QCD Lagrangian only in the limit when the masses of the quarks are zero $m_u, m_d \to 0$. As these quarks have nonzero masses, the chiral symmetry is only approximate and as a consequence the corresponding Goldstone bosons are not strictly massless. That is why pions have masses, although they are the lightest particles among the hadrons.

The phenomenon of spontaneous symmetry breaking is not confined to high energy physics, but appears also frequently in condensed matter physics [4]. For example, when a solid crystallizes from a liquid the translational invariance that is present in the liquid phase is broken to a discrete group of translations that represent the crystal lattice. This symmetry breaking has associated Goldstone bosons that are identified with acoustic phonons, which are the quantum excitation modes of the vibrational degrees of freedom of the lattice.

7.4 The Brout–Englert–Higgs Mechanism

Gauge symmetry seems to prevent a vector field from having a mass. This is obvious once we realize that a term in the Lagrangian like $m^2 A_\mu A^\mu$ is incompatible with gauge invariance.

Certain physical situations, however, seem to require massive vector fields. This became evident during the 1960s in the study of weak interactions. The Glashow model gave a common description of both the electromagnetic and weak forces based on a gauge theory with group $\mathrm{SU}(2) \times \mathrm{U}(1)_Y$. However, in order to reproduce Fermi's four-fermion theory of the β-decay, it was necessary that three of the vector fields involved were massive. Also in condensed matter physics massive vector fields are required to describe certain systems, most notably in superconductivity.

The way out to this situation was found independently by Brout and Englert [5] and by Higgs [6, 7] using the concept of spontaneous symmetry breaking discussed above[4]: if the consistency of the quantum theory requires gauge invariance, this can also be realized à la Nambu–Goldstone. When this happens the full gauge symmetry is not explicitly present in the effective action constructed around the particular vacuum chosen for the theory. This makes possible the existence of mass terms for gauge fields without jeopardizing the consistency of the full theory, which is still invariant under the whole gauge group.

[4] In condensed matter the idea had been previously considered by Anderson [8].

To illustrate the Brout–Englert–Higgs mechanism we study the simplest example, the Abelian Higgs model: a U(1) gauge field coupled to a self-interacting charged complex scalar field φ with Lagrangian

$$\mathscr{L} = -\frac{1}{4}F_{\mu\nu}F^{\mu\nu} + (D_\mu\varphi)^\dagger(D^\mu\varphi) - \frac{\lambda}{4}\left(\varphi^\dagger\varphi - \frac{v^2}{2}\right)^2, \tag{7.78}$$

where the covariant derivative is given in Eq. (4.36). This theory is invariant under the gauge transformations

$$\varphi \longrightarrow e^{i\alpha(x)}\varphi, \quad A_\mu \longrightarrow A_\mu + \partial_\mu\alpha(x). \tag{7.79}$$

The minimum of the potential is defined by the equation $|\varphi| = \frac{v}{\sqrt{2}}$. Thus, there is a continuum of different vacua labelled by the phase of the scalar field. None of these vacua, however, is invariant under the gauge symmetry

$$\langle\varphi\rangle = \frac{v}{\sqrt{2}}e^{i\vartheta_0} \longrightarrow \frac{v}{\sqrt{2}}e^{i\vartheta_0+i\alpha(x)} \tag{7.80}$$

and therefore the symmetry is spontaneously broken.

Let us study now the theory around one of these vacua, for example $\langle\varphi\rangle = \frac{v}{\sqrt{2}}$, by writing the field φ in terms of the excitations around this particular vacuum

$$\varphi(x) = \frac{1}{\sqrt{2}}\left[v + \sigma(x)\right]e^{i\vartheta(x)}. \tag{7.81}$$

The whole Lagrangian is still gauge invariant under (7.79), independently of which vacuum we have chosen. This means that we are at liberty of performing a gauge transformation with parameter $\alpha(x) = -\vartheta(x)$ in order to get rid of the phase in Eq. (7.81). Substituting then $\varphi(x) = \frac{1}{\sqrt{2}}[v + \sigma(x)]$ in Eq. (7.78) we find

$$\begin{aligned}\mathscr{L} = &-\frac{1}{4}F_{\mu\nu}F^{\mu\nu} + \frac{1}{2}e^2v^2A_\mu A^\mu + \frac{1}{2}\partial_\mu\sigma\partial^\mu\sigma - \frac{1}{4}\lambda v^2\sigma^2 \\ &- \frac{1}{\sqrt{2}}\lambda v\sigma^3 - \frac{\lambda}{4}\sigma^4 + \frac{1}{\sqrt{2}}e^2vA_\mu A^\mu\sigma + e^2A_\mu A^\mu\sigma^2.\end{aligned} \tag{7.82}$$

We ask now about the excitation of the theory around the vacuum $\langle\varphi\rangle = v/\sqrt{2}$. There is a real scalar field $\sigma(x)$ with mass squared $\frac{1}{2}\lambda v^2$. What makes the construction interesting is that the gauge field A_μ has acquired a mass given by

$$m_\gamma^2 = e^2v^2. \tag{7.83}$$

What is really remarkable about this way of giving a mass to the photon is that at no point we have given up gauge invariance. The symmetry is only hidden. Therefore in quantizing the theory we can still enjoy all the advantages of having a gauge theory, while at the same time we have managed to generate a mass for the gauge field.

It might look surprising that in the Lagrangian (7.82) we did not find any massless mode. Since the vacuum chosen by the scalar field breaks the single generator of U(1), we would have expected from Goldstone's theorem to have one massless particle. To understand the fate of the missing Goldstone boson we have to revisit the calculation leading to the Lagrangian (7.82). Were we dealing with a global U(1) theory, the Goldstone boson would correspond to excitations of the scalar field along the valley of the potential associated with the phase $\vartheta(x)$. In writing the Lagrangian we managed to get rid of $\vartheta(x)$ using a gauge transformation. With this we shifted the Goldstone mode into the gauge field A_μ. In fact, by identifying the gauge parameter with the Goldstone excitation we have completely fixed the gauge and the Lagrangian (7.82) does not have any residual gauge symmetry.

A massive vector field has three polarizations: two transverse ones $\mathbf{k}\cdot\varepsilon(\mathbf{k},\pm 1) = 0$ with helicities $\lambda = \pm 1$ plus a longitudinal one $\varepsilon_L(\mathbf{k}) \sim \mathbf{k}$. In gauging away the massless Goldstone boson $\vartheta(x)$ we have transformed it into the longitudinal polarization of the massive vector field. In the literature this is usually expressed by saying that the Goldstone mode is "eaten up" by the longitudinal component of the gauge field. One should not forget that, in spite of the fact that the Lagrangian (7.82) looks quite different from the one we started with, we have not lost any degrees of freedom. We started with two polarizations of the photon plus the two degrees of freedom associated with the real and imaginary components of the complex scalar field $\varphi(x)$. After symmetry breaking we ended up with the three polarizations for the massive vector field, plus the degree of freedom represented by the real scalar field $\sigma(x)$.

We can understand the Brout–Englert–Higgs mechanism in the light of our general discussion of gauge symmetry in Chap. 4 (see Sect. 4.7). Remember that there we had considered the set $\mathscr{G}$ of all gauge transformations $g(\mathbf{x}) \in G$ approaching the identity at infinity and the subset $\mathscr{G}_0 \subset \mathscr{G}$ formed by those contractible to the identity. These latter are the ones generated by Gauss' law, $[(\mathbf{D}\cdot\mathbf{E})_A - \rho_A]|\text{phys}\rangle = 0$ where ρ_a represents the matter contribution.

The set of *all* gauge transformations also contains elements $g(\mathbf{x})$ approaching any other element of G as $|\mathbf{x}| \to \infty$. This differs from $\mathscr{G}$ by a copy of the gauge group G at infinity. This is identified as the group of *global* transformations associated with the existence of conserved charges via Noether's theorem. When the gauge symmetry is spontaneously broken, the invariance of the theory under $\mathscr{G}$ is nevertheless preserved, while the invariance under global transformations (i.e. the copy of G at infinity) is broken. Notice that this in no way poses a threat to the consistency of the theory since properties like the decoupling of unphysical states are guaranteed by the fact that Gauss' law is satisfied quantum mechanically. This follows from the invariance under $\mathscr{G}_0$.

In Chap. 10 we will explain why the Abelian Higgs model discussed here can be regarded as a toy model of the Brout–Englert–Higgs mechanism responsible for giving masses to the $W^\pm$ and Z^0 gauge bosons in the standard model. In condensed matter physics the symmetry breaking described by the nonrelativistic version of the Abelian Higgs model can be used to characterize the onset of a superconducting phase in the Ginzburg–Landau and BCS (Bardeen–Cooper–Schrieffer) theories, where the complex scalar field Φ is associated with the Cooper pairs. In this case the parameter

v^2 depends on the temperature. Above the critical temperature T_c, $v^2(T) > 0$ and there is only a symmetric vacuum $\langle \Phi \rangle = 0$. When $T < T_c$ then $v^2(T) < 0$ and symmetry breaking takes place. The onset of a nonzero mass for the photon (7.83) below the critical temperature explains the Meissner effect: the magnetic fields cannot penetrate inside superconductors beyond a distance of order $1/m_\gamma$ (see for example [9] for a review on the subject).

References

1. Mermin, N.D., Wagner, H.: Absence of ferromagnetism or antiferromagnetism in one-dimensional or two-dimensional isotropic Heisenberg models . Phys. Rev. Lett. **17**, 1133 (1966)
2. Hohenberg, P.C.: Existence of long-range order in one and two dimensions. Phys Rev. **158**, 383 (1967)
3. Coleman, S.R.: There are no Goldstone bosons in two-dimensions. Commun. Math. Phys. **31**, 259 (1973)
4. Burgess C.P.: Goldstone and pseudoGoldstone bosons in nuclear, particle and condensed matter physics. Phys. Rept. **330**, 193 (2000, hep-th/9808176)
5. Englert, F., Brout, R.: Broken symmetries and the mass of gauge vector mesons. Phys. Rev. Lett. **13**, 321 (1964)
6. Higgs, P.W.: Broken symmetries and the masses of gauge bosons. Phys. Rev. Lett. **13**, 508 (1964)
7. Higgs, P.W.: Broken symmetries, massless particles and gauge fields. Phys. Lett. B 12, 132 (1964)
8. Anderson, P.W.: Random phase approximation in the theory of superconductivity. Phys. Rev. **112**, 1900 (1958)
9. Annett, J.F.: Superconductivity Superfluidity and Condensates. Oxford University Press, London (2004)

Chapter 8
Renormalization

The computation of quantum corrections to observables in quantum field theory requires making sense of expressions that are formally divergent. In this chapter we are going to show how this is done systematically. The renormalization program has nevertheless a much more profound meaning than just taming infinities. Using concepts borrowed from statistical mechanics we are going to see how the notion of renormalization is linked to the way physics looks like at different scales. These ideas will be further developed in Chap. 12.

8.1 Removing Infinities

From its very early stages, quantum field theory was faced with infinities. They emerged in the calculation of important physical quantities, such as the corrections to the charge of the electron due to the interactions with the radiation field. The way these divergences where handled in the 1940s, starting with Kramers, was physically very much in the spirit of the quantum theory emphasis in observable quantities: the measured magnitude of a physical quantity, such as the electron mass, results from adding the quantum corrections to its unobservable "bare" value. The fact that both of these quantities are divergent is not a problem physically, since only their finite sum is observable. To make things mathematically consistent, the handling of infinities requires the introduction of some regularization procedure cutting off the divergent integrals at some momentum scale Λ. The physical value of an observable $\mathscr{O}_{\text{physical}}$ is then given by

$$\mathscr{O}_{\text{physical}} = \lim_{\Lambda\to\infty} \left[\mathscr{O}(\Lambda)_{\text{bare}} + \Delta\mathscr{O}(\Lambda)_{\hbar}\right], \tag{8.1}$$

where $\Delta\mathscr{O}(\Lambda)_{\hbar}$ represents the regularized quantum corrections.

To make this qualitative discussion more precise we compute the corrections to the electric charge in QED. We consider the process of annihilation of an electron-positron pair to create a muon–antimuon pair $e^-e^+ \to \mu^+\mu^-$. To lowest order in the electric charge e the only diagram contributing is

L. Álvarez-Gaumé and M. Á. Vázquez-Mozo, *An Invitation to Quantum Field Theory*, Lecture Notes in Physics 839, DOI: 10.1007/978-3-642-23728-7_8,

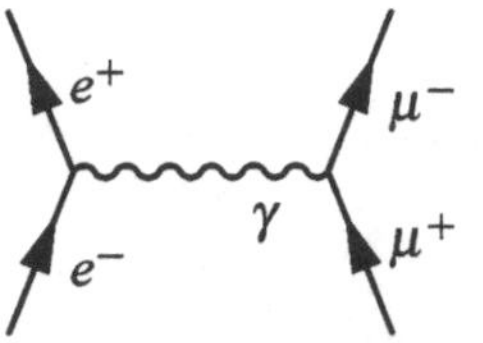

(8.2)

The corrections to order e^4 require the calculation of seven more diagrams

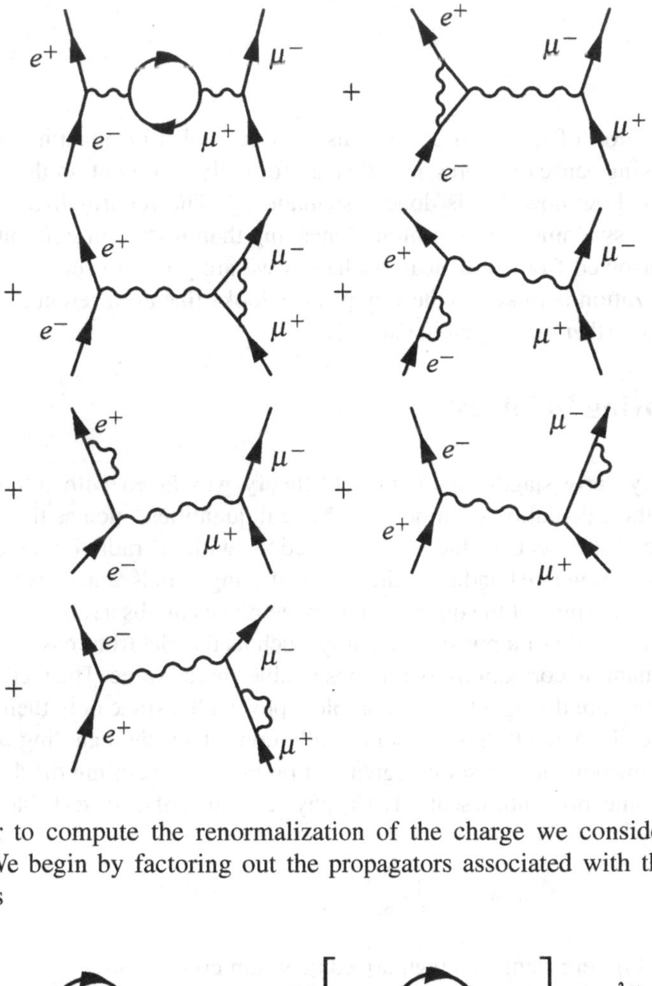

(8.3)

In order to compute the renormalization of the charge we consider the first diagram. We begin by factoring out the propagators associated with the external photon legs

$$= \frac{-i\eta^{\mu\sigma}}{q^2+i\varepsilon}\left[\sigma \quad \lambda\right]\frac{-i\eta^{\lambda\nu}}{q^2+i\varepsilon}, \tag{8.4}$$

where in between brackets we have the *amputated* diagram whose contribution defines the photon polarization tensor

$$\mu \sim\bigcirc\sim \nu \equiv \Pi^{\mu\nu}(q)$$

$$= i^2(-ie)^2(-1)\int \frac{d^4k}{(2\pi)^4}\frac{\mathrm{Tr}\,[(\not{k}+m_e)\gamma^\mu(\not{k}+\not{q}+m_e)\gamma^\nu]}{[k^2-m_e^2+i\varepsilon][(k+q)^2-m_e^2+i\varepsilon]}. \tag{8.5}$$

Physically, this diagram includes the correction to the propagator due to the polarization of the vacuum, i.e. the creation of virtual electron-positron pairs by the propagating photon. The momentum q is the total momentum of the electron-positron pair in the intermediate channel. Notice that the one-loop diagram (8.5) is the Fourier transform of the vacuum expectation value of the time-ordered product of two U(1) gauge currents, namely

$$\Pi^{\mu\nu}(q) = \int d^4x e^{iq\cdot x}\langle 0|T[j^\mu(x)j^\nu(0)]|0\rangle. \tag{8.6}$$

It is instructive to look at the one loop correction to the photon propagator from the point of view of perturbation theory in nonrelativistic quantum mechanics. In each vertex the interaction consists of the annihilation (resp. creation) of a photon and the creation (resp. annihilation) of an electron-positron pair. This can be implemented by the interaction Hamiltonian

$$H_{\mathrm{int}} = e\int d^3x\overline{\psi}\gamma^\mu\psi A_\mu. \tag{8.7}$$

All fields inside the integral can be expressed in terms of the corresponding creation-annihilation operators for photons, electrons and positrons. In quantum mechanics, the change in the wave function to first order in the perturbation H_{int} is given by

$$|\gamma,\mathrm{in}\rangle = |\gamma,\mathrm{in}\rangle_0 + \sum_n \frac{\langle n|H_{\mathrm{int}}|\gamma,\ \mathrm{in}\rangle_0}{E_{\mathrm{in}}-E_n}|n\rangle \tag{8.8}$$

and similarly for $|\gamma,\mathrm{out}\rangle$, where we have denoted symbolically by $|n\rangle$ all the possible states of the electron-positron pair. Since these states are orthogonal to $|\gamma,\ \mathrm{in}\rangle_0, |\gamma,\ \mathrm{out}\rangle_0$, we find to order e^2

$$\begin{aligned}\langle\gamma,\mathrm{in}\,|\gamma',\mathrm{out}\rangle =&\, {}_0\langle\gamma,\mathrm{in}\,|\gamma',\mathrm{out}\,\rangle_0\\ &+\sum_n \frac{{}_0\langle\gamma,\mathrm{in}\,|H_{\mathrm{int}}|n\rangle\langle n|H_{\mathrm{int}}|\gamma',\mathrm{out}\rangle_0}{(E_{\mathrm{in}}-E_n)(E_{\mathrm{out}}-E_n)} + \mathscr{O}(e^4).\end{aligned} \tag{8.9}$$

Hence, we see that the diagram of Eq. (8.4) really corresponds to the order-e^2 correction to the photon propagator $\langle\gamma,\ \text{in}\,|\gamma',\ \text{out}\,\rangle$

$$\longrightarrow \quad {}_0\langle\gamma,\text{in}|\gamma',\text{out}\rangle_0$$

$$\longrightarrow \quad \sum_n \frac{{}_0\langle\gamma,\text{in}|H_{\text{int}}|n\rangle\langle n|H_{\text{int}}|\gamma',\text{out}\rangle_0}{(E_{\text{in}}-E_n)(E_{\text{out}}-E_n)}. \tag{8.10}$$

Once we understand the physical meaning of the Feynman diagram to be computed we proceed to its evaluation. In principle there is no problem in computing the integral in Eq. (8.5) for nonzero values of the electron mass. However since here we are going to be mostly interested in how the divergence of the integral results in an energy scale dependent renormalization of the electric charge, we will set $m_e = 0$. This is something safe to do, since in the case of this diagram we are not inducing new infrared divergences in taking the electron as massless.

To compute the vacuum polarization tensor we are going to exploit what we can expect from gauge symmetry or current conservation. If we contract the external legs of the diagram (8.5) with the polarization tensors of the incoming and outgoing photon $\varepsilon_\mu(q)$ and $\varepsilon'_\nu(q)$, the result must be gauge invariant. That is, the amplitude cannot change under the replacement $\varepsilon_\mu(q) \to \varepsilon_\mu(q)+\lambda q_\mu, \varepsilon'_\mu(q) \to \varepsilon'_\mu(q)+\lambda' q_\mu$, for arbitrary λ and λ'. The consequence is that

$$q_\mu \Pi^{\mu\nu}(q) = 0 = q_\nu \Pi^{\mu\nu}(q). \tag{8.11}$$

This implies the following tensor structure for the polarization tensor

$$\Pi_{\mu\nu}(q) = \left(q^2\eta_{\mu\nu} - q_\mu q_\nu\right)\Pi(q^2). \tag{8.12}$$

Manipulating (8.5) with techniques to be learned in Chap. 12 [using (12.37) and shifting the integration variable], we obtain

$$\Pi(q) = 8e^2 \int_0^1 dx \int \frac{d^4k}{(2\pi)^4} \frac{x(1-x)}{[k^2 - m^2 + x(1-x)q^2 + i\varepsilon]^2}. \tag{8.13}$$

From the representation (8.6) of the polarization tensor we see that the gauge invariance conditions (8.11) implement current conservation.

A more intuitive way to obtain this same result is to think of the diagram in (8.5) as the Fourier transform of the time-ordered correlation function of two gauge currents (8.6). Naively, the conservation of each current implies condition (8.11), and thus the form (8.12) of the polarization tensor. Notice that here we said "naively" because for this to be true we should have a way to compute the correlation function that either preserves gauge invariance (i.e., current conservation) or, if it breaks it, the damage can be fixed without much difficulty.

By looking at the powers of k in the numerator and denominator of the integrand of (8.5) we would conclude that the integral is quadratically divergent. It can be seen, however, that the quadratic divergence does cancel leaving behind only a logarithmic one.[1] In order to handle this divergent integral we have to figure out some procedure to render it finite. This can be done in several ways, but here we choose to cut the integrals off at a high energy scale Λ, where new physics might be at work, $|p| < \Lambda$. This gives the result

$$\Pi(q^2) \simeq \frac{e^2}{12\pi^2} \log\left(\frac{q^2}{\Lambda^2}\right) + \text{ finite terms} . \tag{8.14}$$

As a matter of fact, we have cheated a little bit in this analysis. Regularizing the integral (8.5) using a momentum cutoff does not lead to an expression of the form (8.12). In addition to this piece there is another one proportional to $\Lambda^2\eta_{\mu\nu}$ that spoils gauge invariance. Here we are not very concerned about this term because it can be regarded as an artifact of the chosen regularization. Indeed, in the case of QED there are other regularization methods that preserve gauge invariance, such as dimensional regularization that we will introduce in Chap. 12. In any case the term proportional to Λ^2 could be dealt with by adding an appropriate local counterterm (see Sect. 8.3). Therefore in the following we will pretend that the offending term is absent.

If we want to make sense out of Eq. (8.14), we have to go back to the physical question that led us to compute Eq. (8.4). Our primary motivation was to find the corrections to the annihilation of two electrons into two muons. Including the virtual photon propagation correction, we obtain

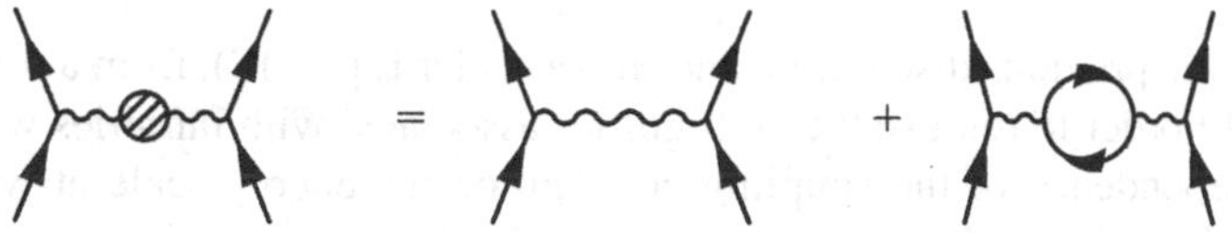

$$\begin{aligned} &= \eta_{\alpha\beta}\left(\overline{v}_e\gamma^\alpha u_e\right)\frac{e^2}{4\pi q^2}\left(\overline{v}_\mu\gamma^\beta u_\mu\right) + \eta_{\alpha\beta}\left(\overline{v}_e\gamma^\alpha u_e\right)\frac{e^2}{4\pi q^2}\Pi(q^2)\left(\overline{v}_\mu\gamma^\beta u_\mu\right) \\ &= \eta_{\alpha\beta}\left(\overline{v}_e\gamma^\alpha u_e\right)\left\{\frac{e^2}{4\pi q^2}\left[1+\frac{e^2}{12\pi^2}\log\left(\frac{q^2}{\Lambda^2}\right)\right]\right\}\left(\overline{v}_\mu\gamma^\beta u_\mu\right). \end{aligned} \tag{8.15}$$

The reader is invited to check that the contribution of the terms proportional to $q_\mu q_\nu$ in (8.12) cancel after using the wave equation for the spinor wave functions. Now let us imagine that in the scattering $e^-e^+ \to \mu^-\mu^+$ we have a center of mass energy μ. From the previous result we can identify the effective charge of the particles at this energy scale $e(\mu)$ as

[1] The change from a quadratically to a logarithmically divergent integral is a consequence of the tensor structure (8.12) of the polarization tensor, and therefore a consequence of gauge invariance.

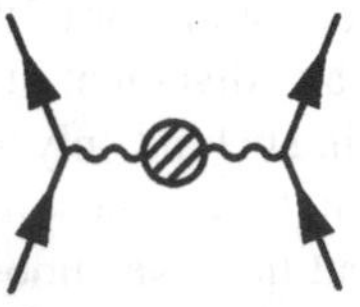

$$= \eta_{\alpha\beta}\left(\overline{v}_e\gamma^\alpha u_e\right)\left[\frac{e(\mu)^2}{4\pi q^2}\right]\left(\overline{v}_\mu\gamma^\beta u_\mu\right). \tag{8.16}$$

This charge, $e(\mu)$, is the physically measurable quantity in our experiment. Now we can make sense of the formally divergent result (8.15) by assuming that the charge appearing in the classical Lagrangian of QED is just a "bare" value that depends on the scale Λ at which we cut off the theory, $e \equiv e_0(\Lambda)$. In order to reconcile (8.15) with the physical results (8.16), we must assume that the dependence of the bare (unobservable) charge $e_0(\Lambda)$ on the cutoff Λ is determined by the identity

$$e(\mu)^2 = e_0(\Lambda)^2\left[1 + \frac{e_0(\Lambda)^2}{12\pi^2}\log\left(\frac{\mu^2}{\Lambda^2}\right)\right]. \tag{8.17}$$

If we still insist in removing the cutoff, $\Lambda \to \infty$, we have to send the bare charge to zero, $e_0(\Lambda) \to 0$, in such a way that the effective coupling has the finite value given by the experiment at the energy scale μ. All observable quantities should be expressed in perturbation theory as a power series in the physical coupling $e(\mu)^2$ and not in terms of the unphysical bare coupling $e_0(\Lambda)$.

8.2 The Beta-Function and Asymptotic Freedom

We can look at the previous discussion, and in particular Eq. (8.17), from a different point of view. In order to remove the ambiguities associated with infinities we have introduced a dependence of the coupling constant on the energy scale at which a process takes place. From the expression of the physical coupling in terms of the bare charge (8.17) we can eliminate the cutoff Λ, whose value after all should not affect the value of physical quantities. Taking into account that we are working in perturbation theory in $e(\mu)^2$, we can express the bare charge $e_0(\Lambda)^2$ in terms of $e(\mu)^2$ as

$$e_0(\Lambda)^2 = e(\mu)^2\left[1 + \frac{e(\mu)^2}{12\pi^2}\log\left(\frac{\mu^2}{\Lambda^2}\right)\right] + \mathscr{O}[e(\mu)^6]. \tag{8.18}$$

This expression allows us to eliminate all dependence in the cutoff in the expression of the effective charge at a scale μ by replacing $e_0(\Lambda)$ in Eq. (8.17) by the one computed using (8.18) at a given reference energy scale μ_0

$$e(\mu)^2 = e(\mu_0)^2\left[1 + \frac{e(\mu_0)^2}{12\pi^2}\log\left(\frac{\mu^2}{\mu_0^2}\right)\right]. \tag{8.19}$$

From this equation we can compute, at this order in perturbation theory, the effective value of the coupling constant at an energy μ, once we know its value at some reference energy scale μ_0. In the case of the electron charge we can use as a reference Thompson's scattering at energies of the order of the electron mass $m_e \simeq 0.5$ MeV, where the value of the electron charge is given by the well known value

$$\alpha(m_e) = \frac{e(m_e)^2}{4\pi} \simeq \frac{1}{137}. \tag{8.20}$$

With this, we can compute $e(\mu)^2$ at any other energy scale by applying Eq. (8.19). In computing the electromagnetic coupling constant at any other scale we must take into account the fact that other charged particles can run in the loop in Eq. (8.15). Suppose, for example, that we want to calculate the fine structure constant at the mass of the Z^0-boson $\mu = m_Z \simeq 92$ GeV. Then, we should include in Eq. (8.19) the effect of other standard model fermions with masses below m_Z. Thus

$$e(m_Z)^2 = e(m_e)^2 \left[1 + \frac{e(m_e)^2}{12\pi^2}\left(\sum_i q_i^2\right)\log\left(\frac{m_Z^2}{m_e^2}\right)\right], \tag{8.21}$$

where q_i is the charge in units of the electron charge of the ith fermionic species running in the loop, and we sum over all fermions with masses below the mass of the Z^0 boson. This expression shows how the electromagnetic coupling grows with energy. To compare with the experimental value of $e(m_Z)^2$ it is not enough to include the effect of fermionic fields, since also the $W^{\pm}$ bosons can run in the loop ($m_W < m_Z$). Taking this into account, as well as threshold effects, the value of the electron charge at the scale m_Z is found to be [1]

$$\alpha(m_Z) = \frac{e(m_Z)^2}{4\pi} \simeq \frac{1}{128.9}. \tag{8.22}$$

This growth of the effective fine structure constant with energy can be understood heuristically by remembering that the effect of the polarization of the vacuum shown in the diagram of Eq. (8.4) amounts to the creation of virtual electron-positron pairs around the location of the charge. These virtual pairs behave as dipoles that, as in a dielectric medium, tend to screen this charge and to decrease its value at long distances (i.e. lower energies).

The variation of the coupling constant with energy is usually given in quantum field theory in terms of the *beta function* defined by

$$\beta(g) = \mu\frac{dg}{d\mu}. \tag{8.23}$$

In the case of QED the beta function can be computed from Eq. (8.19) with the result

$$\beta(e)_{\text{QED}} = \frac{e^3}{12\pi^2}. \tag{8.24}$$

The fact that the coefficient of the leading term in the beta-function is positive gives us the overall behavior of the coupling as we change the scale. Equation (8.24) means that, if we start at an energy where the electric coupling is small enough for our perturbative treatment to be valid, the effective charge grows with the energy scale. This growth of the effective coupling constant with energy means that QED is infrared safe, since the perturbative approximation gives better and better results as we go to lower energies. In fact, since the electron is the lightest electrically charged particle and has a finite nonvanishing mass, the running of the fine structure constant stops at the scale m_e in the well-known value $\frac{1}{137}$. Would other charged fermions with masses below m_e be present in Nature, the effective value of the fine structure constant would run further to lower values at energies below the electron mass.

When we increase the energy scale, $e(\mu)^2$ grows until at some scale the coupling is of order one and the perturbative approximation breaks down. In QED this is known as the problem of the Landau pole but in fact it does not pose any serious threat to the reliability of QED perturbation theory: a calculation including the effect of all standard model fermions shows that the energy scale at which the theory would become strongly coupled is $\Lambda_{\text{Landau}} \simeq 10^{34}$ GeV [2]. However, we expect QED to be unified with other interactions below that scale, and even if this is not the case we will enter the uncharted territory of quantum gravity at energies of the order of 10^{19} GeV.

So much for QED. The next question that one may ask at this stage is whether it is possible to find quantum field theories with a behavior opposite to that of QED, i.e. such that they become weakly coupled at high energies. This is not a purely academic question. In the late 1960s a series of deep inelastic scattering experiments carried out at SLAC showed that the quarks behave essentially as free particles inside hadrons. The apparent problem was that no theory was known at the time that would become free at very short distances: the QED behavior was encountered in all the theories that were analyzed. This posed a very serious problem for quantum field theory as a way to describe subnuclear physics, since it seemed that its predictive power was restricted to electrodynamics but failed when applied to the strong interactions.

This critical time for quantum field theory turned out to be its finest hour. In 1973 David Gross and Frank Wilczek [3] and David Politzer [4] showed that nonabelian gauge theories display the required behavior. For the QCD Lagrangian in Eq. (9.38) the beta function is given by [2]

$$\beta(g) = -\frac{g^3}{16\pi^2}\left(\frac{11}{3}N_c - \frac{2}{3}N_f\right). \tag{8.25}$$

In particular, for real QCD ($N_c = 3$, and N_f equal the number of active flavors) we have that $\beta(g) < 0$. This means that for a weakly coupled theory at an energy scale μ_0 the coupling constant decreases as energy increases $\mu \to \infty$. This explain the

[2] This result has an interesting history. See, for example, [5].

apparent freedom of quarks inside hadrons: when the quarks are very close together their effective color charge tends to zero. This phenomenon is called *asymptotic freedom*.

Asymptotically free theories display a behavior opposite to QED. At high energies their coupling constant approaches zero, whereas at low energies they become strongly coupled (infrared slavery). This features are at the heart of the success of QCD as a theory of the strong interactions, since this is exactly the type of behavior found in quarks: they are quasi-free particles inside the hadrons but the interaction potential between them increases at large distances.

Although asymptotically free theories can be handled in the ultraviolet, they have remarkable properties in the infrared. In the case of QCD they are responsible for color confinement and chiral symmetry breaking (9.52).

In general, the ultraviolet and infrared properties of a theory are controlled by the fixed points of the beta function, i.e. those values of the coupling constant g for which it vanishes

$$\beta(g^*) = 0. \tag{8.26}$$

Using perturbation theory we have seen that for both QED and QCD a fixed point occurs at zero coupling, $g^* = 0$. However, our analysis also showed that the two theories present radically different behavior at high and low energies. From the point of view of the beta function, the difference lies in the energy regime at which the coupling constant approaches its critical value. This is in fact governed by the sign of the beta function around the critical coupling.

If the beta function is negative close to the fixed point (the case of QCD) the coupling tends to its critical value, $g^* = 0$, as the energy is increased. This means that the critical point is *ultraviolet stable*, i.e. it is an attractor as we evolve towards higher energies. If, on the contrary, the beta function is positive (as it happens in QED) the coupling constant approaches the critical value as the energy decreases. This is the case of an *infrared stable* fixed point.

This analysis that we have motivated with the examples of QED and QCD is completely general and can be carried out for any quantum field theory. In Fig. 8.1 we have represented the beta function for a hypothetical theory with three fixed points located at couplings g_1^*, g_2^* and g_3^*. The arrows in the line below the plot represent the evolution of the coupling constant as the energy increases. We learn that $g_1^* = 0$ and g_3^* are ultraviolet stable fixed points, while g_2^* is infrared stable.

In order to understand the high and low energy behavior of a quantum field theory it is crucial to know the structure of the beta functions associated with its couplings. This can be a very difficult task, since perturbation theory only allows the study of the theory around "trivial" fixed points, i.e. those that occur at zero coupling like the case of g_1^* in Fig. 8.1. Any "nontrivial" fixed point occurring in a theory (like g_2^* and g_3^*) cannot be captured in perturbation theory and requires a full nonperturbative analysis.

The lesson to be learned from this discussion is that dealing with the ultraviolet divergences in a quantum field theory has as a consequence the introduction of an energy dependence in the measured value of the coupling constants of the

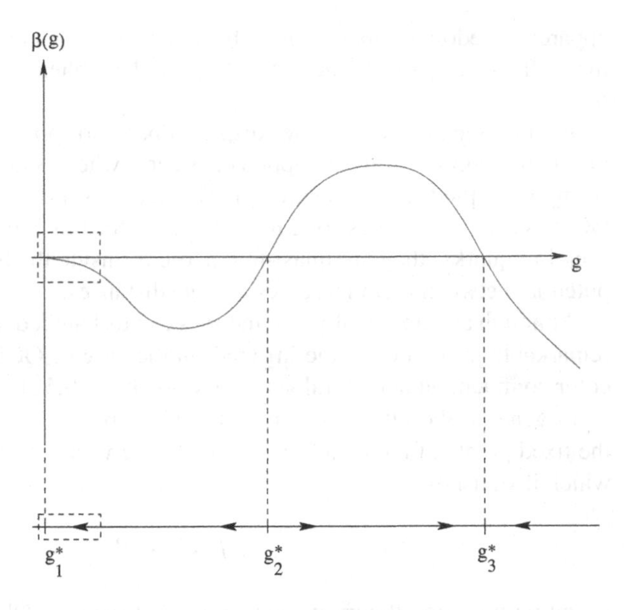

Fig. 8.1 Beta function for a hypothetical theory with three fixed points g_1^*, g_2^* and g_3^*. A perturbative analysis would capture only the regions shown in the *boxes*

theory. This happens even in the case of theories without dimensionful couplings. These theories are scale invariant at the classical level because the action does not contain dimensionful parameters. In this case the running of the coupling constants can be seen as resulting from a quantum breaking of classical scale invariance: different energy scales in the theory are distinguished by different values of the coupling constants. We say that classical scale invariance is an *anomalous* symmetry (see Chap. 9). A heuristic way to understand how the conformal anomaly comes about is to notice that the regularization of an otherwise scale invariant field theory requires the introduction of an energy scale (e.g. a cutoff). In general, the classical invariance cannot be restored after renormalization.

Scale invariance is not completely lost in quantum field theory, however. It is recovered at the fixed points of the beta function where, by definition, the coupling does not run. We consider a scale invariant classical field theory whose field $\phi(x)$ transform under coordinate rescalings as

$$x^\mu \longrightarrow x'^\mu = \lambda x^\mu, \quad \phi(x) \longrightarrow \phi'(x) = \lambda^{-\Delta}\phi(\lambda^{-1}x), \tag{8.27}$$

where Δ is called the canonical scaling dimension of the field. An example of such a theory is a massless ϕ^4 theory in four dimensions

$$\mathscr{L} = \frac{1}{2}\partial_\mu\phi\partial^\mu\phi - \frac{g}{4!}\phi^4, \tag{8.28}$$

where the scalar field has canonical scaling dimension $\Delta = 1$. The Lagrangian density transforms as

$$\mathscr{L} \longrightarrow \lambda^{-4}\mathscr{L}[\phi] \tag{8.29}$$

and the classical action remains invariant [3] under (8.27).

If scale invariance is preserved by quantization, the Green's functions transform as

$$\langle \Omega | T[\phi'(x_1)\ldots\phi'(x_n)]|\Omega\rangle = \lambda^{-n\Delta}\langle \Omega | T[\phi(\lambda^{-1}x_1)\ldots\phi(\lambda^{-1}x_n)]|\Omega\rangle. \quad (8.30)$$

This is what happens in a free theory, whereas in an interacting theory the running of the coupling constant destroys classical scale invariance at the quantum level. In spite of this, at the fixed points of the beta function the Green's functions transform again according to Eq. (8.30) where Δ is replaced by

$$\Delta_{\text{anom}} = \Delta + \gamma^{*}. \quad (8.31)$$

Thus, the canonical scaling dimension of the fields are corrected by γ^{*}, called the *anomalous dimension*. A more detailed discussion of this issue is postponed to Chap. 12.

The previous discussion exhibits some of the high-energy properties of asymptotically free theories like QCD. In the critical theory, the fields have anomalous dimensions different from those in the free theory. These carry the dynamical information about the high-energy behavior.

8.3 A Look at the Systematics of Renormalization

The renormalization program presented in Sect. 8.1 proceeds in two steps. First, the divergences appearing in the calculation of loop diagrams are tamed by introducing a regulator Λ setting an energy scale above which the theory is modified.[4] The second step consists of absorbing the divergences appearing as $\Lambda \to \infty$ in the perturbative calculation of S-matrix amplitudes (or Green's functions) in the parameters of the Lagrangian.

In the particular case of QED this implies a dependence on the regulator of the bare electron charge and mass $e_0(\Lambda)$ and $m_0(\Lambda)$ and also of the global normalization factor of the fields. That is, the "bare" fields $\psi_0(x)$ and $A_{0\mu}(x)$ appearing in the Lagrangian get a dependence on Λ of the form

$$\psi_0(x,\Lambda) = \sqrt{Z_\psi(\Lambda)}\,\psi(x), \quad A_{0\mu}(x,\Lambda) = \sqrt{Z_A(\Lambda)}\,A_\mu(x), \quad (8.32)$$

where $\psi(x)$ and $A_\mu(x)$ are called the renormalized fields and are regulator independent. The dependence on Λ of all bare quantities has to be chosen in such a way that

[3] In a d-dimensional theory the canonical scaling dimensions of the fields coincide with its engineering dimension: $\Delta = \frac{d-2}{2}$ for bosonic fields and $\Delta = \frac{d-1}{2}$ for fermionic ones. For a Lagrangian with no dimensionful parameters classical scale invariance follows then from dimensional analysis.

[4] In the following we denote by Λ any regulator, not necessarily the momentum cutoff used in Sect. 8.1 By convention we consider that the removal of the regulator corresponds to the limit $\Lambda \to \infty$.

the time-ordered Green's functions for the renormalized fields, as computed from the renormalized Lagrangian density

$$\begin{aligned}\mathscr{L}_{\rm ren} =& Z_\psi(\Lambda)\overline{\psi}\left[i\gamma^\mu\partial_\mu - m_0(\Lambda)\right]\psi - \frac{1}{4}Z_A(\Lambda)F_{\mu\nu}F^{\mu\nu}\\ &- e_0(\Lambda)Z_\psi(\Lambda)\sqrt{Z_A(\Lambda)}A_\mu\overline{\psi}\psi,\end{aligned} \tag{8.33}$$

remain finite when $\Lambda \to \infty$. The dependence of Z_ψ, Z_A, e_0 and m_0 on the regulator has to be corrected at each order in perturbation theory.

Since the bare parameters are unphysical we need to identify their physical counterparts that would be measurable in experiments. This we did in Sect. 8.1 by a proper physical interpretation of the sum of two diagrams in Eq. (8.15). We can be more general and instead of considering the one-loop diagram to compute the photon self-energy we can add all *one-particle-irreducible* (1PI) diagrams with two amputated photon external legs. This class of diagrams is defined as those that cannot be split into two disconnected pieces by slicing a single internal line. For example, all diagrams in (8.2) and (8.3) are reducible whereas the one on the left-hand side of Eq. (8.4) is 1PI. We consider the sum

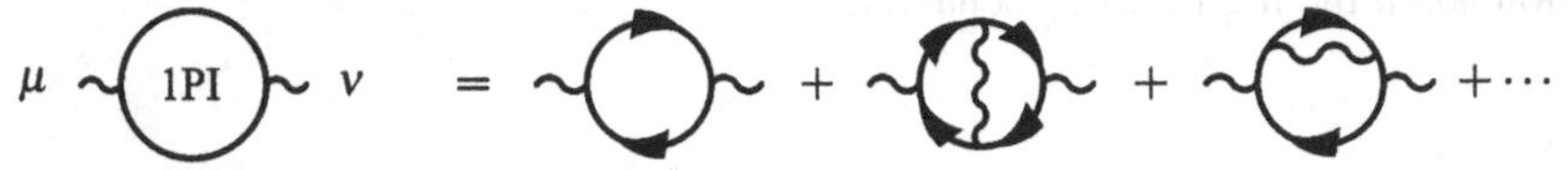

$$= (q^2\eta_\mu - q_\mu q_\nu)\Pi(q^2), \tag{8.34}$$

where the function $\Pi(q^2)$ receives contributions to all orders in perturbation theory. The tensor structure of the previous sum of diagrams is imposed by gauge invariance, as explained in page 148.

What makes the quantity $\Pi(q^2)$ interesting is that the full photon propagator

$$G_{\mu\nu}(q^2) = \int d^4x e^{iq\cdot x}\langle\Omega|T\left[A_\mu(x)A_\nu(0)\right]|\Omega\rangle \tag{8.35}$$

can be written in terms of it. The diagrammatic expansion of $G_{\mu\nu}(q^2)$ has the form

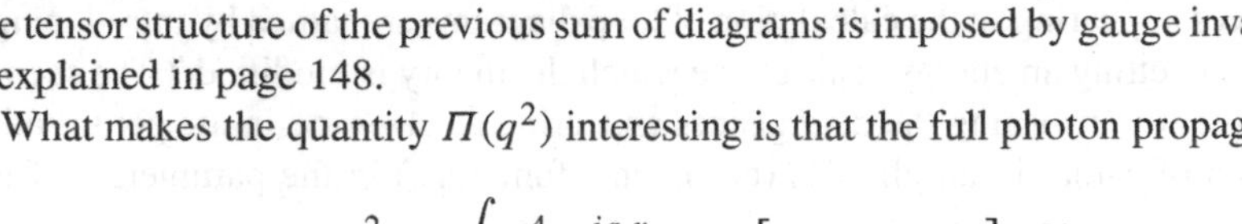

(8.36)

Using the simple identity

$$\left(\delta^\mu_\alpha - \frac{q^\mu q_\alpha}{q^2}\right)\left(\delta^\alpha_\nu - \frac{q^\alpha q_\nu}{q^2}\right) = \delta^\mu_\nu - \frac{q^\mu q_\nu}{q^2} \tag{8.37}$$

it is not difficult to show that (8.36) reduces to a geometric series whose sum is

$$G_{\mu\nu}(q^2) = \frac{-i}{q^2[1 - \Pi(q^2)]}\left(\eta_{\mu\nu} - \frac{q_\mu q_\nu}{q^2}\right) - i\frac{q_\mu q_\nu}{q^4}. \tag{8.38}$$

Using the Lagrangian (8.33) we compute the full propagator $G_{\mu\nu}(q^2, \Lambda)_0$ of the bare gauge field

$$G_{\mu\nu}(q^2, \Lambda)_0 = \frac{-i\eta_{\mu\nu}}{q^2[1 - \Pi_0(q^2, \Lambda)]} \tag{8.39}$$

Here we have removed the terms in the full propagator proportional to $q_\mu q_\nu$ since they vanish once contracted with the fermion lines. The divergence of $\Pi_0(q^2, \Lambda)$ can be absorbed in the normalization factor of the bare field. To see this we notice that

$$G_{\mu\nu}(q^2, \Lambda)_0 = Z_A(\Lambda) G_{\mu\nu}(q^2), \tag{8.40}$$

where the Green's function on the right-hand side is that of the renormalized photon field and therefore remains finite as $\Lambda \to \infty$. We choose $Z_A(\Lambda)$ satisfying the condition

$$\lim_{\Lambda\to\infty} Z_A(\Lambda)[1 - \Pi_0(q^2, \Lambda)] < \infty. \tag{8.41}$$

This does not determine uniquely $Z_A(\Lambda)$. To fix the ambiguity we impose a *renormalization condition*. For example, we demand that the renormalized Green's function $G_{\mu\nu}(q^2)$ behaves close to the pole in the same way as the free photon propagator

$$G_{\mu\nu}(q^2) \sim \frac{-i\eta_{\mu\nu}}{q^2} \quad \text{as} \quad q^2 \to 0. \tag{8.42}$$

This fixes the wave function renormalization to be

$$Z_A(\Lambda) = \frac{1}{1 - \Pi_0(0, \Lambda)}. \tag{8.43}$$

We should bear in mind that the denominator on the right-hand side of this expression does not vanish. This follows from the condition that the photon remains massless to all orders in perturbation theory. Hence, (8.39) should have a single pole at $q^2 = 0$, and we get the condition $\Pi_0(0, \Lambda) \neq 1$.

The calculation of other physical parameters can be done along similar lines. In the case of the electron mass we start by summing the contributions of all 1PI corrections that defines the fermion self-energy $\Sigma_{ab}(\not{p})$

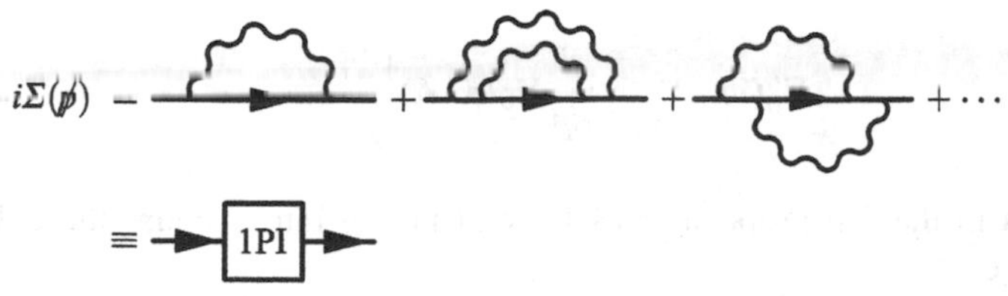

$i\Sigma(\not{p})$ = + + + ···

≡ 1PI

(8.44)

Similarly to the photon case, the whole perturbative expansion of the full fermion propagator $S_{ab}(p)$ can be formally obtained by iterating the insertion of self-energy blobs in the fermion line

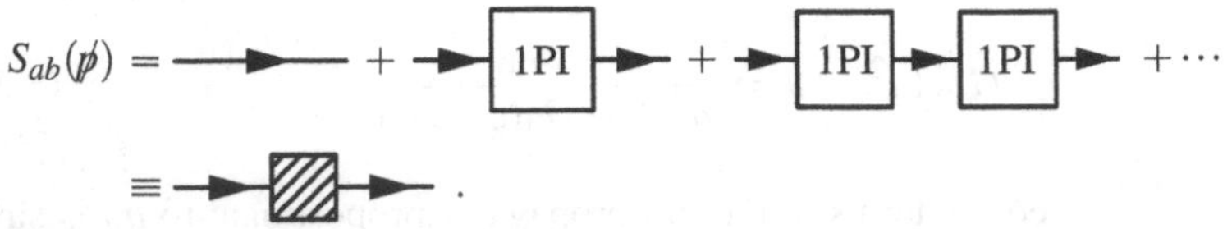

$$S_{ab}(\not{p}) = \longrightarrow + \rightarrow\boxed{\text{1PI}}\rightarrow + \rightarrow\boxed{\text{1PI}}\rightarrow\boxed{\text{1PI}}\rightarrow + \cdots$$
$$\equiv \rightarrow\blacksquare\rightarrow . \tag{8.45}$$

The resulting geometric series yields

$$S(\not{p}) = \frac{i}{\not{p} - m} \sum_{n=0}^{\infty} \left[\frac{1}{\not{p} - m} \Sigma(\not{p}) \right]^n = \frac{i}{\not{p} - m - \Sigma(\not{p})}. \tag{8.46}$$

In a free fermion theory the mass of the particle can be identified as the pole in the propagator

$$S(\not{p}) = \frac{i}{\not{p} - m}. \tag{8.47}$$

We extend this definition of the physical fermion mass to QED. Working with the Lagrangian (8.33) the complete propagator for the bare fermion field reads

$$S_0(\not{p}, \Lambda) = \frac{i}{\not{p} - m_0(\Lambda) - \Sigma_0(\not{p}, \Lambda)}. \tag{8.48}$$

The physical mass is identified with the value $\not{p} = m$ at which the denominator of the full propagator vanishes

$$m = m_0(\Lambda) + \Sigma_0(\not{p}, \Lambda)\Big|_{\not{p}=m}. \tag{8.49}$$

This gives the dependence of the bare mass on the regulator Λ.

With this we do not get rid of all infinities since the fermion self-energy can have a divergent piece linear in the momentum p. To deal with this problem we write the following expansion around the physical fermion mass m

$$\Sigma_0(\not{p}, \Lambda) = \Sigma_0(m, \Lambda) + (\not{p} - m)\Sigma_0'(m, \Lambda) + (\not{p} - m)^2 \tilde{\Sigma}_0(\not{p}, \Lambda), \tag{8.50}$$

where

$$\Sigma_0'(m, \Lambda) = \frac{d}{d\not{p}} \Sigma_0(\not{p}, \Lambda)\Big|_{\not{p}=m}. \tag{8.51}$$

Plugging this in the full propagator (8.48) and taking into account the definition (8.49) we have

$$S_0(\not{p}, \Lambda) = \frac{i}{(\not{p} - m)[1 - \Sigma_0'(m, \Lambda) - (\not{p} - m)\tilde{\Sigma}(\not{p}, \Lambda)]}. \tag{8.52}$$

The divergence of $\Sigma_0'(m, \Lambda)$ as $\Lambda \to \infty$ can now be absorbed in the normalization of the bare field. Indeed, the bare propagator is written in terms of the renormalized one as

$$S_0(\not{p}, \Lambda) = Z_\psi(\Lambda) S(\not{p}), \tag{8.53}$$

where $Z_\psi(\Lambda)$ is chosen to satisfy

$$\lim_{\Lambda \to \infty} Z_\psi(\Lambda) \left[1 - \Sigma_0'(m, \Lambda) - (\not{p} - m)\tilde{\Sigma}(\not{p}, \Lambda) \right] < \infty. \tag{8.54}$$

To fix the freedom in choosing the field normalization we use a renormalization condition similar to the one for the photon propagator and demand that the renormalized fermion propagator satisfies

$$S(\not{p}) \sim \frac{i}{\not{p} - m} \quad \text{when} \quad \not{p} \to m. \tag{8.55}$$

With this we find

$$Z_\psi(\Lambda) = \frac{1}{1 - \Sigma_0'(m, \Lambda)}. \tag{8.56}$$

We still have to express the bare charge $e_0(\Lambda)$ in terms of renormalized parameters. This requires to know how the fermion-photon interaction is corrected by quantum effects. These corrections are contained in the 1PI diagrams with one photon and two fermion lines

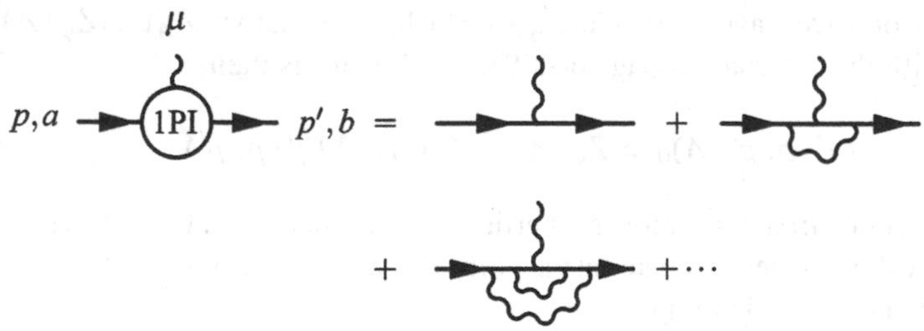

$$\equiv \Gamma^\mu_{ab}(p, p'). \tag{8.57}$$

We compute these diagrams in the regularized theory with the Lagrangian (8.33)

$$\Gamma^\mu(p, p'; \Lambda)_0 = -ie_0(\Lambda) \left[\gamma^\mu + \Lambda_0^\mu(\not{p}, \not{p}'; \Lambda) \right], \tag{8.58}$$

where the function $-ie(\Lambda)\Lambda_0^{\mu}(\not{p},\not{p}';\Lambda)$ contains the contributions of all 1PI *loop* diagrams in Eq. (8.57)

To define the renormalized charge e we point out that at tree level the electric charge is read from the vertex $-ie\gamma^{\mu}$. We use this as a guiding principle and define the renormalized coupling in terms of the *renormalized* 1PI vertex function using the renormalization condition

$$\lim_{\not{p},\not{p}'\to m} \Gamma^{\mu}(p,p') = -ie\gamma^{\mu}. \tag{8.59}$$

We only need to express $\Gamma^{\mu}(p,p')$ in terms of the bare function $\Gamma^{\mu}(p,p';\Lambda)_0$ computed in (8.58). Were we dealing with the complete Green's function

$$\begin{aligned} &G^{\mu}_{ab}(\not{p},\not{p}';\Lambda)_0(2\pi)^4\delta^{(4)}(p+p'+q) \\ &= \int d^4x_1 d^4x_2 d^4x_3 e^{ip\cdot x_1+ip'\cdot x_2+iq\cdot x_3}\langle\Omega|T[\psi_{0a}(x_1)\overline{\psi}_{0b}(x_2)A_0^{\mu}(x_3)]|\Omega\rangle, \end{aligned} \tag{8.60}$$

we would only need to multiply by the corresponding field renormalizations

$$G^{\mu}(\not{p},\not{p}';\Lambda)_0 = \sqrt{Z_A(\Lambda)}Z_{\psi}(\Lambda)G^{\mu}(\not{p},\not{p}'), \tag{8.61}$$

as dictated by (8.32). This Green's function can be expressed in terms of the exact propagators and the 1PI vertex (8.57) as shown in the following "skeleton" diagram

$G^{\mu}(p,p';\Lambda)_0 =$ 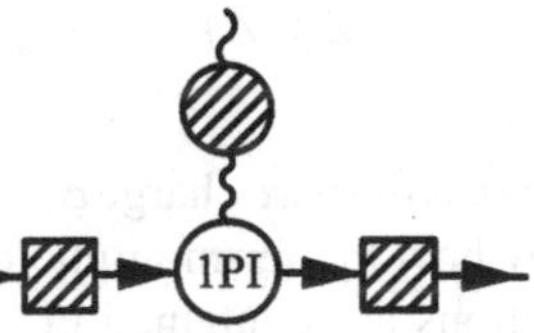 1PI (8.62)

It is now clear that in order to write the 1PI bare vertex function in terms of the renormalized one we have to divide Eq. (8.61) by the factor $[Z_A(\Lambda)Z_{\psi}(\Lambda)^2]^{-1}$ associated with the external propagators. The final result is then

$$\Gamma^{\mu}(p,p';\Lambda)_0 = Z_{\psi}(\Lambda)^{-1}Z_A(\Lambda)^{-\frac{1}{2}}\Gamma^{\mu}(p,p'). \tag{8.63}$$

Using the Lorentz transformation properties of the function $\Lambda^{\mu}(\not{p},\not{p}';\Lambda)_0$, it is possible to show that when evaluated at the point $\not{p}=\not{p}'=m$ we have $\Lambda_0^{\mu}(m,m;\Lambda) = \gamma^{\mu}\Lambda_0(m,m;\Lambda)$. Thus, defining

$$Z_1(\Lambda) = \frac{1}{1+\Lambda_0(m,m;\Lambda)} \tag{8.64}$$

we find the renormalized coupling to be

$$e = e_0(\Lambda)\frac{Z_{\psi}(\Lambda)\sqrt{Z_A(\Lambda)}}{Z_1(\Lambda)}. \tag{8.65}$$

The form of the renormalized coupling we have found can in fact be simplified due to the following identity

$$\Sigma_0'(m, \Lambda) = -\Lambda_0(m, m; \Lambda) \quad \Longrightarrow \quad Z_1(\Lambda) = Z_\psi(\Lambda), \tag{8.66}$$

valid to all orders in QED whenever the theory is regularized in a way preserving gauge invariance (see Ref. [1–15] in Chap. 1). Taking this into account, Eq. (8.65) gives

$$e^2 = e_0(\Lambda)^2 Z_A(\Lambda) = \frac{e_0(\Lambda)^2}{1 - \Pi_0(0, \Lambda)}. \tag{8.67}$$

Physically, the renormalized charge e is identified with the physical coupling at low transferred momentum, as can be seen from the diagram describing the interaction of an electron and a muon by the interchange of a full photon propagator

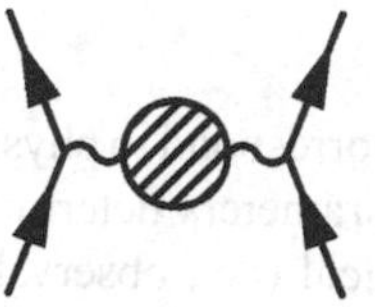

$$= \frac{e_0(\Lambda)^2}{4\pi} (\overline{v}_e \gamma^\mu u_e) \frac{\eta_{\mu\nu}}{q^2[1 - \Pi_0(q^2, \Lambda)]} (\overline{v}_\mu \gamma^\nu u_\mu). \tag{8.68}$$

In fact, the identity (8.66) guarantees that the charge renormalization is universal and independent of the fermion species.

All divergences in QED can be handled order by order in the bare coupling using the renormalization procedure that we just overviewed. Using the relation between the bare and renormalized quantities derived previously, it is possible to express every physical quantity, such as S-matrix amplitudes or the effective charge at different energy scales, solely in terms of the renormalized parameters e and m.

A very practical way of implementing the renormalization program systematically to all orders is to use renormalized perturbation theory. This means that instead of using the bare couplings as expansion parameters we use the renormalized ones that are cutoff independent. In the case of QED the starting point is the action written in terms of the renormalized fields, mass and charge

$$\mathscr{L} = \overline{\psi}\left(i\gamma^\mu \partial_\mu - m\right)\psi - \frac{1}{4} F_{\mu\nu}F^{\mu\nu} - eA_\mu \overline{\psi}\gamma^\mu \psi. \tag{8.69}$$

The divergences appearing in the computation of loop diagrams from this action are dealt with in the following way: for each divergent 1PI diagram we add a *counterterm* to the action such that the new vertex induced by this counterterm cancels the divergence.

This can be done systematically to each order of the perturbative expansion in powers of the renormalized couplings. By construction, after adding the counterterms to (8.69) the Green's functions calculated using the renormalized Lagrangian

$$\mathscr{L}_{\text{ren}} = \mathscr{L} + \mathscr{L}_{\text{ct}} \tag{8.70}$$

are finite in the limit where the cutoff is removed, $\Lambda \to \infty$. For QED, the counterterm Lagrangian has the form

$$\mathscr{L}_{\text{ct}} = iA(\Lambda)\overline{\psi}\gamma^{\mu}\partial_{\mu}\psi - mB(\Lambda)\overline{\psi}\psi - \frac{1}{4}C(\Lambda)F_{\mu\nu}F^{\mu\nu} - eD(\Lambda)A_{\mu}\overline{\psi}\psi. \quad (8.71)$$

Adding this to (8.69) and comparing with the form of the renormalized Lagrangian given in (8.33) we find the bare mass, charge and field renormalizations in terms of the counterterm couplings

$$\begin{aligned} m_0(\Lambda) &= m\frac{1+B(\Lambda)}{1+A(\Lambda)}, \\ e_0(\Lambda) &= e\frac{1+D(\Lambda)}{[1+A(\Lambda)]\sqrt{1+C(\Lambda)}} \\ Z_{\psi}(\Lambda) &= 1+A(\Lambda), \\ Z_A(\Lambda) &= 1+C(\Lambda). \end{aligned} \quad (8.72)$$

In general, the renormalized parameters m and e do not have to correspond to physical values of the mass and the electric charge. They are finite parameters determined by the renormalization conditions in terms of which all physical (i.e., observable) quantities are expressed. All these issues will become clear in Chap. 12, where we will study the one-loop renormalization of an interacting scalar field theory in some detail using the techniques of renormalized perturbation theory.

QED, and in general Yang-Mills theories, belong to a class of quantum field theories called renormalizable. This means that the operators appearing in the renormalized Lagrangian are exactly the same ones as those of the classical action. In other words, the counterterms needed to cancel the divergences in the Green's functions have the same structure as the operators already present in the original Lagrangian [cf. Eqs. (8.69) and (8.71)]. This is not necessarily the case for other theories where the elimination of the divergences at higher orders in perturbation theory requires the introduction of new operators to absorb them in their couplings. When the number of new couplings grows with the order of perturbation theory we say that the quantum field theory is nonrenormalizable. Until the 1970s it was believed that nonrenormalizability would render a theory inconsistent. Nowadays, however, we know that nonrenormalizable theories are perfectly consistent and can be used to compute observables at energies below the natural scale of the theory. We will have more to say about effective field theories in Sect. 8.5 and in Chap. 12.

8.4 Renormalization in Statistical Mechanics

In spite of its successes, the renormalization procedure presented above could still be seen as some kind of prescription or recipe to get rid of the divergences in an ordered way. This discomfort about renormalization was expressed in occasions by

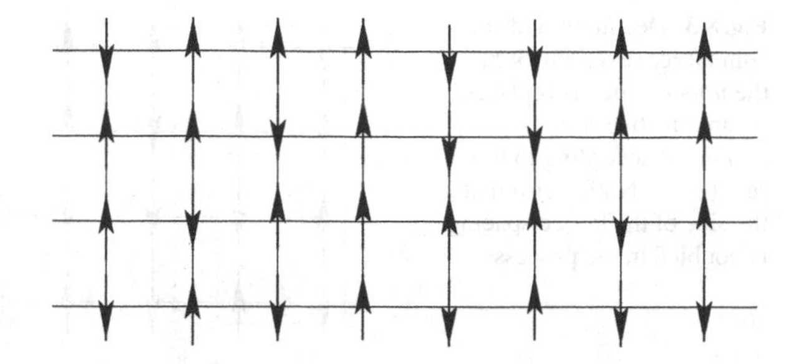

Fig. 8.2 Systems of spins in a two-dimensional square lattice

comparing it with "sweeping the infinities under the rug". After the work of Ken Wilson [6–8], the process of renormalization is now understood in a very profound way as a procedure to incorporate the effects of physics at high energies by modifying the value of the parameters that appear in the Lagrangian.

Wilson's ideas are both simple and profound and consist of thinking about quantum field theory as the analog of a thermodynamical description of a statistical system. To be more precise, let us consider an Ising spin system in a two-dimensional square lattice as the one depicted in Fig. 8.2. In terms of the spin variables $s_i = \pm\frac{1}{2}$, where i labels the lattice site, the Hamiltonian of the system is given by

$$H = -J \sum_{\langle i,j\rangle} s_i s_j, \tag{8.73}$$

where $\langle i, j\rangle$ indicates that the sum extends over nearest neighbors and J is the coupling constant between neighboring spins (no interaction with an external magnetic field is considered). The starting point to study the statistical mechanics of this system is the partition function defined as

$$\mathscr{Z} = \sum_{\{s_i\}} e^{-\beta H}, \tag{8.74}$$

where the sum is over all possible configurations of the spins and $\beta = \frac{1}{T}$ is the inverse temperature. For $J > 0$ the Ising model presents spontaneous magnetization below a critical temperature T_c, in any dimension higher than one. Away from this temperature correlations between spins decay exponentially at large distances

$$\langle s_i s_j\rangle \sim e^{-\frac{|x_{ij}|}{\xi}}, \tag{8.75}$$

with $|x_{ij}|$ the distance between the spins located in the ith and jth sites of the lattice. This expression serves as a definition of the correlation length ξ setting the characteristic length scale at which spins can influence each other by their interaction through their nearest neighbors.

Suppose now that we are interested in a macroscopic description of this spin system. We can capture the relevant physics by integrating out the physics at short

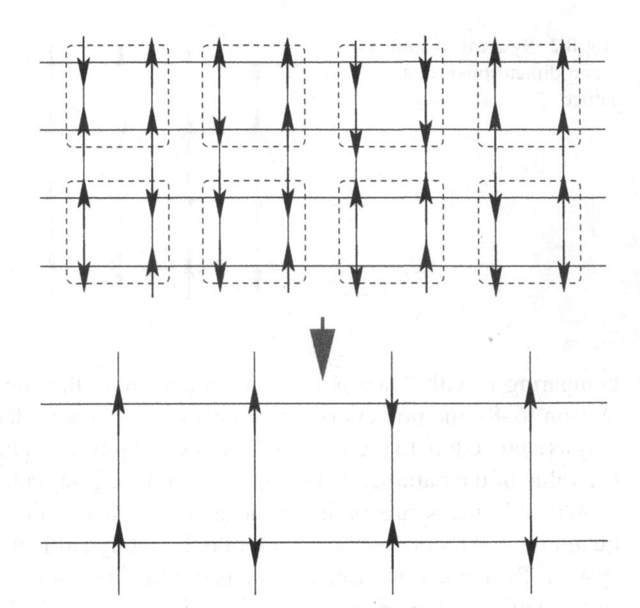

Fig. 8.3 Decimation of the spin lattice. Each block in the *upper lattice* is replaced by an effective spin computed according to the rule (8.77). Notice also that the size of the lattice spacing is doubled in the process

scales. A way in which this can be done was proposed by Leo Kadanoff [9] and consists of dividing our spin system in spin-blocks like the ones showed in Fig. 8.3. Now, we can construct another spin system where each spin-block of the original lattice is replaced by an effective spin calculated according to some rule from the spins contained in each block B_a

$$\{s_i : i \in B_a\} \longrightarrow s_a^{(1)}. \tag{8.76}$$

For example we can define the effective spin associated with the block B_a by taking the majority rule with an additional prescription in case of a draw

$$s_a^{(1)} = \frac{1}{2}\,\text{sign}\left(\sum_{i\in B_a} s_i\right), \tag{8.77}$$

where we have used the sign function, $\text{sign}\,(x) \equiv \frac{x}{|x|}$, with the additional definition $\text{sign}\,(0) = 1$. This procedure is called decimation and leads to a new spin system with a double lattice space.

The idea now is to rewrite the partition function (8.74) only in terms of the new effective spins $s_a^{(1)}$. We start by splitting the sum over spin configurations into two nested sums, one over the spin blocks and the other over the spins within each block

$$\mathscr{Z} = \sum_{\{s\}} e^{-\beta H[s_i]} = \sum_{\{s^{(1)}\}} \sum_{\{s\in B_a\}} \delta\left[s_a^{(1)} - \text{sign}\left(\sum_{i\in B_a} s_i\right)\right] e^{-\beta H[s_i]}. \tag{8.78}$$

The interesting point is that the sum over spins inside each block can be written as the exponential of a new effective Hamiltonian depending only on the effective spins, $H^{(1)}[s_a^{(1)}]$

$$\sum_{\{s\in B_a\}} \delta\left[s_a^{(1)} - \text{sign}\left(\sum_{i\in B_a} s_i\right)\right] e^{-\beta H[s_i]} = e^{-\beta H^{(1)}[s_a^{(1)}]}. \tag{8.79}$$

The new Hamiltonian is of course more complicated

$$H^{(1)} = -J^{(1)} \sum_{\langle i,j\rangle} s_i^{(1)} s_j^{(1)} + \cdots \tag{8.80}$$

where the dots stand for other interaction terms between the effective block spins. The new terms appear because in the process of integrating out short distance physics we induce interactions between the new effective degrees of freedom. For example, the interaction between the spin block variables $s_i^{(1)}$ will not in general be restricted to nearest neighbors in the new lattice. The important point is that we have managed to rewrite the partition function solely in terms of this new (renormalized) spin variables $s^{(1)}$ interacting through a new Hamiltonian $H^{(1)}$

$$\mathscr{Z} = \sum_{\{s^{(1)}\}} e^{-\beta H^{(1)}[s_a^{(1)}]}. \tag{8.81}$$

We can think about the space of all possible Hamiltonians for our statistical system including all kinds of possible couplings between the individual spins compatible with the symmetries of the system. If we denote by $\mathscr{R}$ the decimation operation, it defines a map in the space of Hamiltonians

$$\mathscr{R} : H \to H^{(1)}. \tag{8.82}$$

At the same time the operation $\mathscr{R}$ replaces a lattice with spacing a by another one with double spacing $2a$. As a consequence, the correlation length in the new lattice measured in units of the lattice spacing is divided by two, $\mathscr{R} : \xi \to \frac{\xi}{2}$.

Now we can iterate the operation $\mathscr{R}$ an indefinite number of times. Eventually we might reach a Hamiltonian $H_\star$ that is not further modified by the operation $\mathscr{R}$

$$H \xrightarrow{\mathscr{R}} H^{(1)} \xrightarrow{\mathscr{R}} H^{(2)} \xrightarrow{\mathscr{R}} \ldots \xrightarrow{\mathscr{R}} H_\star. \tag{8.83}$$

The fixed point Hamiltonian $H_\star$ is *scale invariant* because it does not change as $\mathscr{R}$ is performed. As a consequence of this invariance, the correlation length of the system at the fixed point does not change under $\mathscr{R}$. This fact is compatible with the transformation $\xi \to \frac{\xi}{2}$ only if $\xi = 0$ or $\xi = \infty$. Here we will focus in the case of nontrivial fixed points with infinite correlation length.

The space of Hamiltonians can be parametrized by specifying the values of the coupling constants associated with all possible interaction terms between individual spins of the lattice. If we denote by $\mathscr{O}_a[s_i]$ these (possibly infinite) interaction terms, the most general Hamiltonian for the spin system under study can be written as

$$H[s_i] = \sum_{a=1}^{\infty} \lambda_a \mathscr{O}_a[s_i], \tag{8.84}$$

where $\lambda_a \in \mathbb{R}$ are the coupling constants for the corresponding operators. These constants can be thought of as coordinates in the space of all Hamiltonians. Therefore the operation $\mathscr{R}$ defines a transformation in the set of coupling constants

$$\mathscr{R} : \lambda_a \longrightarrow \lambda_a^{(1)}. \tag{8.85}$$

For example, in our case we started with a Hamiltonian in which only one of the coupling constants is different from zero (say $\lambda_1 = -J$). As a result of the decimation $\lambda_1 \equiv -J \to -J^{(1)}$ while some of the originally vanishing coupling constants will take nonzero values. Of course, for the fixed point Hamiltonian the coupling constants do not change under the scale transformation $\mathscr{R}$.

Physically, the transformation $\mathscr{R}$ integrates out short distance physics. The consequence for physics at long distances is that we have to replace our Hamiltonian by a new one with different values for the coupling constants. That is, our ignorance of the details of the physics going on at short distances result in a *renormalization* of the coupling constants of the Hamiltonian describing the long range physical processes. It is important to stress that although $\mathscr{R}$ is sometimes called a renormalization group transformation in fact this is a misnomer. Transformations between Hamiltonians defined by $\mathscr{R}$ do not form a group: since these transformations proceed by integrating out degrees of freedom at short scales they cannot be inverted.

In statistical mechanics fixed points under renormalization group transformations with $\xi = \infty$ are associated with phase transitions. From our previous discussion we can conclude that the space of Hamiltonians is divided into regions corresponding to the basins of attraction of the different fixed points. We can ask ourselves now about the stability of those fixed points. Suppose we have a statistical system described by a fixed-point Hamiltonian $H_\star$ and we perturb it by changing the coupling constant associated with an interaction term $\mathscr{O}$. This is equivalent to replace $H_\star$ by the perturbed Hamiltonian

$$H = H_\star + \delta\lambda \mathscr{O}, \tag{8.86}$$

where $\delta\lambda$ is a perturbation of the coupling constant corresponding to $\mathscr{O}$ (we can also consider perturbations in more than one coupling constant). Thinking of the λ_a's as coordinates in the space of all Hamiltonians this corresponds to moving slightly away from the position of the fixed point.

The question to decide now is in which direction the renormalization group flow will take the perturbed system. Working to first order in $\delta\lambda$ there are three possibilities:

- The renormalization group flow takes the system back to the fixed point. In this case the corresponding interaction $\mathscr{O}$ is called *irrelevant*.

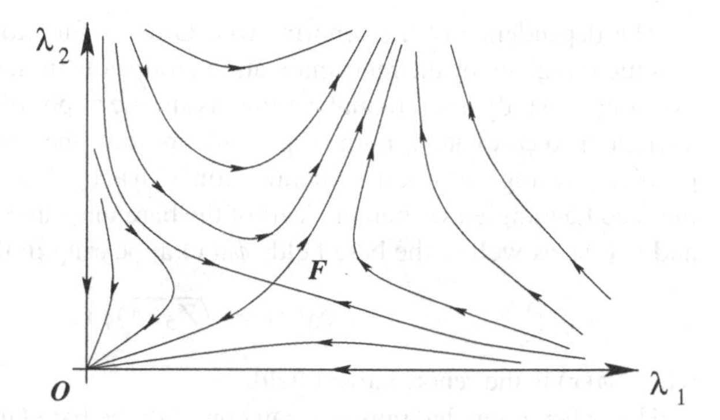

Fig. 8.4 Example of a renormalization group flow

- $\mathscr{R}$ takes the system away from the fixed point. If this is what happens the interaction is called *relevant*.
- It is possible that the perturbation does not take the system away from the fixed point at first order in $\delta\lambda$. In this case the interaction is said to be *marginal* and it is necessary to go to higher orders in $\delta\lambda$ in order to decide whether the system moves to or away the fixed point, or whether we have a family of fixed points.

We can picture the action of the renormalization group transformation as a flow in the space of coupling constants. In Fig. 8.4 we have depicted an example of such a flow in the case of a system with two coupling constants λ_1 and λ_2. In this example we find two fixed points, one at the origin O and another at F for a finite value of the couplings. The arrows indicate the direction in which the renormalization group flow acts. The free theory at $\lambda_1 = \lambda_2 = 0$ is a stable fix point since any perturbation $\delta\lambda_1, \delta\lambda_2 > 0$ makes the theory flow back to the free theory at long distances. On the other hand, the fixed point F is stable with respect to certain perturbations (along the line with incoming arrows) whereas for any other perturbations the system flows either to the free theory at the origin or to a theory with infinite values for the couplings.

8.5 The Renormalization Group in Quantum Field Theory

In the renormalization program in quantum field theory a key role is played by the renormalization conditions. It is through them that the renormalized parameters are related to the bare ones. In the case of QED that we have analyzed in Sect. 8.3, we have used what is called on-shell renormalization in which the renormalized parameters are defined by evaluating the corresponding Green's functions for on-shell values of the external momenta.

In carrying out the renormalization of QED, we could have defined the renormalized charge and mass at any other value of the momentum (physical or unphysical) $p^2 = \mu^2$. This is called a change in the renormalization scheme. It would have lead to different values of the renormalized parameters and Green's functions, although all physical quantities would be independent of the chosen renormalization scheme.

The dependence of the renormalized Green's functions on the renormalization scheme μ is given by the renormalization group equation, or Callan–Symanzik equation, that we study next. To make things as simple as possible, we consider a theory of a single field $\phi(x)$ and ignore all possible indices (they are mostly irrelevant for the problem). Doing perturbative quantization with a regulator Λ we work with the renormalized Lagrangian written in terms of the bare mass and coupling constants $m_0(\Lambda)$ and $g_0(\Lambda)$, as well as the bare fields $\phi_0(x)$ appearing in the classical Lagrangian

$$\phi_0(x) = \sqrt{Z_\phi(\Lambda)}\phi(x), \tag{8.87}$$

where $\phi(x)$ is the renormalized field.

Using Feynman diagrams we can compute the bare Green's functions

$$\begin{aligned} &G_n(p_1, \ldots, p_n; \Lambda)_0 (2\pi)^4 \delta^{(4)}(p_1 + \cdots + p_n) \\ &\quad = \int d^4x_1 \ldots d^4x_n e^{ip_1 \cdot x_1 + \cdots + ip_n \cdot x_n} \langle \Omega | T[\phi_0(x_1) \ldots \phi_0(x_n)] | \Omega \rangle \end{aligned} \tag{8.88}$$

order by order in perturbation theory in the bare coupling constant $g_0(\Lambda)$. The renormalized Green's function $G(p_1, \ldots, p_n)$, on the other hand, is regulator independent. It only depends on the renormalized quantities and the energy scale μ at which the renormalization conditions are implemented. From the relation (8.87) between the bare and renormalized fields we find the following identity

$$G(p_1, \ldots, p_n; \Lambda)_0 = Z_\phi(\Lambda)^{\frac{n}{2}} G(p_1, \ldots, p_n; \mu). \tag{8.89}$$

In the following we focus on the 1PI Green's functions $\Gamma_n(p_1, \ldots, p_n)$, obtained by summing the all 1PI diagrams contributing to the corresponding amplitude. Any other Green's function can be computed in terms of them (see the example of QED studied in Sect. 8.3). The relation between the bare and renormalized 1PI functions can be obtained from (8.89) taking into account that in passing from the Green's functions to the 1PI functions we have to remove the contribution from the external propagators, each one contributing a factor of $Z_\phi(\Lambda)$. This leads to

$$\Gamma_n(p_1, \ldots, p_n; \Lambda)_0 = Z_\phi(\Lambda)^{-\frac{n}{2}} \Gamma_n(p_1, \ldots, p_n; \mu). \tag{8.90}$$

The function on the left-hand side of this equation depends on the regulator Λ both explicitly and through the bare parameters $m_0(\Lambda)$ and $g_0(\Lambda)$. The renormalized function, on the other hand, does not depend on Λ but only on the renormalized parameters m and g as well as on the energy scale μ.

We can find now how $\Gamma_n(p_1, \ldots, p_n; \mu)$ changes when we change the scale μ. Keeping fixed the bare mass $m_0(\Lambda)$ and coupling constant $g_0(\Lambda)$ while varying μ results in a change both in the renormalized parameters m and g and of the field renormalization $Z_\phi(\Lambda)$. Remembering that the left-hand side of (8.90) is independent of μ

$$\mu \frac{d}{d\mu} \left[Z_\phi(\Lambda)^{-\frac{n}{2}} \Gamma_n(p_1, \ldots, p_n; \mu) \right] = 0, \tag{8.91}$$

and using the μ-dependence of m, g and Z_ϕ we arrive at the Callan–Symanzik equation

$$\left[\mu\frac{\partial}{\partial\mu}+\beta(g)\frac{\partial}{\partial g}+\gamma_m(g)\frac{\partial}{\partial m}-n\gamma(g)\right]\Gamma_n(p_1,\ldots,p_n;\mu)=0, \tag{8.92}$$

where we have defined the functions

$$\begin{aligned}\beta(g)&=\mu\frac{\partial g}{\partial\mu},\\ \gamma_m(g)&=\mu\frac{\partial m}{\partial\mu},\\ \gamma(g)&=\frac{1}{2}\mu\frac{\partial}{\partial\mu}\log Z_\phi.\end{aligned} \tag{8.93}$$

Two of this functions already appeared in Sect. 8.2: the beta function $\beta(g)$, that governs the evolution of the coupling with the energy, and the anomalous dimension $\gamma(g)$.

The application to quantum field theory of the idea of the renormalization group in statistical mechanics introduced in Sect. 8.4 leads to a profound understanding of what renormalizing a quantum field theory means in physical terms. Consider a theory with a number of fields ϕ_a defined by a Lagrangian

$$\mathscr{L}[\phi_a]=\mathscr{L}_0[\phi_a]+\sum_i g_i\mathscr{O}_i[\phi_a], \tag{8.94}$$

where $\mathscr{L}_0[\phi_a]$ is the kinetic part and the g_i's are the coupling constants associated with the operators $\mathscr{O}_i[\phi_a]$. In order to make sense of the quantum theory we introduce a momentum cutoff Λ. In principle, we include all operators $\mathscr{O}_i$ compatible with the symmetries of the theory.

In Sect. 8.2 we learned how in the cases of QED and QCD the value of the coupling constants changed with the scale from their values at the scale Λ. We can understand this behavior along the lines of the analysis presented for the Ising model. If we would like to compute the effective dynamics of the theory at an energy scale $\mu < \Lambda$, we only have to integrate out all physical modes with energies between the cutoff Λ and the scale of interest μ. This is analogous to what we did in the Ising model by replacing the original spins by the block spins. In the case of field theory the effective action $S[\phi_a,\mu]$ at scale μ can be written in the language of functional integration as

$$e^{iS[\phi'_a,\mu]}=\int_{\mu<p<\Lambda}\prod_a\mathscr{D}\phi_a e^{iS[\phi_a,\Lambda]} \tag{8.95}$$

Here $S[\phi_a,\Lambda]$ is the action at the cutoff scale

$$S[\phi_a, \Lambda] = \int d^4x \left\{ \mathscr{L}_0[\phi_a] + \sum_i g_i(\Lambda) \mathscr{O}_i[\phi_a] \right\} \tag{8.96}$$

and the functional integral in Eq. (8.95) is carried out only over the field modes with momenta in the range $\mu < p < \Lambda$. The action resulting from integrating out the physics at the intermediate scales between Λ and μ depends not on the original field variable ϕ_a but on some renormalized field ϕ'_a. At the same time, the couplings $g_i(\mu)$ differ from their values at the cutoff scale $g_i(\Lambda)$. This is analogous to what we learned in the Ising model: by integrating out short distance physics we ended up with a new Hamiltonian depending on renormalized effective spin variables and with renormalized values for the coupling constants. Therefore the resulting effective action at scale μ can be written as

$$S[\phi'_a, \mu] = \int d^4x \left\{ \mathscr{L}_0[\phi'_a] + \sum_i g_i(\mu) \mathscr{O}_i[\phi'_a] \right\}. \tag{8.97}$$

This Wilsonian interpretation of renormalization sheds light to what in Sect. 8.1 might have looked just a smart way to get rid of the infinities. The running of the coupling constant with the energy scale can be understood instead as a way of incorporating into an effective action at scale μ the effects of field excitations at higher energies $E > \mu$.

As in statistical mechanics, there are also quantum field theories that are fixed points of the renormalization group flow, i.e. whose coupling constants do not change with the scale. We have encountered them already in Sect. 8.2 when studying the properties of the beta function. The most trivial example of such theories are massless free quantum field theories, but there are also examples of scale invariant, four-dimensional interacting quantum field theories. We can ask the question of what happens when a scale invariant theory is perturbed with some operator. In general, the perturbed theory is not scale invariant anymore but we may wonder whether the perturbed theory flows at low energies towards or away the fixed point theory.

In quantum field theory this can be decided by looking at the canonical dimension $D_{\mathscr{O}}$ of the operator $\mathscr{O}[\phi_a]$ used to perturb the theory at the fixed point. In four dimensions the three possibilities are:

- $D_{\mathscr{O}} > 4$: irrelevant perturbation. The running of the coupling constants takes the theory back to the fixed point.
- $D_{\mathscr{O}} < 4$: relevant perturbation. At low energies the theory flows away from the scale-invariant theory.
- $D_{\mathscr{O}} = 4$: marginal deformation. The direction of the flow cannot be decided only on dimensional grounds.

As an example, let us consider first a massless fermion theory perturbed by a four-fermion interaction term

$$\mathscr{L} = i\overline{\psi}\partial\!\!\!/\psi - \frac{1}{M^2}(\overline{\psi}\psi)^2. \tag{8.98}$$

This is indeed a perturbation by an irrelevant operator, since in four-dimensions $D_\psi = \frac{3}{2}$. Interactions generated by the extra term are suppressed at low energies since typically their effects are weighted by the dimensionless factor E^2/M^2, where E is the energy scale of the process. This means that as we try to capture the relevant physics at lower and lower energies, the effect of the perturbation is weaker and weaker rendering in the infrared limit $E \to 0$ again a free theory. Hence, the irrelevant perturbation in (8.98) makes the theory flow back to the fixed point.

On the other hand, relevant operators dominate the physics at low energies. This is the case, for example, of a mass term. As we lower the energy the mass becomes more important, and once the energy goes below the mass of the field its dynamics is completely dominated by the mass term. This is, for example, how Fermi's theory of weak interactions emerges from the standard model at energies below the mass of the $W^\pm$ boson

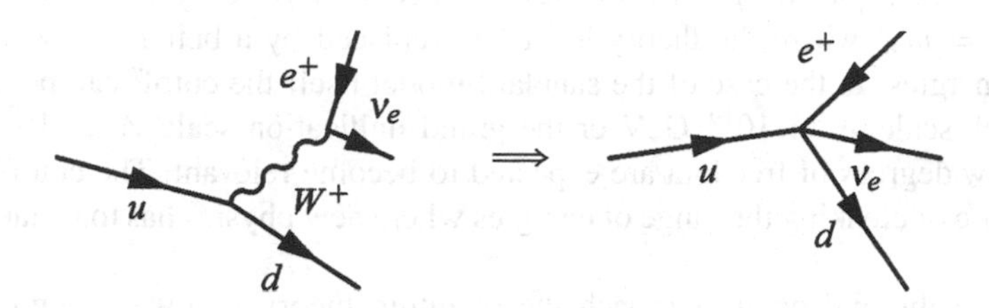

At energies below $m_W = 80.4$ GeV the dynamics of the W^+ boson is dominated by its mass term and therefore becomes nonpropagating, giving rise to the effective four-fermion Fermi theory.

To summarize our discussion so far, we found that while relevant operators dominate the dynamics in the infrared, taking the theory away from the fixed point, irrelevant perturbations become suppressed in the same limit. Finally, we consider the effect of marginal operators. As an example we take the interaction term in massless QED, $\mathscr{O} = \overline{\psi}\gamma^\mu\psi A_\mu$. Taking into an account that in $d = 4$ the dimension of the electromagnetic potential is $[A_\mu] = 1$, the operator $\mathscr{O}$ is a marginal perturbation. In order to decide whether the fixed point theory

$$\mathscr{L}_0 = -\frac{1}{4}F_{\mu\nu}F^{\mu\nu} + i\overline{\psi}\partial\!\!\!/\psi \tag{8.99}$$

is restored at low energies or not we need to study the perturbed theory in more detail. This we have done in Sect. 8.1 where we learned that the effective coupling in QED decreases at low energies. Then we conclude that the perturbed theory flows towards the fixed point in the infrared.

As an example of a marginal operator with the opposite behavior, we write the Lagrangian for a SU(N_c) gauge theory, $\mathscr{L} = -\frac{1}{4}F^A_{\mu\nu}F^{A\,\mu\nu}$, as

$$\mathscr{L} = -\frac{1}{4}\left(\partial_\mu A^A_\nu - \partial_\nu A^A_\mu\right)\left(\partial^\mu A^{A\nu} - \partial^\nu A^{A\mu}\right) - 4g f^{ABC} A^A_\mu A^B_\nu \partial^\mu A^{C\nu} + g^2 f^{ABC} f^{ADE} A^B_\mu A^C_\nu A^{D\mu} A^{E\nu} \equiv \mathscr{L}_0 + \mathscr{O}_g, \tag{8.100}$$

i.e. a marginal perturbation of the free theory described by $\mathscr{L}_0$, which is obviously a fixed point under renormalization group transformations. Unlike QED, the full theory is asymptotically free, so the coupling constant grows at low energies. This implies that the operator $\mathscr{O}_g$ becomes more and more important in the infrared, and the theory flows away the fixed point in this limit.

It is very important to notice here that in the Wilsonian view the cutoff is not necessarily regarded as just some artifact to remove infinities but it has a physical origin. For example in the case of Fermi's theory of β-decay there is a natural cutoff $\Lambda = m_W$ where the theory has to be replaced by a better behaved theory at high energies. In the case of the standard model itself the cutoff can be taken at the Planck scale $\Lambda \simeq 10^{19}$ GeV or the grand unification scale $\Lambda \simeq 10^{16}$ GeV, where new degrees of freedom are expected to become relevant. The cutoff serves the purpose of cloaking the range of energies where new physics has to be taken into account.

Since in the Wilsonian approach the quantum theory is always defined with a physical cutoff, there is no fundamental difference between renormalizable and nonrenormalizable theories. A renormalizable field theory, like the standard model, can generate nonrenormalizable operators at low energies such as the effective four-fermion interaction of Fermi's theory. They are not sources of any trouble if we are interested in the physics at scales much below the cutoff, $E \ll \Lambda$, since their contribution to the amplitudes is suppressed by powers of E/Λ. A more detailed analysis of effective field theories will be presented in Chap. 12.

References

1. Eidelman, S. et al.: Review of particle physics. Phys. Lett. B **592**, 1 (2004). http://pdg.lbl.gov
2. Yndurain, F.J.: Landau Poles, Violations of Unitarity and a Bound on the Top Quark Mass. In: Akhoury, R., de Wit, B., van Nieuwenhuizen, P., Veltman, H. (eds.) Gauge Theories—Past and Future. World Scientific, Singapore (1992)
3. Gross D, J., Wilczek, F.: Ultraviolet behavior of nonabelian gauge theories. Phys. Rev. Lett. **30**, 1343 (1973)
4. Politzer H, D.: Reliable perturbative results for strong interations?. Phys. Rev. Lett. **30**, 1346 (1973)
5. Hoddeson, L., Brown, L., Riordan, M., Dresden, M. (eds.) The Rise of the Standard Model. Particle Physics in the 1960s and 1970s. Cambridge University Press, New York (1997)
6. Wilson K, G.: Renormalization group and critical phenomena 1. Renormalization group and the Kadanoff scaling picture. Phys. Rev. B **4**, 3174 (1971)

7. Wilson K, G.: Renormalization group and critical phenomena 2. Phase space cell analysis of critical behavior. Phys. Rev. B **4**, 3184 (1971)
8. Wilson K, G.: The renormalization group and critical phenomena. Rev. Mod. Phys. **55**, 583 (1983)
9. Kadanoff, LP.: Scaling Laws for Ising Models Near T_c. Physics **2**, 263 (1966)

Chapter 9
Anomalies

So far we did not worry about how the classical symmetries of a theory are carried over to the quantum theory. We have implicitly assumed that classical symmetries are preserved in the process of quantization.

This is not necessarily the case. As we have seen in the previous chapter, quantizing an interacting field theory is a very involved process requiring regularization and renormalization. Sometimes, it does not matter how hard we try, there is no way for a classical symmetry to survive quantization. When this happens one says that the theory has an *anomaly* (for a review see [1]). It is important to avoid the misconception that anomalies appear due to a bad choice of the way a theory is regularized in the process of quantization. When we talk about anomalies we mean a classical symmetry that *cannot* be realized in the quantum theory, no matter how smart we are in choosing the regularization procedure.

In Chap. 8 we have already encountered an example of an anomaly: the quantum breaking of classical scale invariance reflected in the running of the coupling constants with the energy. In the following we focus on other examples of anomalies, this time associated with the global and local symmetries of the classical theory.

9.1 A Toy Model for the Axial Anomaly

Probably the best known examples of anomalies appear when we consider axial symmetries. In a theory of two Weyl spinors $u_{\pm}$

$$\mathscr{L} = i\overline{\psi}\partial\!\!\!/\psi = iu_{+}^{\dagger}\sigma_{+}^{\mu}\partial_{\mu}u_{+} + iu_{-}^{\dagger}\sigma_{-}^{\mu}\partial_{\mu}u_{-} \quad \text{with} \quad \psi = \begin{pmatrix} u_{+} \\ u_{-} \end{pmatrix} \tag{9.1}$$

the Lagrangian is invariant under two types of global U(1) transformations. In the first one both chiralities transform with the same phase, this is a *vector* transformation:

$$\mathrm{U}(1)_{\mathrm{V}} : u_{\pm} \longrightarrow e^{i\alpha}u_{\pm}, \tag{9.2}$$

L. Álvarez-Gaumé and M. Á. Vázquez-Mozo, *An Invitation to Quantum Field Theory*, Lecture Notes in Physics 839, DOI: 10.1007/978-3-642-23728-7_9,

whereas in the second, the axial U(1), the signs of the phases are different for the two chiralities

$$\mathrm{U(1)_A} : u_\pm \longrightarrow e^{\pm i\alpha} u_\pm. \tag{9.3}$$

Using Noether's theorem, there are two conserved currents, a vector current

$$J^\mu_\mathrm{V} = \overline{\psi}\gamma^\mu\psi = u_+^\dagger \sigma_+^\mu u_+ + u_-^\dagger \sigma_-^\mu u_- \quad \Longrightarrow \quad \partial_\mu J^\mu_\mathrm{V} = 0 \tag{9.4}$$

and an axial vector current

$$J^\mu_\mathrm{A} = \overline{\psi}\gamma^\mu\gamma_5\psi = u_+^\dagger \sigma_+^\mu u_+ - u_-^\dagger \sigma_-^\mu u_- \quad \Longrightarrow \quad \partial_\mu J^\mu_\mathrm{A} = 0. \tag{9.5}$$

The theory described by the Lagrangian (9.1) can be coupled to the electromagnetic field. The resulting classical theory is still invariant under the vector and axial U(1) symmetries (9.2) and (9.3). Surprisingly, upon quantization it turns out that the conservation of the axial vector current (9.5) is spoiled by quantum effects

$$\partial_\mu J^\mu_\mathrm{A} \sim \hbar\,\mathbf{E}\cdot\mathbf{B}. \tag{9.6}$$

To understand more clearly how this result comes about, we study first a simple model in two dimensions that captures the relevant physics involved in the four-dimensional case [2]. We work in a two-dimensional Minkowski space with coordinates $(x^0, x^1) \equiv (t, x)$ and where the spatial direction is compactified to a circle S^1 with length L. In this setup we consider a fermion coupled to a classical electromagnetic field. Notice that in our two-dimensional world the field strength $\mathscr{F}_{\mu\nu}$ has only one independent component that corresponds to the electric field, $\mathscr{F}_{01} \equiv -\mathscr{E}$ (in two dimensions there are no magnetic fields!).

To write the Lagrangian for the spinor field we need to find a representation of the algebra of γ-matrices

$$\{\gamma^\mu, \gamma^\nu\} = 2\eta^{\mu\nu} \quad \text{with} \quad \eta = \begin{pmatrix} 1 & 0 \\ 0 & -1 \end{pmatrix}. \tag{9.7}$$

In two dimensions the dimension of the representation of the γ-matrices is 2. In fact, remembering the anticommutation relation of the Pauli matrices $\{\sigma_i, \sigma_j\} = 2\delta_{ij}$ is not very difficult to come up with the following representation

$$\gamma^0 \equiv \sigma_1 = \begin{pmatrix} 0 & 1 \\ 1 & 0 \end{pmatrix}, \quad \gamma^1 \equiv i\sigma_2 = \begin{pmatrix} 0 & 1 \\ -1 & 0 \end{pmatrix}. \tag{9.8}$$

This is a chiral representation since the matrix γ_5 is diagonal[1]

$$\gamma_5 \equiv -\gamma^0\gamma^1 = \begin{pmatrix} 1 & 0 \\ 0 & -1 \end{pmatrix}. \tag{9.9}$$

---

[1] In any even number of dimensions γ_5 is defined to satisfy the conditions $(\gamma_5)^2 = \mathbf{1}$ and $\{\gamma_5, \gamma^\mu\} = 0$.

Writing a two-component Dirac spinor ψ as

$$\psi = \begin{pmatrix} u_+ \\ u_- \end{pmatrix} \tag{9.10}$$

and defining as usual the projectors $P_\pm = \frac{1}{2}(\mathbf{1} \pm \gamma_5)$, we find that the components $u_\pm$ of ψ are respectively right- and left-handed Weyl spinors in two dimensions.

Once we have a representation of the γ-matrices we can write the Dirac equation. Expressed in terms of the components $u_\pm$ of the Dirac spinor, we have

$$(\partial_0 - \partial_1)u_+ = 0, \quad (\partial_0 + \partial_1)u_- = 0. \tag{9.11}$$

The general solution of these equations can be immediately written as

$$u_+ = u_+(x^0 + x^1), \quad u_- = u_-(x^0 - x^1). \tag{9.12}$$

Hence $u_\pm$ are two wave packets moving along the spatial dimension respectively to the left (u_+) and to the right (u_-). Notice that according to our convention the left-moving u_+ is a right-handed spinor (positive helicity) whereas the right-moving u_- is a left-handed spinor (negative helicity).

If we insist in interpreting (9.11) as the wave equation for two-dimensional Weyl spinors, we find the following properly normalized wave functions for free particles with well defined energy-momentum $p^\mu = (E, p)$

$$v_\pm^{(E)}(x^0 \pm x^1) = \frac{1}{\sqrt{L}} e^{-iE(x^0 \pm x^1)} \quad \text{with} \quad p = \mp E. \tag{9.13}$$

As it is always the case with a relativistic wave equation, we have found both positive and negative energy solutions. For $v_+^{(E)}$, since $E = -p$, we see that the solutions with positive energy are those with negative momentum $p < 0$, whereas the negative energy solutions are plane waves with $p > 0$. For the left-handed spinor u_- the situation is reversed. Besides, since the spatial direction is compact with length L the momentum p is quantized according to

$$p = \frac{2\pi n}{L}, \quad n \in \mathbb{Z}. \tag{9.14}$$

The spectrum of the theory is represented in Fig. 9.1.

Knowing the spectrum of the theory the next step is to obtain the vacuum. As with the Dirac equation in four dimensions, we identify the ground state of the theory with the one where all states with $E \leqslant 0$ are filled (see Fig. 9.2.). Exciting a particle in the Dirac sea produces a positive energy fermion plus a hole that is interpreted as an antiparticle. This gives us the key on how to quantize the theory. In the expansion of the operator $u_\pm$ in terms of the modes (9.13) we associate positive energy states with annihilation operators, whereas the states with negative energy are associated with creation operators for the corresponding antiparticle

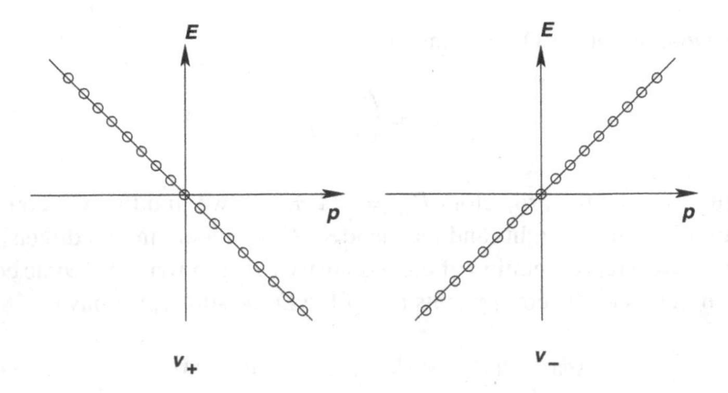

Fig. 9.1 Spectrum of the massless two-dimensional Dirac field. We denote by $v_{\pm}$ the states with dispersion relation $E = \mp p$

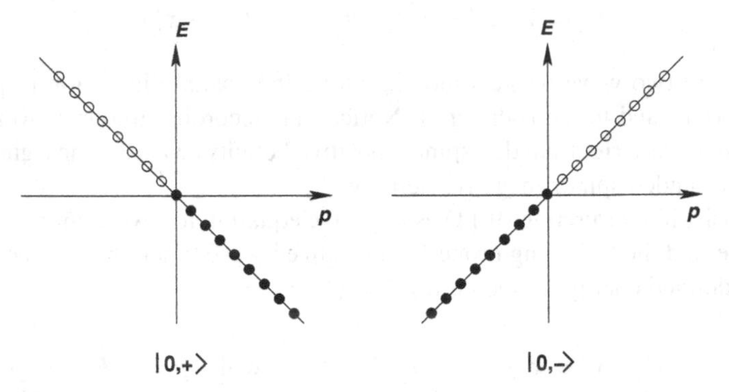

Fig. 9.2 The two branches in the vacuum of the theory. The solid points represent the filled negative energy states

$$u_{\pm}(x) = \sum_{E>0} \left[a_{\pm}(E) v_{\pm}^{(E)}(x) + b_{\pm}^{\dagger}(E) v_{\pm}^{(E)}(x)^* \right]. \tag{9.15}$$

The operator $a_{\pm}(E)$ annihilates a particle with positive energy E and momentum $\mp E$, and $b_{\pm}^{\dagger}(E)$ creates out of the vacuum an antiparticle with positive energy E and spatial momentum $\mp E$. In the Dirac sea picture the operator $b_{\pm}(E)^{\dagger}$ is originally an annihilation operator for a state of the sea with negative energy $-E$. As in four dimensions, the problem of the negative energy states is solved by interpreting annihilation operators for negative energy states as creation operators for the corresponding antiparticle with positive energy (and vice versa). The operators appearing in the expansion of $u_{\pm}$ in Eq. (9.15) satisfy the usual fermionic algebra

$$\{a_{\lambda}(E), a_{\lambda'}^{\dagger}(E')\} = \{b_{\lambda}(E), b_{\lambda'}^{\dagger}(E')\} = \delta_{E,E'}\delta_{\lambda\lambda'}, \tag{9.16}$$

where we have introduced the label $\lambda, \lambda' = \pm$. In addition, $a_{\lambda}(E)$, $a_{\lambda}^{\dagger}(E)$ anticommute with $b_{\lambda'}(E')$, $b_{\lambda'}^{\dagger}(E')$.

The Lagrangian of the theory

$$\mathscr{L} = iu_+^\dagger(\partial_0 + \partial_1)u_+ + iu_-^\dagger(\partial_0 - \partial_1)u_- \tag{9.17}$$

is invariant under both the $U(1)_V$ transformations shown in Eq. (9.2), and $U(1)_A$ of Eq. (9.3). The corresponding Noether currents are

$$J_V^\mu = \begin{pmatrix} u_+^\dagger u_+ + u_-^\dagger u_- \\ -u_+^\dagger u_+ + u_-^\dagger u_- \end{pmatrix}, \quad J_A^\mu = \begin{pmatrix} u_+^\dagger u_+ - u_-^\dagger u_- \\ -u_+^\dagger u_+ - u_-^\dagger u_- \end{pmatrix}. \tag{9.18}$$

The associated conserved charges are given by

$$Q_V \equiv \int_0^L dx^1 J_V^0 = \int_0^L dx^1 \left(u_+^\dagger u_+ + u_-^\dagger u_-\right), \tag{9.19}$$

for the vector current, and

$$Q_A \equiv \int_0^L dx^1 J_A^0 = \int_0^L dx^1 \left(u_+^\dagger u_+ - u_-^\dagger u_-\right) \tag{9.20}$$

for the vector axial one. Using the orthonormality relations for the modes $v_\pm^{(E)}(x)$

$$\int_0^L dx^1 v_\pm^{(E)}(x)^* v_\pm^{(E')}(x) = \delta_{E,E'}, \tag{9.21}$$

the conserved charges can be explicitly computed as

$$\begin{aligned} Q_V &= \sum_{E>0}\left[a_+^\dagger(E)a_+(E) - b_+^\dagger(E)b_+(E) + a_-^\dagger(E)a_-(E) - b_-^\dagger(E)b_-(E)\right], \\ Q_A &= \sum_{E>0}\left[a_+^\dagger(E)a_+(E) - b_+^\dagger(E)b_+(E) - a_-^\dagger(E)a_-(E) + b_-^\dagger(E)b_-(E)\right]. \end{aligned} \tag{9.22}$$

From these expressions we see how Q_V counts the net fermion number, i.e. the number of particles minus antiparticles, independently of their helicity. The axial charge Q_A, on the other hand, counts the net number of positive minus negative helicity states. In the case of the vector current we have subtracted a formally divergent vacuum contribution to the charge (the "charge of the Dirac sea").

In the free theory there is of course no problem with the conservation of either Q_V or Q_A, since the occupation numbers do not change. What we want to study is the effect of coupling the theory to the electric field $\mathscr{E}$. We work in the gauge $\mathscr{A}^0 = 0$. Instead of solving the problem exactly we are going to use the following trick: we simulate the electric field by adiabatically varying in a long time τ_0 the

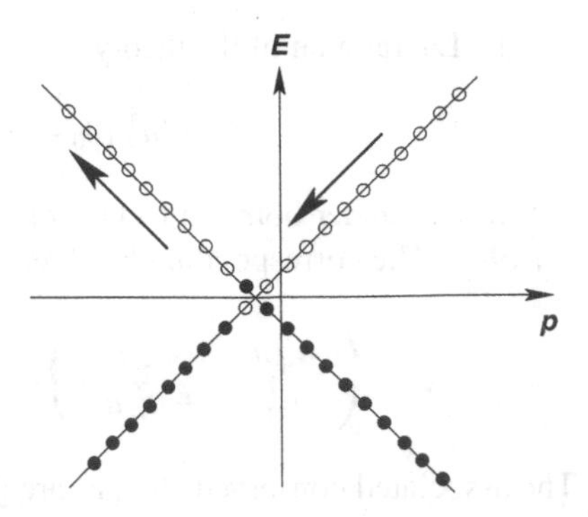

Fig. 9.3 Effect of the electric field on the vacuum shown in Fig. 9.2. Some of the occupied negative energy states in the brach ν_+ acquires positive energy, while the same number of empty positive energy states in the branch ν_- shift to negative energy and become holes in the Dirac sea

vector potential $\mathscr{A}^1$ from zero value to $\mathscr{E}\tau_0$. From our discussion in Chap. 4 (see Sect. 4.1) we know that the effect of the electromagnetic coupling in the theory is a shift in the momentum according to

$$p \longrightarrow p - e\mathscr{A}^1, \tag{9.23}$$

where e is the charge of the fermions. Since we assumed that the vector potential varies adiabatically, we can take it to be approximately constant at each time.

We have to understand the effect on the vacuum depicted in Fig. 9.2 of switching on the vector potential. Increasing adiabatically $\mathscr{A}^1$ results, according to Eq. (9.23), in decreasing the momentum of the state. What happens to the energy depends on whether we consider states with dispersion relation $E = -p$ (the branch ν_+) or $E = p$ (the branch ν_-).

The result is that the two branches move as shown in Fig. 9.3. Thus, some of the negative energy states of the ν_+ branch acquire positive energy while the same number of the empty positive energy states of the other branch ν_- become empty negative energy states. Physically, this means that the external electric field $\mathscr{E}$ creates a number of particle-antiparticle pairs out of the vacuum.

We have to count the number of such pairs created by the electric field after a time τ_0. This is given by

$$N = \frac{L}{2\pi} e\mathscr{E}\tau_0. \tag{9.24}$$

To get this expression we have divided the shift of the spectrum $e\mathscr{E}\tau_0$ by the separation between energy levels given by $\frac{2\pi}{L}$ [cf. Eq. (9.14)]. The value of the charges at the time τ_0 are

$$\begin{aligned} Q_V(\tau_0) &= (N-0) + (0-N) = 0, \\ Q_A(\tau_0) &= (N-0) - (0-N) = 2N. \end{aligned} \tag{9.25}$$

We conclude that the coupling to the electric field produces a violation in the conservation of the axial charge per unit time given by

$$\dot{Q}_{\rm A} = \frac{e}{\pi}\mathscr{E}L. \tag{9.26}$$

This result translates into a nonconservation of the axial vector current

$$\partial_\mu J^\mu_{\rm A} = \frac{e\hbar}{\pi}\mathscr{E}, \tag{9.27}$$

where we have restored $\hbar$ to make clear that we are dealing with a quantum effect. In addition, the fact that $\Delta Q_{\rm V} = 0$ guarantees that the vector current remains conserved also quantum mechanically, $\partial_\mu J^\mu_{\rm V} = 0$.

9.2 The Triangle Diagram

We have just studied a two-dimensional example of the Adler-Bell-Jackiw axial anomaly [3, 4]. We have presented a heuristic analysis consisting of studying the coupling of a two-dimensional massless fermion to an external classical electric field to compute the violation in the conservation of the axial vector current due to quantum effects.

This suggests an alternative, more sophisticated way to compute the axial anomaly. Gauge invariance requires that the fermion couples to the external gauge field through the vector current $J^\mu_{\rm V}$ via a term in the Lagrangian

$$\mathscr{L} = i\overline{\psi}\partial\!\!\!/\psi + eJ^\mu_{\rm V}\mathscr{A}_\mu, \tag{9.28}$$

where $\mathscr{A}_\mu(x)$ represents the classical external gauge field. To decide whether the axial vector current is conserved quantum mechanically we compute the vacuum expectation value

$$\langle \partial_\mu J^\mu_{\rm A}(x)\rangle_{\mathscr{A}}, \tag{9.29}$$

where the subscript indicates the expectation value is computed in the vacuum of the theory coupled to the external field. This quantity can be evaluated in powers of $\mathscr{A}_\mu$ using either the operator formalism or functional integrals. The first nonvanishing term is

$$\langle \partial_\mu J^\mu_{\rm A}(x)\rangle_{\mathscr{A}} = ie\int d^2y\,\partial_\mu C^{\mu\nu}(y)\mathscr{A}_\nu(x-y), \tag{9.30}$$

where

$$C^{\mu\nu}(x) = \langle 0|T\left[J^\mu_{\rm A}(x)J^\nu_{\rm V}(0)\right]|0\rangle = \quad \text{[diagram: fermion loop with } J^\mu_{\rm A} \text{ and } J^\nu_{\rm V} \text{ insertions]} \tag{9.31}$$

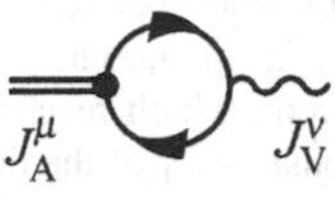

In this correlation function the state $|0\rangle$ represents the Fock space vacuum of the free fermion theory. It can be evaluated using Wick's theorem. The Feynman diagram summarizes the Wick contractions required to compute the time-ordered correlation function of the two currents

$$C^{\mu\nu}(x) = \langle 0|\overline{\psi}\gamma^{\mu}\gamma_5\psi(x)\overline{\psi}\gamma^{\nu}\psi(0)|0\rangle. \tag{9.32}$$

We have concluded that the axial anomaly is controlled by the quantity $\partial_\mu C^{\mu\nu}(x)$. In computing the anomaly we have to impose the conservation of the vector current. This is crucial, since the gauge invariance of the theory depends upon it.[2] Doing this, one arrives at the result

$$\langle \partial_\mu J^{\mu}_{\text{A}}\rangle_{\mathscr{A}} = -\frac{e\hbar}{2\pi}\varepsilon^{\nu\sigma}\mathscr{F}_{\nu\sigma}, \tag{9.33}$$

with $\varepsilon^{01} = -\varepsilon^{10} = 1$ and $\mathscr{F}_{\mu\nu}$ is the field strength of the external gauge field. It is immediate to check that the diagramatic calculation renders the same result (9.27) obtained in the previous section using a more heuristic argumentation.

The calculation of the axial anomaly can be also carried out in four dimensions along the same lines. Again, we have to compute the vacuum expectation value of the axial vector current coupled to an external classical gauge field $\mathscr{A}_\mu$. Now, however, the first nonvanishing contribution comes from the term quadratic in the external gauge field, namely

$$\langle \partial_\mu J^{\mu}\rangle_{\mathscr{A}} = -\frac{e^2}{2}\int d^4y_1 d^4y_2 \partial^{(x)}_{\mu} C^{\mu\nu\sigma}(x,y)\mathscr{A}_\nu(x-y_1+y_2)\mathscr{A}_\sigma(x-y_2), \tag{9.34}$$

where now

$$C^{\mu\nu\sigma}(x,y) = \left\langle 0\right| T\left[J^{\mu}_{\text{A}}(x)J^{\nu}_{\text{V}}(y)J^{\sigma}_{\text{V}}(0)\right]\left|0\right\rangle. \tag{9.35}$$

This correlation function can be computed diagrammatically as

$$C^{\mu\nu\sigma}(x,y) = \Big[\ \cdots\ \Big]_{\text{symmetric}} \tag{9.36}$$

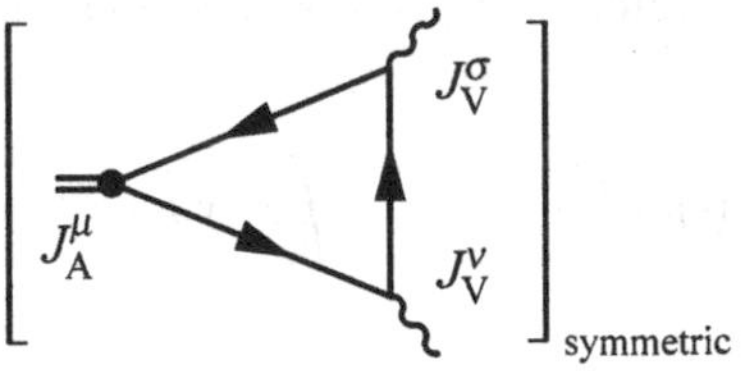

J^{μ}_{A} J^{σ}_{V} J^{ν}_{V}

[2] In fact there is a tension between the conservation of the vector an axial vector currents. The calculation of the diagram shown in Eq. (9.31) can be carried out imposing the conservation of the axial vector current, which results in an anomaly for the vector current. Since this would be disastrous for the consistency of the theory, we choose the other alternative.

This is the celebrated triangle diagram. The subscript indicates that, in fact, $C^{\mu\nu\sigma}$ is given by two triangle diagrams with the two photon external legs interchanged. This is the result of Bose symmetry and can be explicitly checked by performing the Wick contractions in the correlation function (9.35).

The evaluation of the integral in the right-hand side of (9.34) is complicated by the presence of divergences that have to be regularized. As in the two-dimensional case, the conservation of the vector currents has to be imposed. The calculation gives the following anomaly for the axial vector current [3, 4]

$$\langle \partial_\mu J^\mu_{\mathrm{A}} \rangle_{\mathscr{A}} = -\frac{e^2}{16\pi^2} \varepsilon^{\mu\nu\sigma\lambda} \mathscr{F}_{\mu\nu} \mathscr{F}_{\sigma\lambda}. \tag{9.37}$$

This result has very important consequences in the physics of strong interactions as we will see in the next section.

We have paid attention to the axial anomaly in two and four dimensions. Chiral fermions exists in all even-dimensional space-times and, as a matter of fact, the axial vector current has an anomaly in all even-dimensional space-times. More precisely, if the dimension of the space-time is $d = 2k$, with $k = 1, 2, \ldots$, the anomaly is given by a one-loop diagram with one axial current and k vector currents, i.e. a $(k+1)$-gon. For example, in 10 dimensions the axial anomaly comes from the following hexagon diagram

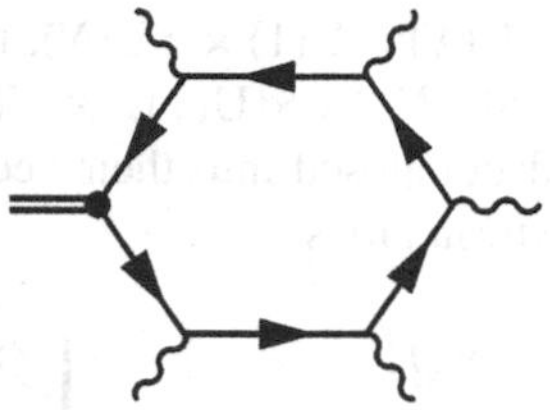

As in the four-dimensional case, Bose symmetry and the conservation of all vector currents has to be imposed.

9.3 Chiral Symmetry in QCD

Our knowledge of the physics of strong interactions is based on the theory of Quantum Chromodynamics (QCD) introduced in Sect. 5.3 (see also [5–7] for reviews). Here we will consider a slightly more general version with an arbitrary number of colors and flavors: a nonabelian gauge theory with gauge group $\mathrm{SU}(N_c)$ coupled to a number N_f of quarks. These are spin-$\frac{1}{2}$ particles Q_i^f labelled by the color and flavor quantum numbers $i = 1, \ldots, N_c$ and $f = 1, \ldots, N_f$. The interaction between them is mediated by the $N_c^2 - 1$ gauge bosons, the gluons A^A_μ, with $A = 1, \ldots, N_c^2 - 1$. Let us recall that in the real world $N_c = 3$ and $N_f = 6$, corresponding to the six quarks: up (u), down (d), charm (c), strange (s), top (t) and bottom (b).

For reasons that will be clear later we work in the limit of vanishing quark masses[3] $m_f \to 0$. In this case the QCD Lagrangian is given by

$$\mathscr{L}_{\rm QCD} = -\frac{1}{4}F^A_{\mu\nu}F^{A\mu\nu} + \sum_{f=1}^{N_f}\left(i\overline{Q}^f_L \not{D} Q^f_L + i\overline{Q}^f_R \not{D} Q^f_R\right), \tag{9.38}$$

where the subscripts L and R indicate respectively left and right-handed spinors,

$$Q^f_{L,R} \equiv \frac{1}{2}(1 \pm \gamma_5)Q^f, \tag{9.39}$$

and the field strength $F^A_{\mu\nu}$ and covariant derivative D_μ are respectively defined in Eqs. (4.52) and (4.46). Apart from the gauge symmetry, this Lagrangian is also invariant under a global $\mathrm{U}(N_f)_L \times \mathrm{U}(N_f)_R$ acting on the flavor indices and defined by

$$\mathrm{U}(N_f)_L : \begin{cases} Q^f_L \to \sum_{f'=1}^{N_f} (U_L)_{ff'} Q^{f'}_L \\ Q^f_R \to Q^f_R \end{cases} \qquad \mathrm{U}(N_f)_R : \begin{cases} Q^f_L \to Q^f_L \\ Q^r_R \to \sum_{f'=1}^{N_f} (U_R)_{ff'} Q^{f'}_R \end{cases} \tag{9.40}$$

with $U_L, U_R \in \mathrm{U}(N_f)$. Since U(*N*) = U(1) × SU(*N*), this global symmetry group can be written as $\mathrm{SU}(N_f)_L \times \mathrm{SU}(N_f)_R \times \mathrm{U}(1)_L \times \mathrm{U}(1)_R$. The abelian subgroup $\mathrm{U}(1)_L \times \mathrm{U}(1)_R$ can be now decomposed into their vector $\mathrm{U}(1)_B$ and axial $\mathrm{U}(1)_{\rm A}$ subgroups defined by the transformations

$$\mathrm{U}(1)_B : \begin{cases} Q^f_L \to e^{i\alpha} Q^f_L \\ Q^f_R \to e^{i\alpha} Q^f_R \end{cases} \qquad \mathrm{U}(1)_{\rm A} : \begin{cases} Q^f_L \to e^{i\alpha} Q^f_L \\ Q^f_R \to e^{-i\alpha} Q^f_R \end{cases}$$

According to Noether's theorem, associated with these two abelian symmetries we have two conserved currents:

$$J^\mu_{\rm V} = \sum_{f=1}^{N_f} \overline{Q}^f \gamma^\mu Q^f, \quad J^\mu_{\rm A} = \sum_{f=1}^{N_f} \overline{Q}^f \gamma^\mu \gamma_5 Q^f. \tag{9.41}$$

The conserved charge associated with the vector current $J^\mu_{\rm V}$ is the baryon number counting the number of quarks minus the number of antiquarks.

The nonabelian part of the global symmetry, group $\mathrm{SU}(N_f)_L \times \mathrm{SU}(N_f)_R$ can also be decomposed into its vector and axial factors, $\mathrm{SU}(N_f)_{\rm V} \times \mathrm{SU}(N_f)_{\rm A}$, defined by the following transformations of the quarks fields

[3] In the real world this makes sense only for the up and down, and perhaps the strange quarks.

$$
\mathrm{SU}(N_f)_{\mathrm{V}}: \begin{cases} Q_L^f \to \sum\limits_{f'=1}^{N_f} U_{ff'} Q_L^{f'} \\ \\ Q_R^f \to \sum\limits_{f'=1}^{N_f} U_{ff'} Q_R^{f'} \end{cases} \qquad \mathrm{SU}(N_f)_{\mathrm{A}}: \begin{cases} Q_L^f \to \sum\limits_{f'=1}^{N_f} U_{ff'} Q_L^{f'} \\ \\ Q_R^f \to \sum\limits_{f'=1}^{N_f} U_{ff'}^{-1} Q_R^{f'} \end{cases} \tag{9.42}
$$

where U is a $\mathrm{SU}(N_f)$ matrix. Again, the application of Noether's theorem shows the existence of the following nonabelian conserved charges

$$
J_{\mathrm{V}}^{I\mu} \equiv \sum_{f,f'=1}^{N_f} \overline{Q}^f \gamma^\mu (T^I)_{ff'} Q^{f'},
$$

$$
J_{\mathrm{A}}^{I\mu} \equiv \sum_{f,f'=1}^{N_f} \overline{Q}^f \gamma^\mu \gamma_5 (T^I)_{ff'} Q^{f'}. \tag{9.43}
$$

To summarize, we have shown that the initial flavor chiral symmetry of the QCD Lagrangian (9.38) can be decomposed according to

$$
\mathrm{U}(N_f)_L \times \mathrm{U}(N_f)_R = \mathrm{SU}(N_f)_{\mathrm{V}} \times \mathrm{SU}(N_f)_{\mathrm{A}} \times \mathrm{U}(1)_B \times \mathrm{U}(1)_{\mathrm{A}}. \tag{9.44}
$$

Up to now we have worked with the classical Lagrangian. The question to address next is which part of the classical global symmetry is preserved in the quantum theory.

As argued in Sect. 9.1, the conservation of the axial vector currents J_{A}^μ and $J_{\mathrm{A}}^{A\mu}$ can in principle be spoiled by an anomaly. In the case of the abelian axial current J_{A}^μ the relevant quantity to compute is the correlation function

$$
C^{\mu\nu\sigma}(x,x') \equiv \langle 0| T\left[J_{\mathrm{A}}^\mu(x) j_{\mathrm{gauge}}^{A\,\nu}(x') j_{\mathrm{gauge}}^{B\sigma}(0)\right] |0\rangle
$$

$$
= \sum_{f=1}^{N_f} \Big[\quad \Big]_{\text{symmetric}} \tag{9.45}
$$

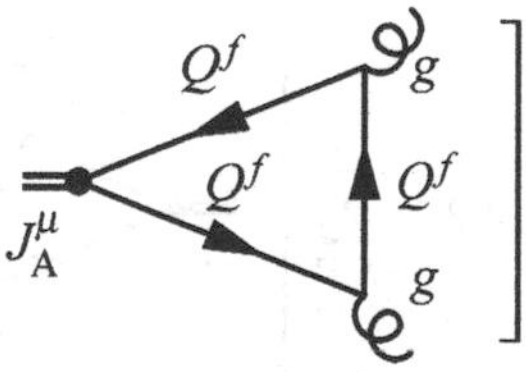

Q^f g Q^f Q^f J_{A}^μ g

Here $j_{\mathrm{gauge}}^{A\mu}$ is the nonabelian conserved current coupling to the gluon field

$$
j_{\mathrm{gauge}}^{A\mu} \equiv \sum_{f=1}^{N_f} \overline{Q}^f \gamma^\mu \tau^A Q^f, \tag{9.46}
$$

where, to avoid confusion with the generators of the global symmetry, we have denoted by τ^A the generators of the gauge group $\mathrm{SU}(N_c)$. The anomaly can be read now from $\partial_\mu^{(x)} C^{\mu\nu\sigma}(x, x')$. If we impose Bose symmetry with respect to the interchange of the two outgoing gluons and the conservation of the vector currents, we find that the axial abelian global current has an anomaly given by[4]

$$\partial_\mu J^\mu_{\mathrm{A}} = -\frac{g^2 N_f}{32\pi^2} \varepsilon^{\mu\nu\sigma\lambda} F^A_{\mu\nu} F^A_{\sigma\lambda}. \tag{9.47}$$

In the case of the nonabelian axial global symmetry $\mathrm{SU}(N_f)_{\mathrm{A}}$ the calculation of the anomaly is made as above. The result, however, is quite different since in this case we conclude that the nonabelian axial vector current $J^{A\mu}_{\mathrm{A}}$ is not anomalous. This can be easily seen by noticing that associated with the axial vector current vertex we have a generator T^I of $\mathrm{SU}(N_f)$, whereas for the two gluon vertices we have the generators τ^A of the gauge group $\mathrm{SU}(N_c.)$ Therefore, the triangle diagram is proportional to the group-theory factor

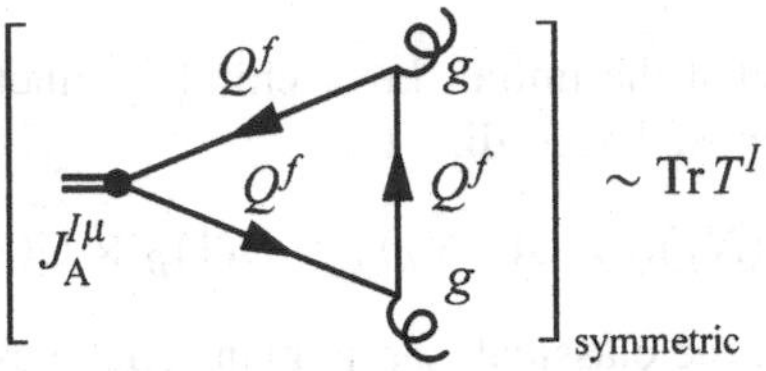

$$\left[J^{I\mu}_{\mathrm{A}} \quad Q^f \quad Q^f \quad Q^f \quad g \quad g \right]_{\text{symmetric}} \sim \mathrm{Tr}\, T^I\, \mathrm{Tr}\{\tau^A, \tau^B\} = 0 \tag{9.48}$$

vanishing because the generators of $\mathrm{SU}(N_f)$ are traceless.

From here we could be tempted to conclude that the nonabelian axial symmetry $\mathrm{SU}(N_f)_{\mathrm{A}}$ is nonanomalous. However this is not the whole story, since quarks are charged particles that also couple to photons. Thus there is a second potential source of an anomaly coming from the the one-loop triangle diagram coupling $J^{I\mu}_{\mathrm{A}}$ to two photons

$$\langle 0|T\left[J^{I\mu}_{\mathrm{A}}(x) j^\nu_{\mathrm{em}}(x') j^\sigma_{\mathrm{em}}(0)\right]|0\rangle = \sum_{f=1}^{N_f} \left[J^{I\mu}_{\mathrm{A}} \quad Q^f \quad Q^f \quad Q^f \quad \gamma \quad \gamma \right]_{\text{symmetric}} \tag{9.49}$$

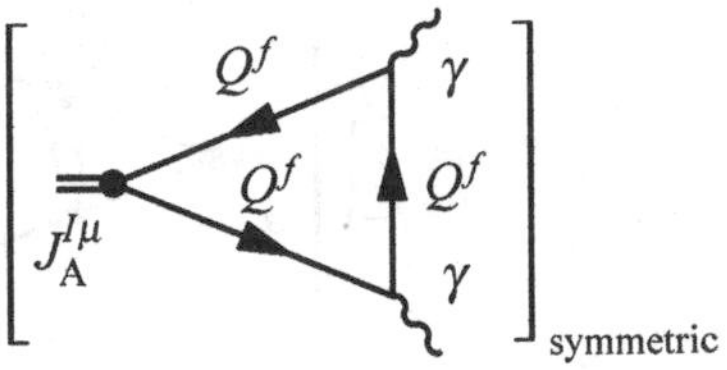

where j^μ_{em} is the electromagnetic current

$$j^\mu_{\mathrm{em}} = \sum_{f=1}^{N_f} q_f \overline{Q}^f \gamma^\mu Q^f, \tag{9.50}$$

---

[4] The normalization of the generators T^I of the global $\mathrm{SU}(N_f)$ is given by $\mathrm{Tr}(T^I T^J) = \frac{1}{2}\delta^{IJ}$.

with q_f the electric charge of the f-th quark flavor. A calculation of the diagram in (9.49) shows the existence of the Adler-Bell-Jackiw anomaly given by

$$\partial_\mu J^{I\mu}_{\rm A} = -\frac{N_c}{16\pi^2}\left[\sum_{f=1}^{N_f}(T^I)_{ff}q_f^2\right]\varepsilon^{\mu\nu\sigma\lambda}F_{\mu\nu}F_{\sigma\lambda}, \tag{9.51}$$

where $F_{\mu\nu}$ is the field strength of the electromagnetic field coupling to the quarks. The only chance for the anomaly to cancel is that the factor between brackets in this equation be identically zero.

Before proceeding let us summarize the results found so far. Due to the presence of anomalies the axial part of the global chiral symmetry, $\mathrm{SU}(N_f)_{\rm A}$ and $\mathrm{U}(1)_{\rm A}$, are not realized quantum mechanically in general. We found that $\mathrm{U}(1)_{\rm A}$ is always affected by an anomaly. However, the right-hand side of the anomaly equation (9.47) is a total derivative, so the anomalous character of $J^\mu_{\rm A}$ docs not cxplain the absence of $\mathrm{U}(1)_{\rm A}$ multiplets in the hadron spectrum, since a new current can be constructed which is conserved. In addition, the nonexistence of candidates for an associated Nambu-Goldstone boson with the right quantum numbers indicates that $\mathrm{U}(1)_{\rm A}$ is not spontaneously broken either, so it has to be explicitly broken somehow. This is the so-called U(1)-problem solved by 't Hooft [8], who showed how the contribution from instantons describing quantum transitions between vacua with topologically nontrivial gauge field configurations results in an explicit breaking of this symmetry.

Due to the dynamics of the $\mathrm{SU}(N_c)$ gauge theory, the axial nonabelian symmetry is spontaneously broken due to the presence at low energies of a vacuum expectation value for the fermion bilinear $\overline{Q}^f Q^f$

$$\langle 0|\overline{Q}^f Q^f|0\rangle \neq 0 \qquad \text{(no summation in } f!). \tag{9.52}$$

This nonvanishing vacuum expectation value for the quark bilinear breaks chiral invariance spontaneously to the vector subgroup $\mathrm{SU}(N_f)_{\rm V}$, so the only subgroup of the original global symmetry that is realized in the full theory at low energy is

$$\mathrm{U}(N_f)_L \times \mathrm{U}(N_f)_R \longrightarrow \mathrm{SU}(N_f)_{\rm V} \times \mathrm{U}(1)_B. \tag{9.53}$$

Associated with this breaking, Nambu–Goldstone bosons should appear with the quantum numbers of the broken nonabelian currents. For example, in the case of QCD the Nambu–Goldstone bosons associated with the spontaneous symmetry breaking induced by the vacuum expectation values $\langle\overline{u}u\rangle$, $\langle\overline{d}d\rangle$ and $\langle(\overline{u}d-\overline{d}u)\rangle$ have been identified as the pions π^0, $\pi^\pm$. These bosons are not exactly massless due to the nonvanishing mass of the u and d quarks. Since the global chiral symmetry is already slightly broken by mass terms in the Lagrangian, the associated Goldstone bosons also have masses although they are very light compared to the masses of other hadrons.

In order to have a better physical understanding of the role of anomalies in the physics of the strong interactions we particularize our analysis to the case of real QCD. Since the u and d quarks are much lighter than the other four flavors, QCD at low energies can be well described by including only these two flavors and ignoring heavier quarks. In this approximation, from our previous discussion we know that the low energy global symmetry of the theory is $\mathrm{SU(2)_V} \times \mathrm{U(1)}_B$, where now the vector group $\mathrm{SU(2)_V}$ is the well-known isospin symmetry. The axial $\mathrm{U(1)_A}$ current is anomalous due to Eq. (9.47) with $N_f = 2$. In the case of the nonabelian axial symmetry $\mathrm{SU(2)_A}$, taking into account that $q_u = \frac{2}{3}e$ and $q_d = -\frac{1}{3}e$ and that the three generators of SU(2) can be written in terms of the Pauli matrices as $T^K = \frac{1}{2}\sigma_K$ we find

$$\sum_{f=u,d} (T^1)_{ff} q_f^2 = \sum_{f=u,d} (T^2)_{ff} q_f^2 = 0,$$
$$\sum_{f=u,d} (T^3)_{ff} q_f^2 = \frac{e^2}{6}. \tag{9.54}$$

Therefore $J_{\mathrm{A}}^{3\mu}$ is anomalous.

The anomaly in the axial vector current $J_{\mathrm{A}}^{3\mu}$ has an important physical consequence. As we learned in Chap. 5 the flavor wave function of the neutral pion π^0 is given by

$$|\pi^0\rangle = \frac{1}{\sqrt{2}} \left(|\bar{u}u\rangle - |\bar{d}d\rangle \right). \tag{9.55}$$

The isospin quantum numbers of $|\pi^0\rangle$ are those of $J_{\mathrm{A}}^{3\mu}$. In fact, the correspondence goes even further. The divergence of the axial vector current $\partial_\mu J_{\mathrm{A}}^{3\mu}$ has precisely the same quantum numbers as the pion. This means that, properly normalized, it can be identified as the operator creating a pion π^0 out of the vacuum

$$|\pi^0\rangle \sim \partial_\mu J_{\mathrm{A}}^{3\mu} |0\rangle. \tag{9.56}$$

This leads to the physical interpretation of the triangle diagram (9.49) with $J_{\mathrm{A}}^{3\mu}$ as the one loop contribution to the decay of a neutral pion into two photons

$$\pi^0 \longrightarrow 2\gamma. \tag{9.57}$$

This is an interesting piece of physics. In 1967 Sutherland and Veltman [9, 10] presented a calculation, using current algebra techniques, according to which the decay of the pion into two photons should be suppressed. This however contradicted the experimental evidence showing the existence of such a decay. The way out to this paradox, as pointed out in [3, 4], is the axial anomaly. What happens is that the current algebra analysis overlooks the ambiguities associated with the regularization of divergences in quantum field theory. A QED evaluation of the triangle diagram leads to a divergent integral that has to be regularized. It is in this process that the

Adler-Bell-Jackiw axial anomaly appears resulting in a nonvanishing value for the $\pi^0 \to 2\gamma$ amplitude.[5]

9.4 Gauge Anomalies

The existence of anomalies associated with global currents does not necessarily mean difficulties for the theory. On the contrary, as we saw in the case of the axial anomaly, its existence provides a solution of the Sutherland–Veltman paradox and an explanation of the electromagnetic decay of the pion. The situation is very different when we deal with local symmetries. A quantum mechanical violation of gauge symmetry leads to many problems, from lack of renormalizability to nondecoupling of negative norm states. This is because the presence of an anomaly in the theory implies that the Gauss' law constraint $\mathbf{D} \cdot \mathbf{E}_A = \rho_A$ cannot be consistently implemented in the quantum theory. As a consequence, states that classically were eliminated by the gauge symmetry become propagating in the quantum theory, thus spoiling the consistency of the theory.

Anomalies in a gauge symmetry can be expected only in chiral theories where left and right-handed fermions transform in different representations of the gauge group. Physically, the most interesting example of such theories is the electroweak sector of the standard model where, for example, left handed fermions transform as doublets under SU(2) whereas right-handed fermions are singlets. On the other hand, QCD is free of gauge anomalies since both left- and right-handed quarks transform in the fundamental representation of SU(3).

We consider the Lagrangian

$$\mathscr{L} = -\frac{1}{4} F^{A\mu\nu} F^A_{\mu\nu} + i \sum_{i=1}^{N_+} \overline{\psi}^i_+ \not{D}^{(+)} \psi^i_+ + i \sum_{j=1}^{N_-} \overline{\psi}^j_- \not{D}^{(-)} \psi^j_-, \tag{9.58}$$

where the chiral fermions $\psi^i_\pm$ transform according to the representations $\tau^A_{i,\pm}$ of the gauge group G ($A = 1, \ldots, \dim G$). The covariant derivatives $D^{(\pm)}_\mu$ are, as usual, defined by

$$D^{(\pm)}_\mu \psi^i_\pm = \partial_\mu \psi^i_\pm - i g_{\rm YM} A^A_\mu \tau^A_\pm \psi^i_\pm. \tag{9.59}$$

The anomaly is determined by the parity-violating part of the triangle diagram with three external gauge bosons, summed over all chiral fermion species running in the loop. All three vertices in the diagram include a projector P_+ or P_- and the parity-violating terms are identified as those containing a single γ_5. Splitting the gauge current into its vector and axial vector part, we conclude that the gauge anomaly comes from the triangle diagram with one axial and two vector gauge currents

---

5 An early computation of the triangle diagram for the electromagnetic decay of the pion was made by Steinberger in [11].

$$\langle 0|T\left[j_{\mathrm{A}}^{A\mu}(x)j_{\mathrm{V}}^{B\nu}(x')j_{\mathrm{V}}^{C\sigma}(0)\right]|0\rangle = \Bigg[$$ 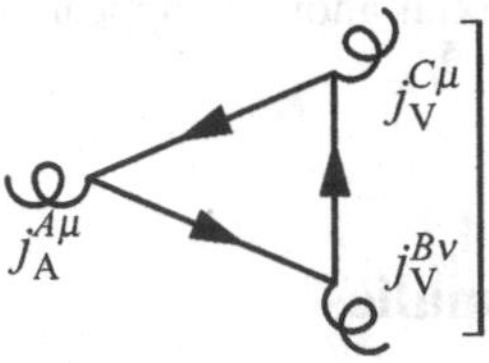 $$\Bigg]_{\text{symmetric}} \tag{9.60}$$

where $j_{\mathrm{V}}^{A\mu}$ and $j_{\mathrm{A}}^{A\mu}$ are given by

$$j_{\mathrm{V}}^{A\mu} = \sum_{i=1}^{N_+} \overline{\psi}_+^i \tau_+^A \gamma^\mu \psi_+^i + \sum_{j=1}^{N_-} \overline{\psi}_-^j \tau_-^A \gamma^\mu \psi_-^j,$$
$$j_{\mathrm{A}}^{A\mu} = \sum_{i=1}^{N_+} \overline{\psi}_+^i \tau_+^A \gamma^\mu \psi_+^i - \sum_{i=1}^{N_-} \overline{\psi}_-^j \tau_-^A \gamma^\mu \psi_-^j. \tag{9.61}$$

Luckily, we do not have to compute the whole diagram in order to find an anomaly cancellation condition. It is enough if we calculate the overall group theoretical factor. In the case of the diagram in Eq. (9.60) for each fermion species running in the loop this factor is equal to

$$\mathrm{Tr}\left[\tau_{i,\pm}^A\{\tau_{i,\pm}^B, \tau_{i,\pm}^C\}\right], \tag{9.62}$$

where the sign $\pm$ corresponds respectively to the generators of the representations of the gauge group for the left and right-handed fermions. Hence, the anomaly cancellation condition reads

$$\sum_{i=1}^{N_+} \mathrm{Tr}\left[\tau_{i,+}^A\{\tau_{i,+}^B, \tau_{i,+}^C\}\right] - \sum_{j=1}^{N_-} \mathrm{Tr}\left[\tau_{j,-}^A\{\tau_{j,-}^B, \tau_{j,-}^C\}\right] = 0. \tag{9.63}$$

Knowing this we can proceed to check the anomaly cancellation in the standard model SU(3)$\times$ SU(2)$\times$ U(1)$_Y$. Left handed fermions (both leptons and quarks) transform as doublets with respect to the SU(2) factor whereas the right-handed components are singlets. The charge with respect to the U(1)$_Y$ part, the weak hypercharge Y, is determined by the Gell-Mann–Nishijima formula

$$Q = T_3 + Y, \tag{9.64}$$

where Q is the electric charge of the corresponding particle and T_3 is the eigenvalue with respect to the third generator of the SU(2) group in the corresponding representation: $T_3 = \frac{1}{2}\sigma_3$ for the doublets and $T_3 = 0$ for the singlets. For the first family of quarks (u, d) and leptons (e, ν_e) we have the following field content

$$\text{quarks:} \quad \begin{pmatrix} u^i \\ d^i \end{pmatrix}_{L,\frac{1}{6}} \qquad u^i_{R,\frac{2}{3}} \qquad d^i_{R,-\frac{1}{3}}$$
$$\text{leptons:} \quad \begin{pmatrix} \nu_e \\ e \end{pmatrix}_{L,-\frac{1}{2}} \qquad e_{R,-1} \tag{9.65}$$

where $i = 1, 2, 3$ labels the color quantum number and the subscript indicates the value of the weak hypercharge Y. Denoting the representations of SU(3)× SU(2) × U(1)$_Y$ by $(n_c, n_w)_Y$, with n_c and n_w the representations of SU(3) and SU(2) respectively and Y the hypercharge, the matter content of the standard model consists of a three family replication of the representations

$$\begin{aligned} \text{left-handed fermions:} \quad & (3,2)^L_{\frac{1}{6}} \quad (1,2)^L_{-\frac{1}{2}} \\ \text{right-handed fermions:} \quad & (3,1)^R_{\frac{2}{3}} \quad (3,1)^R_{-\frac{1}{3}} \quad (1,1)^R_{-1}. \end{aligned} \tag{9.66}$$

In computing the triangle diagram we have 10 possibilities depending on which factor of the gauge group SU(3)× SU(2)× U(1)$_Y$ appears in each vertex:

SU(3)^3	SU(2)^3	U(1)^3
SU(3)^2 SU(2)	SU(2)^2 U(1)	
SU(3)^2 U(1)	SU(2) U(1)^2	
SU(3) SU(2)^2		
SU(3) SU(2) U(1)		
SU(3) U(1)^2		

It is easy to verify that some of them do not give rise to anomalies. For example, the anomaly for the SU(3)^3 case cancels because left and right-handed quarks transform in the same representation. In the case of SU(2)^3 the cancellation happens term by term using the Pauli matrices identity $\sigma_j\sigma_k = \delta_{jk} + i\varepsilon_{jk\ell}\sigma_\ell$ leading to

$$\operatorname{Tr}\left[\sigma_i\{\sigma_j, \sigma_k\}\right] = 2\,(\operatorname{Tr}\sigma_i)\,\delta_{jk} = 0. \tag{9.67}$$

The hardest condition comes from the three U(1)'s. In this case the absence of anomalies within a single family is guaranteed by the nontrivial identity

$$\begin{aligned} \sum_{\text{left}} Y_+^3 - \sum_{\text{right}} Y_-^3 &= 3 \times 2 \times \left(\frac{1}{6}\right)^3 + 2 \times \left(-\frac{1}{2}\right)^3 - 3 \times \left(\frac{2}{3}\right)^3 \\ &\quad - 3 \times \left(-\frac{1}{3}\right)^3 - (-1)^3 = \left(-\frac{3}{4}\right) + \left(\frac{3}{4}\right) = 0. \end{aligned} \tag{9.68}$$

It is remarkable that the anomaly exactly cancels between leptons and quarks. Notice that this result holds even if a right-handed sterile neutrino is added since such a particle is a singlet under the whole standard model gauge group and therefore does not contribute to the triangle diagram. We see how the matter content of the standard model conspires to yield a consistent quantum field theory.

In all our discussion of anomalies we only considered the computation of one-loop diagrams. It might happen that higher loop orders impose additional conditions. Fortunately this is not so: the Adler–Bardeen theorem [12] guarantees that the axial anomaly only receives contributions from one loop diagrams. Therefore, once anomalies are canceled (if possible) at one loop we know that there will be no new conditions coming from higher-loop diagrams in perturbation theory.

The Adler–Bardeen theorem, however, only applies in perturbation theory. It is nonetheless possible that nonperturbative effects can result in the quantum violation of a gauge symmetry. This is precisely the case pointed out by Witten [13] with respect to the SU(2) gauge symmetry of the standard model. In this case the problem lies in the nontrivial topology of the gauge group SU(2). The invariance of the theory with respect to non-trivial gauge transformations requires the number of fermion doublets to be even. It is again remarkable that the family structure of the standard model makes this anomaly cancel

$$3 \times \begin{pmatrix} u \\ d \end{pmatrix}_L + 1 \times \begin{pmatrix} \nu_e \\ e \end{pmatrix}_L = 4 \text{ SU(2)-doublets}, \tag{9.69}$$

where the factor of 3 comes from the number of colors.

References

1. Álvarez-Gaumé, L.: An introduction to anomalies. In: Velo, G., Wightman, A.S. (eds) Fundamental Problems of Gauge Field Theory, Plenum Press, London (1986)
2. Jackiw, R.: Topological investigations of quantized gauge theories. In: Treiman, S.B., Jackiw, R., Zumino, B., Witten, E. (eds) Current Algebra and Anomalies, Princeton, NJ (1985)
3. Adler, S.: Axial-vector vertex in spinor electrodynamics. Phys. Rev. **177**, 2426 (1969)
4. Bell, J.S., Jackiw, R.: A PCAC puzzle: $\pi^0 \to 2\gamma$ in the sigma model. Nuovo Cim. A **60**, 47 (1969)
5. Ynduráin, F.J.: The Theory of Quark and Gluon Interactions. Springer, New York (1999)
6. Dissertori, G., Knowles, I., Schmelling, M.: Quantum Chromodynamics: High Energy Experiments and Theory. Oxford, New York (2003)
7. Narison, S.: QCD as a Theory of Hadrons: From Partons to Confinement. Cambridge University Press, Cambridge (2004)
8. 't Hooft, G.: How the instantons solve the U(1) problem. Phys. Rept. **142**, 357 (1986)
9. Sutherland, D.G.: Current algebra and some nonstrong mesonic decays. Nucl. Phys. B **2**, 433 (1967)
10. Veltman, M.J.G.: Theoretical aspects of high-energy neutrino interactions. Proc. R. Soc. A **301**, 107 (1967)
11. Steinberger, J.: On the use of substraction fields and the lifetimes of some types of meson decay. Phys. Rev. **76**, 1180 (1949)
12. Adler, S.L., Bardeen, W.A.: Absence of higher order corrections in the anomalous axial vector divergence equation. Phys. Rev. **182**, 1517 (1969)
13. Witten, E.: An SU(2) anomaly. Phys. Lett. B **117**, 324 (1982)

Chapter 10
The Origin of Mass

The time has come to finally address a central problem left pending in the discussion of the standard model carried out in Chap. 5: how particle masses can be generated preserving gauge invariance. We apply the Brout–Englert–Higgs mechanism introduced in Chap. 7 to solve the problem of mass in the electroweak theory.

We will see, however, that this is not the end of the story. The masses generated by spontaneous symmetry breaking in the standard model cannot account for the mass of protons and neutrons, and therefore for most of the mass we see around us, including our own. We will see that its origin is a purely quantum mechanical effect in QCD.

10.1 The Masses in the Standard Model

We are finally ready to give a solution to the double problem that we left unsolved in Chap. 5. First, the chiral nature of the electroweak interaction forbade writing mass terms for the quark and lepton fields, while we know for sure that electrons, muons and other particles are massive. Secondly, the phenomenology of weak decays indicated that this interaction should be mediated by massive gauge bosons, something that at face value is impossible to reconcile with gauge invariance.

Chapter 7 has provided the crucial hint on how this problem can be cured: by breaking the $\mathrm{SU}(2) \times \mathrm{U}(1)_Y$ gauge symmetry spontaneously to the electromagnetic $\mathrm{U}(1)$ one could give mass to three of the gauge bosons mediating the electroweak interaction leaving a massless photon behind. To do so we have to introduce a new field, the Higgs field, transforming under the electroweak gauge group and whose vacuum expectation value breaks it properly. Since we are not interested in breaking Lorentz invariance, the field has to be a scalar.

To find the transformation of the Higgs field under the gauge group we take into account that, in acquiring its vacuum expectation value, it should also give mass to the matter fields. To see how this can be done we go back for a moment to the Abelian Higgs model discussed in Chap. 7 [see Eq. (7.78)]. We add a massless fermion ψ

L. Álvarez-Gaumé and M. Á. Vázquez-Mozo, *An Invitation to Quantum Field Theory*,
Lecture Notes in Physics 839, DOI: 10.1007/978-3-642-23728-7_10,

and couple it to the complex scalar field $\varphi(x)$ introducing the Yukawa coupling term

$$\mathscr{L}_{\text{Yukawa}} = -c\varphi\overline{\psi}\psi, \tag{10.1}$$

where c is a real constant. Upon symmetry breaking, this term in the Lagrangian takes the form

$$\mathscr{L}_{\text{Yukawa}} = -\frac{cv}{\sqrt{2}}\overline{\psi}\psi - \frac{c}{\sqrt{2}}\sigma\overline{\psi}\psi. \tag{10.2}$$

The first term gives a Dirac mass $m_\psi = \frac{1}{\sqrt{2}}cv$ to the fermion $\psi(x)$, while the second one couples it to the scalar field $\sigma(x)$.

This shows the way to solve the problem of giving mass to fermions coupling to gauge fields in a chiral way without breaking gauge invariance. In Chap. 5 we learned that in the standard model the left-handed fermions transform as doublets under the SU(2) factor of the gauge group, whereas the right-handed components are singlets. Then, the gauge invariance of the Yukawa couplings indicates that the Higgs field has to be a SU(2) doublet

$$\mathbf{H} = \begin{pmatrix} H^+ \\ H^0 \end{pmatrix}, \tag{10.3}$$

where H^+ and H^0 are complex scalar fields. Taking this into account we add to the standard model Lagrangian the piece

$$\mathscr{L}^{(\ell)}_{\text{Yukawa}} = -\sum_{i,j=1}^{3}\left(C^{(\ell)}_{ij}\overline{\mathbf{L}}^i\mathbf{H}\ell^j_R + C^{(\ell)*}_{ji}\overline{\ell}^i_R\mathbf{H}^\dagger\mathbf{L}^j\right), \tag{10.4}$$

invariant under SU(2) gauge transformations. Here $C^{(\ell)}_{ij}$ are dimensionless coupling constants and we have used the notation introduced in Table 5.1. The Yukawa couplings have been constructed in such a way that neutrinos do not get Dirac masses.

The masses of the quarks are generated by Yukawa couplings similar to the ones already written for the leptons. One important difference, however, lies in the fact that now we want to give mass to the two components of the left-handed SU(2) doublets. To achieve this we need to couple the fermions not only to the Higgs doublet $\mathbf{H}$ but also to its "charge conjugate"

$$\tilde{\mathbf{H}} \equiv i\sigma_2\mathbf{H}^* = \begin{pmatrix} H^{0*} \\ -H^{+*} \end{pmatrix}. \tag{10.5}$$

From the identity

$$(i\sigma_2)e^{-i\mathbf{a}\cdot\frac{\boldsymbol{\sigma}^*}{2}} = e^{i\mathbf{a}\cdot\frac{\boldsymbol{\sigma}}{2}}(i\sigma_2), \tag{10.6}$$

it follows that the conjugated Higgs field $\tilde{\mathbf{H}}$ also transforms as a SU(2) doublet. Then, the Dirac masses of the quark fields can be obtained from the following Yukawa couplings

$$\mathscr{L}^{(q)}_{\text{Yukawa}} = -\sum_{i,j=1}^{3}\left(C^{(q)}_{ij}\overline{\mathbf{Q}}^i\mathbf{H}D^j_R + C^{(q)*}_{ji}\overline{D}^i_R\mathbf{H}^\dagger\mathbf{Q}^j\right)$$
$$-\sum_{i,j=1}^{3}\left(\tilde{C}^{(q)}_{ij}\overline{\mathbf{Q}}^i\tilde{\mathbf{H}}U^j_R + \tilde{C}^{(q)*}_{ji}\overline{U}^i_R\tilde{\mathbf{H}}^\dagger\mathbf{Q}^j\right). \tag{10.7}$$

The notation also follows Table 5.2.

We have constructed an interaction term between the Higgs field and the fermions demanding invariance under SU(2) gauge transformations. It is easy to see that the Yukawa couplings (10.4) and (10.7) are invariant also under the $\mathrm{U}(1)_Y$ gauge symmetry factor provided the Higgs field is assigned the weak hypercharge $Y(\mathbf{H}) = \frac{1}{2}$. The Gell–Mann–Nishijima formula then implies that

$$Q(H^+) = 1, \quad Q(H^0) = 0, \tag{10.8}$$

thus justifying our notation.

To implement symmetry breaking we have to add the following term to the standard model Lagrangian

$$\mathscr{L}_{\text{Higgs}} = (D_\mu\mathbf{H})^\dagger D^\mu\mathbf{H} - V(\mathbf{H},\mathbf{H}^\dagger), \tag{10.9}$$

where D_μ is the corresponding $\mathrm{SU}(2)\times\mathrm{U}(1)_Y$ covariant derivative (see Sect. 5.4). The potential has to be wisely chosen in such a way that spontaneous symmetry breaking takes place and solves our problems with the particle masses in a satisfactory way. In fact, gauge invariance and the condition that the theory is renormalizable (see Chap. 8) imply that the Higgs potential should be of the form

$$V(\mathbf{H},\mathbf{H}^\dagger) = \frac{\lambda}{4}\left(\mathbf{H}^\dagger\mathbf{H} - \frac{v^2}{2}\right)^2. \tag{10.10}$$

The system exhibits spontaneous symmetry breaking if $v^2 > 0$. Then, the theory has a degenerate family of vacua defined by $\mathbf{H}^\dagger\mathbf{H} = \frac{1}{2}v^2$.

The only surviving gauge symmetry in the electroweak sector at low energies is the U(1) invariance of QED. This means that this symmetry is realized à la Wigner–Weyl and therefore the vacuum has zero electric charge. Taking into account Eq. (10.8) this means that we are forced to take[1]

$$\langle\mathbf{H}\rangle = \begin{pmatrix} 0 \\ \frac{1}{\sqrt{2}}v \end{pmatrix}. \tag{10.11}$$

Since $Y(\mathbf{H}) = \frac{1}{2}$ this vacuum expectation value breaks not only SU(2) but also $\mathrm{U}(1)_Y$. It however preserves the electromagnetic U(1) and therefore implements correctly the symmetry breaking pattern, $\mathrm{SU}(2)\times\mathrm{U}(1)_Y \to \mathrm{U}(1)$.

[1] It can be shown that, by appropriate gauge transformations, any other vacuum expectation value can always be brought to this form.

Following the example of the Abelian Higgs model, the fluctuations around this vacuum can be parametrized as [cf. Eq. (7.81)]

$$\mathbf{H}(x) = \frac{1}{\sqrt{2}} e^{i\mathbf{a}(x)\cdot\frac{\sigma}{2}} \begin{pmatrix} 0 \\ v + h(x) \end{pmatrix}. \tag{10.12}$$

There are four different fields associated with these fluctuations, here denoted by $\mathbf{a}(x)$ and $h(x)$. The factor $e^{i\mathbf{a}(x)\cdot\frac{\sigma}{2}}$ represents the action of the three broken generators,[2] and can be eliminated by a SU(2) gauge transformation. This removes the three would-be Nambu–Goldstone bosons $\mathbf{a}(x)$ that are transmuted into the longitudinal components of the massive gauge bosons W^+, W^- and Z^0. The remaining propagating degree of freedom $h(x)$ is the neutral scalar whose elementary excitation is known as the Higgs boson. Inserting

$$\mathbf{H}(x) = \frac{1}{\sqrt{2}} \begin{pmatrix} 0 \\ v + h(x) \end{pmatrix} \tag{10.13}$$

in $V(\mathbf{H}, \mathbf{H}^\dagger)$ and expanding the result in powers of the field $h(x)$, the mass of the Higgs particle is found to be

$$m_{\mathrm{H}} = v\sqrt{\frac{\lambda}{2}}. \tag{10.14}$$

The dimensionless coupling λ governs the self-interaction of the Higgs bosons.

Substituting Eq. (10.13) in the Yukawa couplings (10.4), we find, at low energies, the following lepton mass terms

$$\mathscr{L}^{(\ell)}_{\text{mass}} = -(\overline{e}_L, \overline{\mu}_L, \overline{\tau}_L) M^{(\ell)} \begin{pmatrix} e_R \\ \mu_R \\ \tau_R \end{pmatrix} + \text{h.c.} \tag{10.15}$$

We notice that no mass term for the neutrinos is generated through the Brout–Englert–Higgs mechanism, so neutrino masses have to be explained in some other way. On the other hand, for the quarks we find

$$\mathscr{L}^{(q)}_{\text{mass}} = -(\overline{d}_L, \overline{s}_L, \overline{b}_L) M^{(q)} \begin{pmatrix} d_R \\ s_R \\ b_R \end{pmatrix} - (\overline{u}_L, \overline{c}_L, \overline{t}_L) \tilde{M}^{(q)} \begin{pmatrix} u_R \\ c_R \\ t_R \end{pmatrix} + \text{h.c.} \tag{10.16}$$

and the mass matrices are given by

---

[2] It might seem strange that, apparently, we have included only the action of the SU(2) generators on the vacuum. As a matter of fact, this is not the case. What happens is that the electromagnetic U(1) remains unbroken and therefore $Q_{\text{vac}} = 0$. Then, using the Gell–Mann–Nishijima relation, the action of the weak hypercharge generator Y on the vacuum can be written in terms of the generators of SU(2) as $Y = -2T_3 = -\sigma_3$.

$$M_{ij}^{(\ell,q)} = \frac{1}{\sqrt{2}} v C_{ij}^{(\ell,q)}, \quad \tilde{M}_{ij}^{(q)} = \frac{1}{\sqrt{2}} v \tilde{C}_{ij}^{(q)}, \tag{10.17}$$

with $C_{ij}^{(\ell,q)}$ and $\tilde{C}_{ij}^{(q)}$ the strength of the Yukawa couplings defining general complex 3×3 matrices. We notice as well that the mass scale of all charged fermion is set by the Higgs vacuum expectation value v.

So far we have written the standard model Lagrangian in terms of fields with well defined transformations under the gauge group (this we call flavor eigenstates). Now, however, there is no a priori reason for the mass matrices in (10.15) and (10.16) to be diagonal. This means that the corresponding propagators are not diagonal and therefore the different flavor eigenstates mix with each other as they propagate. In order to quantize the theory, however, it is more convenient to work with fields whose propagators, at low energies, are diagonal and therefore have well-defined masses. These fields are constructed by noticing that a general complex matrix can always be diagonalized by a biunitary transformation. More precisely, this means that there are unitary matrices $V_{L,R}^{(\ell,q)}$, $\tilde{V}_{L,R}^{(q)}$ such that

$$V_L^{(\ell)\dagger} M^{(\ell)} V_R^{(\ell)} = \begin{pmatrix} m_e & 0 & 0 \\ 0 & m_\mu & 0 \\ 0 & 0 & m_\tau \end{pmatrix} \tag{10.18}$$

for the leptons, whereas for the quarks we have

$$V_L^{(q)\dagger} M^{(q)} V_R^{(q)} = \begin{pmatrix} m_d & 0 & 0 \\ 0 & m_s & 0 \\ 0 & 0 & m_b \end{pmatrix}, \quad \tilde{V}_L^{(q)\dagger} \tilde{M}^{(q)} \tilde{V}_R^{(q)} = \begin{pmatrix} m_u & 0 & 0 \\ 0 & m_c & 0 \\ 0 & 0 & m_t \end{pmatrix}. \tag{10.19}$$

In view of this, we define the *mass eigenstate* quark fields as[3]

$$\begin{pmatrix} u'_{L,R} \\ c'_{L,R} \\ t'_{L,R} \end{pmatrix} = \tilde{V}_{L,R}^{(q)\dagger} \begin{pmatrix} u_{L,R} \\ c_{L,R} \\ t_{L,R} \end{pmatrix}, \quad \begin{pmatrix} d'_{L,R} \\ s'_{L,R} \\ b'_{L,R} \end{pmatrix} = V_{L,R}^{(q)\dagger} \begin{pmatrix} d_{L,R} \\ s_{L,R} \\ b_{L,R} \end{pmatrix}, \tag{10.20}$$

and similarly for the charged lepton fields, this time using the matrices $V_{L,R}^{(\ell)}$. By construction, the propagators are diagonal when expressed in terms of the new fields. The couplings with the gauge fields, on the other hand, can get a dependence on the unitary matrices involved in the diagonalization of the mass matrices. To see how this dependence comes about we look, for example, at the quark charged current coupling to the W^+ bosons

$$j_+^\mu = (\overline{u}_L, \overline{c}_L, \overline{t}_L)\gamma^\mu \begin{pmatrix} d_L \\ s_L \\ b_L \end{pmatrix} = (\overline{u}'_L, \overline{c}'_L, \overline{t}'_L)\gamma^\mu \tilde{V}_L^{(q)\dagger} V_L^{(q)} \begin{pmatrix} d'_L \\ s'_L \\ b'_L \end{pmatrix} \tag{10.21}$$

[3] Our notation at this point differs from the usual one in the literature in that we use primed fields to indicate the mass eigenstates. The reason to use this notation is to avoid cluttering the equations with primes both in this chapter and in Chap. 5.

A similar calculation for the neutral quark current shows that it does not depend on the unitary matrices relating flavor to mass eigenstates. This means that at tree level there are no flavor changing neutral currents (FCNC), as a consequence of the quantum numbers of the three families. This is the tree-level version of the Glashow–Iliopoulos–Maiani (GIM) mechanism that works for complete families.

We have shown that the couplings of the quarks to the $W^{\pm}$ bosons mix the different mass eigenstates. This mixing is given by the 3×3 matrix

$$V \equiv \tilde{V}_L^{(q)\dagger} V_L^{(q)}, \tag{10.22}$$

called the Cabibbo–Kobayashi–Maskawa (CKM) quark mixing matrix. It is immediate to check that this matrix is unitary and therefore in general complex. In Chap. 11 we will see that this has important physical consequences.

We analyze next the leptonic sector. The charged lepton-neutrino current is

$$j_+^{\mu} = (\overline{\nu}_{eL}, \overline{\nu}_{\mu L}, \overline{\nu}_{\tau L})\gamma^{\mu} \begin{pmatrix} e_L \\ \mu_L \\ \tau_L \end{pmatrix} = (\overline{\nu}_{e,L}, \overline{\nu}_{\mu,L}, \overline{\nu}_{\tau,L})\gamma^{\mu} V_L^{(\ell)} \begin{pmatrix} e'_L \\ \mu'_L \\ \tau'_L \end{pmatrix}. \tag{10.23}$$

Were the neutrino massless, the matrix $V_L^{(\ell)}$ could be reabsorbed in a redefinition of the neutrino fields without making the propagator nondiagonal. We know, however, that the neutrinos are massive and the only question is whether their mass terms are of Dirac, Majorana or a mixture of both. In either case one has to redefine the neutrino fields to diagonalize their mass matrix and this results in the introduction of a second CKM matrix in the leptonic sector.

Higgs Couplings

Having learned how fermion masses are generated, we would like to know how these states couple to the Higgs field itself. This is important because these couplings determine both how the Higgs particle can be produced in a scattering experiment and also what its decay signatures are. Looking at the terms linear in $h(x)$ in (10.4), we find that the Higgs boson couples to the *charged* mass eigenstates $f = (e', \mu', \tau', u', d', c', s', t', b')$ according to the vertices

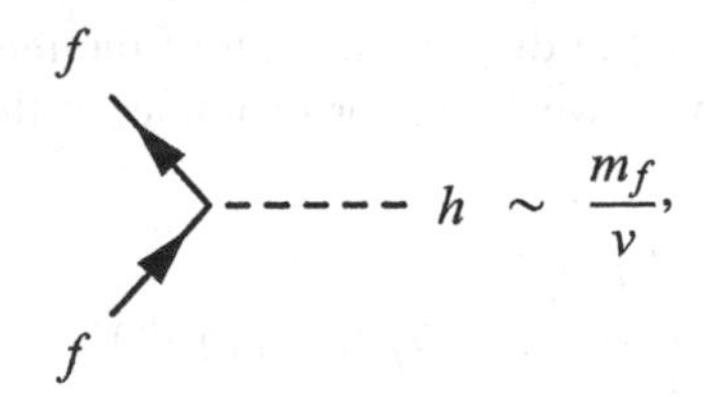

$$\sim \frac{m_f}{v},$$

where m_f is the mass of the charged fermion. Thus, the Higgs-fermion couplings are suppressed by the ratio between the fermion masses and the vacuum expectation value of the Higgs field.

The masses of the gauge fields $W^{\pm}$ and Z^0 and their couplings to the Higgs are obtained by expanding $\mathbf{H}(x)$ around the vacuum in the covariant derivative terms in Eq. (10.9). For the masses one finds

$$m_W = \frac{1}{2} g v, \quad m_Z = \frac{g v}{2 \cos \theta_w}, \tag{10.24}$$

with g the electroweak coupling constant and θ_w the weak mixing angle (see Chap. 5). As for the coupling of the vector bosons to the Higgs field, the terms linear in $h(x)$ give rise to the following interaction vertices

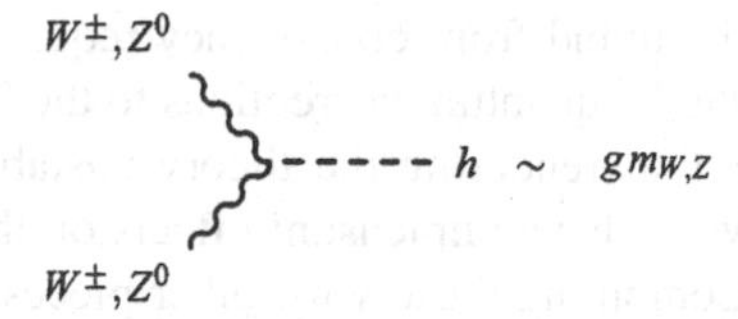

In addition, the theory contains also vertices that couple two vector bosons to two Higgs fields, as well as self-interaction vertices with three and four Higgs bosons. They can be found, for example, in Ref. [4–8] of Chap. 5.

The implementation of symmetry breaking has resulted in the introduction of a new energy scale, the Higgs vacuum expectation value v, and a number of dimensionless couplings: the Higgs self-interaction λ, and the Yukawa couplings for leptons and quarks, $C_{ij}^{(\ell)}$, $C_{ij}^{(q)}$ and $\tilde{C}_{ij}^{(q)}$. In fact, the Higgs vacuum expectation value v is related to the Fermi coupling constant G_F introduced in Chap. 5. Using the relation between v and the mass of the W boson (10.24) we find

$$G_F = \frac{1}{\sqrt{2} v^2}. \tag{10.25}$$

Since G_F can be measured, for example, from muon decay we learn that the Higgs vacuum expectation value is

$$v \approx 246\,\text{GeV}. \tag{10.26}$$

Once the value of the only energy scale v is determined, one can use the relations (10.17) to fix the Yukawa couplings for quarks and leptons from measurement of the mass matrices for the different matter fields. With this, however, we still get no information about the value of the Higgs self-coupling constant λ, or equivalently, the Higgs boson mass m_H. This is the last standard model parameter that remains to be measured and the Higgs boson the last particle of the model to be detected.

What makes the Higgs field so elusive? First, our ignorance of the value of the Higgs mass makes its detection difficult because it is not possible to know a priori "where" to look for it. Depending on the value of m_H different channels have to be considered for the production of this particle. A second aspect is that the Higgs boson couples to other standard model particles with a strength proportional to their masses. Thus, its coupling to light fermions is very small and to produce Higgs particles one must begin by producing heavy fermions, $W^{\pm}$, or Z^0 vector bosons in large quantities. The situation is complicated by the fact that many decay channels of the Higgs boson produce signals that are quite common also in other standard model processes not involving Higgs particles (or, in technical jargon, they have "large backgrounds").

This however does not mean that we do not know anything about the Higgs mass. The Higgs particle enters in the calculation of higher order corrections to standard model processes and bounds to m_H can be found by comparing these calculations with the precision measurements carried out at the Large Electron Positron (LEP) collider, running at CERN between the years 1989 and 2000. Additional bounds for m_H can also be found from consistency requirements. For example, if the Higgs boson is too light the quantum corrections to the Higgs coupling constant λ could make it negative, thus rendering the theory unstable. On the other side, a too-heavy Higgs boson would have unpleasant effects on the good behavior of the theory at high energies. Combining these with other pieces of information a likely range for the Higgs mass can be obtained depending on the energy scale Λ up to which we consider the standard models to describe the physics correctly [1]. Taking, for example, $\Lambda \sim 1$ TeV one finds

$$50\,\text{GeV} \lesssim m_H \lesssim 800\,\text{GeV}, \tag{10.27}$$

while if $\Lambda \sim 10^{16}$ GeV the range narrows to

$$130\,\text{GeV} \lesssim m_H \lesssim 180\,\text{GeV}. \tag{10.28}$$

Searches for Higgs boson are currently underway at both the Tevatron at Fermilab and the LHC at CERN. Particularly promising channels are the decay of the Higgs into two photons or into two Z^0, that in turn decay into a couple of lepton–antilepton pairs:

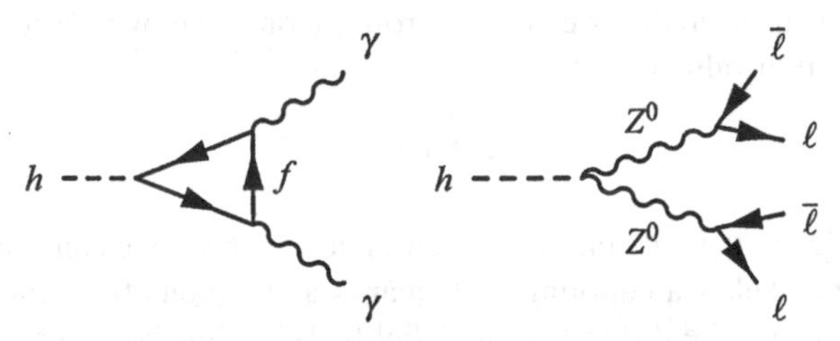

The first process would give a distinctive signature for a Higgs with mass $m_H \lesssim 150\,\text{GeV}$, whereas the second would be important in the regime $m_H \gtrsim 2m_Z$.

Remarks on Symmetry Breaking in the Standard Model

The Higgs sector of the standard model cannot be regarded as a mere attachment to it, as just a smart "trick" intended to circumvent the conflict between masses and gauge invariance. There are more fundamental reasons to think that the Higgs particle, or something very similar, should be there. It is an experimental fact that the $W^{\pm}$ and

Z^0 bosons are massive and therefore have longitudinal components that have been detected.

If we only worry about giving masses to the vector bosons and fermions, it is clear that freezing the field $h(x)$ in Eq. (10.12) suffices. For all practical purposes the theory we obtain has massive $W^{\pm}$ and Z^0 bosons and massive fermions, but no elementary Higgs scalar. So long as we work at low enough energies, this may be a reasonably good phenomenological description.

This naive Higgsless standard model has problems: scattering amplitudes involving the longitudinal components of the gauge bosons behaves badly as the energy approaches the scale $v \sim m_{W,Z}/g$. The amplitudes grow so fast with the energy as to be incompatible with something as basic as the conservation of probability. This problem is automatically solved by including a neutral scalar field in the theory that couples to the massive gauge bosons and fermions in precisely the same way as the Higgs particle does. But this is not the only possibility.

We illustrate this point in more detail using the example of a SU(2) massive gauge field coupled to a pair of chiral doublets Ψ_L, Ψ_R transforming as

$$\Psi_L(x) \longrightarrow g(x)\Psi_L(x), \quad \Psi_R(x) \longrightarrow \Psi_R(x), \tag{10.29}$$

where $g(x)$ belongs to the fundamental representation of SU(2). The Lagrangian

$$\begin{aligned}\mathscr{L} = &-\frac{1}{2}\mathrm{Tr}\left(F_{\mu\nu}F^{\mu\nu}\right) + M^2\mathrm{Tr}\left(A_\mu A^\mu\right) + i\overline{\Psi}_L \not{D}\Psi_L + i\overline{\Psi}_R \not{D}\Psi_R \\ &- m\left(\overline{\Psi}_L\Psi_R + \overline{\Psi}_R\Psi_L\right)\end{aligned} \tag{10.30}$$

is not gauge invariant due to the presence of mass terms for the gauge and fermion fields. Gauge invariance can be "restored" using a trick originally due to Stückelberg [3] (see [4] for a review). We introduce a scalar field $U(x)$, called the Stückelberg field, taking values in the gauge group and transforming under SU(2) as $U(x) \to g(x)U(x)$. The Lagrangian

$$\begin{aligned}\mathscr{L} = &-\frac{1}{2}\mathrm{Tr}\left(F_{\mu\nu}F^{\mu\nu}\right) - \frac{M^2}{g_{\mathrm{YM}}^2}\mathrm{Tr}\left[(U^\dagger D_\mu U)(U^\dagger D^\mu U)\right] \\ &+ i\overline{\Psi}_L \not{D}\Psi_L + i\overline{\Psi}_R \not{D}\Psi_R - m\left(\overline{\Psi}_L U\Psi_R + \overline{\Psi}_R U^\dagger \Psi_L\right)\end{aligned} \tag{10.31}$$

is gauge invariant. Using this gauge freedom we can set $U(x) = \mathbf{1}$ and recover the original Lagrangian (10.30). In this picture the breaking of gauge invariance in the massive theory can be seen as resulting from gauge fixing. In the process, the field $U(x)$ becomes the longitudinal component of the massive vector field.

Replacing (10.30) by (10.31) does not solve our ultraviolet problems. The theory is still ill-defined at energies of order M/g_{YM} and should be completed by specifying the dynamics of $U(x)$ at high energies. Here we are faced with various alternatives. One of them is the Brout–Englert–Higgs mechanism presented: a gauge invariant potential implementing symmetry breaking is added

$$V(U^\dagger U) = \frac{\lambda}{4}\left(\frac{M}{g_{\rm YM}}\right)^4\left[\frac{1}{2}{\rm Tr}(U^\dagger U) - 1\right]^2, \tag{10.32}$$

and the field $U(x)$ is linearized around the vacuum

$$U(x) = U_0(x)\left[1 + \frac{g_{\rm YM}}{M}h(x)\right], \tag{10.33}$$

where $U_0(x) \in {\rm SU}(2)$ and $h(x)$ is the Higgs field of mass $m_H^2 = 2\lambda M^2/g_{\rm YM}^2$. At energies below m_H the Higgs field is frozen, $U(x) \simeq U_0(x)$, and the Stückelberg Lagrangian (10.31) provides a reliable phenomenological description.

This linear realization is the simplest, and historically the first one used. Many other scenarios have been proposed as alternative ultraviolet cures of the mass generation mechanism. Among them, technicolor, where $U(x)$ is a bound state (analogous to the pion) of a set of strongly coupled new fermions. There is a large collection of alternatives to the standard Higgs mechanism (for a clear exposition see [4]), however they all share the same mechanism of giving masses to the vector bosons by absorbing the relevant Nambu–Goldstone bosons. This is reasonable, the masses of the $W^\pm$ and Z^0 bosons are infrared properties of the theory and their origin is not necessarily related to the high energy fate of the "Higgs"-mode.

This discussion should help clarifying the statement contained in the closing paragraph of Sect. 5.5. The Lagrangian (10.30) can be used to describe the physics of a nonabelian massive gauge field chirally coupled to massive fermions, as long as we restrict our attention to energies below the mass scales of the problem. In this regime, the absence of gauge invariance is no big deal. As the reader has repeatedly been reminded along the book, gauge invariance is not a real symmetry but rather a redundancy. The point of Stückelberg's trick is to "fake" this redundancy, allowing to write a formally gauge invariant Lagrangian.

The situation is different if we aim at constructing a theory whose predictions can be trusted to arbitrary high energies, in the spirit of good old QED.[4] In this case gauge invariance is a crucial ingredient for consistency. The Brout–Englert–Higgs mechanism provides a renormalizable, gauge invariant ultraviolet completion of the massive low energy theory. Historically, this explains the enormous effect the proof of renormalizability of spontaneously broken gauge theories by 't Hooft and Veltman [5–8] had on the acceptance of the Glashow–Weinberg–Salam theory.

10.2 Quark Masses

The previous presentation might have led to the mistaken conclusion that the Brout-Englert-Higgs mechanism settles once and for all the problem of accounting for the masses of the subatomic particles. The only task left is the experimental measurement

[4] Let us forget for the moment about the presence of the Landau pole.

of the quark and lepton masses that in turn determine the value of the Yukawa couplings $C_{ij}^{(\ell,q)}$.

This idea works indeed for the leptons. Since they exist as asymptotic states, their masses can be unambiguously determined, and with them the corresponding parameters in the Lagrangian. The complication comes with the quarks. As they cannot be pulled out of the hadrons their masses cannot be measured directly.

One definition of the quark masses is provided by the nonrelativistic quark model. Here the hadrons are considered to be the bound states of a quark–antiquark pair (mesons) or three quarks (baryons). The mass of the hadron can be written in terms of the masses of its constituents plus the corresponding binding energy

$$M_{\text{meson}} = m_q + m_{\overline{q}} + \Delta E_{q\overline{q}}$$
$$M_{\text{baryon}} = m_{q1} + m_{q2} + m_{q3} + \Delta E_{qqq}. \tag{10.34}$$

As quarks are considered to be nonrelativistic in the bound state, the binding energy is subleading with respect to the quark masses, $\Delta E \ll m_q$. In fact, it can be modelled as

$$\Delta E_{q\overline{q}} = \frac{4a}{m_q m_{\overline{q}}} \mathbf{s}_q \cdot \mathbf{s}_{\overline{q}}, \quad \Delta E_{qqq} = 4a' \sum_{i<j}^{3} \frac{1}{m_{q_i} m_{q_j}} \mathbf{s}_{q_i} \cdot \mathbf{s}_{q_j}, \tag{10.35}$$

where a, a' are undetermined numerical constants and $\mathbf{s}_q$ is the quark spin operator. Their products are numbers that depend on the total spin S of the system. This is easy to see in the case of the quark–antiquark bound state, where

$$\mathbf{s}_q \cdot \mathbf{s}_{\overline{q}} = \frac{1}{2}\left[S(S+1) - \frac{3}{2} \right], \tag{10.36}$$

with $S = 0, 1$ the spin of the corresponding meson.

At first sight the ansatz (10.35) for the binding energy might look surprising. It has the form of the hyperfine splitting of the hydrogen atom that we know is a small perturbation to the energy levels determined by the Coulomb interaction. This, however, is not the case for the quark bound states. In the hydrogen atom the smallness of the hyperfine splitting is due to the fact that the corresponding term in the Hamitonian comes suppressed by a factor $\frac{m_e}{m_p} \simeq 0.0004$. In the case of the quark system the factor in front of this term is of order one and therefore its contribution is expected to be of the same order as the quark-quark potential. Due to this we can parametrize our ignorance about the latter in terms of the numerical parameters a and a'.

Using (10.35) the spectrum of hadrons can be fit to get m_q, a and a'. The masses thus obtained are the so-called *constituent quark masses*. They make up a large fraction of the mass of the hadron. For example, in the case of the u and d quarks their constituent masses have the values

$$m_u \simeq m_d \simeq 310\,\text{MeV}, \tag{10.37}$$

about $\frac{1}{3}$ of the proton mass.

Despite the success of the nonrelativistic quark model in accounting for certain properties of hadrons such as their masses and magnetic moments, the constituent masses of the quarks cannot be identified with the mass parameter appearing in the standard model Lagrangian in its broken phase. For historical reasons these parameters are called the *current-algebra quark masses*.

In fact, there is experimental evidence showing that there is much more stuff inside hadrons than the nonrelativistic quark model picture shows. The most compelling comes from the deep inelastic scattering of leptons off protons already described in Chap. 5. In these experiments it is possible to measure the distribution function of the proton momentum among the constituents of the hadron, collectively called partons. The remarkable thing is that about 50% of the total momentum is carried by constituents that do not participate in the electroweak interactions! These have to be identified with virtual gluons responsible for the interaction between the quarks. The remaining proton momentum is shared between the quarks responsible for the quantum numbers such as charge, spin and isospin of the hadron (called the *valence quarks*), and virtual quark–antiquark pairs (*sea quarks*).

With this picture of the hadron interior in mind, constituent quarks can be seen as effective "quasiparticles" resulting from the dressing of the valence quarks by the QCD interaction. This heuristic idea, that would explain the success of the nonrelativistic quark model, is unfortunately too hard to make quantitative due to computational difficulties.

10.3 Λ_{QCD} and the Hadron Masses

The techniques described in Chap. 8 can be used to calculate the beta function in perturbation theory. The running coupling constant can then be formally written in terms of a single dimensionful integration constant as

$$\Lambda = \mu \exp\left[-\int^{g(\mu)} \frac{dx}{\beta(x)}\right]. \tag{10.38}$$

We observe that, whenever the beta function is nonvanishing, quantum corrections generate a characteristic energy scale. This happens even when the classical Lagrangian contains no dimensionful parameters, a phenomenon called dimensional transmutation. It is important to keep in mind that the dynamically generated scale Λ is an integration constant and therefore has to be fixed experimentally. This is related to the fact that quantum field theory only determines the rate of change of the coupling constants with the energy through the renormalization group functions (8.93). Fixing the numerical values of the couplings requires measurements at a reference scale.

In the case of QCD, using the value of the one-loop beta function (8.25) particularized to the case of three colors ($N_c = 3$)

$$\beta(g) = -\frac{g^3}{48\pi^2}(33 - N_f), \tag{10.39}$$

we find the QCD energy scale to be

$$\Lambda_{\rm QCD} = \mu e^{-\frac{24\pi}{(33-N_f)}\frac{1}{g(\mu)^2}}. \tag{10.40}$$

The strong coupling constant can be written in terms of it as

$$g(\mu)^2 = \frac{24\pi^2}{(33 - N_f)\log\left(\frac{\mu}{\Lambda_{\rm QCD}}\right)}. \tag{10.41}$$

The physical meaning of $\Lambda_{\rm QCD}$ becomes clear: it sets the energy scale at which the theory becomes strongly coupled. Notice that the divergence of the coupling constant at $\mu = \Lambda_{\rm QCD}$ following from the one loop computation cannot be taken literally. When the coupling constant grows the perturbative approximation used to compute the beta function (10.39) breaks down.

In the context of the physics of hadrons, $\Lambda_{\rm QCD}$ determines the characteristic size of a hadron. Indeed, the theory becomes strongly coupled when the hadron constituents are at distances larger than $\Lambda_{\rm QCD}^{-1}$, setting thus the length scale inside which quarks are confined.

One of the big problems in QCD is to calculate the mass of particles such as the proton and the neutron in terms of the mass parameters of the quarks. The reason why this problem is difficult lies in the fact that the valence quarks u and d have masses that are much smaller than the natural scale of the theory, $\Lambda_{\rm QCD}$. To see this we estimate the kinetic energy of these quarks using Heisenberg's uncertainty principle. Since they are confined inside a hadron of typical size $\Lambda_{\rm QCD}^{-1}$, the uncertainty in their momenta can be estimated to be $\Delta p \sim \Lambda_{\rm QCD}$. Moreover, using isotropy we can assume that the average momentum of the quarks is equal to zero, $\langle \mathbf{p} \rangle = 0$. Then, $(\Delta p)^2 = \langle \mathbf{p}^2 \rangle$ and we finally conclude that

$$\langle \mathbf{p}^2 \rangle \sim \Lambda_{\rm QCD}^2. \tag{10.42}$$

So far we have not made any hypothesis as to the mass of the quarks. Let us now assume that we are dealing with *light quarks*. They are defined as those whose masses satisfy $m_q \ll \Lambda_{\rm QCD}$. This is the case of the u and d quarks that make up most of the matter that we see around us. In this case, Eq. (10.42) can be recast as

$$\langle \mathbf{p}^2 \rangle \gg m_q^2 \quad (q = u, d). \tag{10.43}$$

This means that light quarks inside hadrons are relativistic. What is more important, Eq. (10.42) implies that the typical energy of these quarks if of order $\Lambda_{\rm QCD}$ and therefore we are in regime where QCD is strongly coupled.

There are two conclusions to be extracted from this discussion. The first is that we have found the reason behind the technical problems in calculating the masses of hadrons such as protons or neutrons from first principles: we would have to deal with a theory in a regime where perturbation theory does not work. Hence, we have to resort to numerical approaches such as lattice field theory.

The second lesson we learn is that the Brout–Englert–Higgs mechanism actually contributes very little to explaining the mass we see around us. In fact, most of the mass of an atom comes from the nucleus (from about 99.95% for hydrogen to 99.9997% for uranium) that is made of protons and neutrons. What we have argued is that the quark mass parameters m_q generated by electroweak spontaneous symmetry breaking contribute very little to the mass of these hadrons: most of the mass of protons and neutrons, and therefore of the world we see, come from Λ_{QCD}.

That all difficulties in computing hadron masses come from having light quarks can be seen in a toy model due to Howard Georgi [9]. He imagines a world essentially identical to our own but with a single crucial difference: the masses of the u and d quarks satisfy

$$m_u \simeq m_d \simeq \frac{1}{3} m_{\text{proton}} \gg \Lambda_{\text{QCD}}. \tag{10.44}$$

Therefore $\langle \mathbf{p}^2 \rangle \ll m_q^2$ and the quarks can be treated nonrelativistically. Thus, the typical energy of the processes inside the proton is m_q, and the condition (10.44) implies that the theory at this scale is weakly coupled. Tuning m_q/Λ_{QCD} we can even make

$$\alpha_s(m_q) \equiv \frac{g(m_q)^2}{4\pi} \simeq \frac{1}{137}. \tag{10.45}$$

This sets $\Lambda_{\text{QCD}} \sim 10^{-42} m_q$.

Given all this, it should be possible to study the bound state of the three quarks in the proton using the techniques of atomic physics. Since the theory is in a coupling regime where perturbation theory can be used, the static potential between the quarks is obtained from the diagram where the two quarks interchange a gluon. In fact we do not even have to compute the diagram. It suffices to compare the corresponding processes in QCD and QED

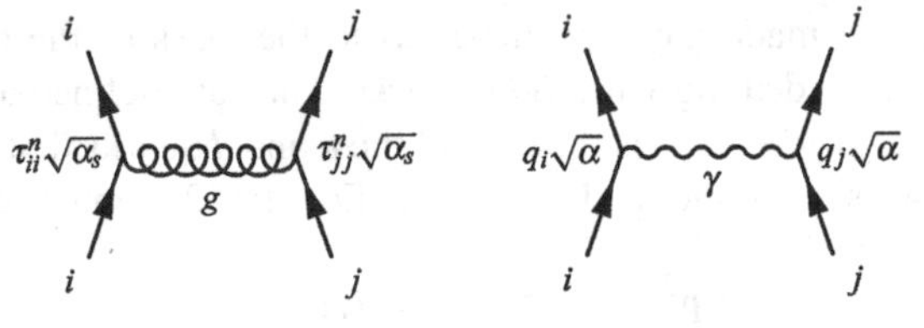

where q_i, q_j are the charges of the corresponding quarks. A look at the Feynman rules for nonabelian Yang–Mills theories listed in Chap. 6 shows that the only difference between the contribution of the two previous diagrams comes from the presence of the SU(3) generators in the vertices of the former. This means that the first diagram is obtained from the second by the replacement

$$\alpha q_i q_j \longrightarrow \alpha_s(m_q) \sum_{n=1}^{8} \tau^n_{ii} \tau^n_{jj}, \tag{10.46}$$

where $\tau^n = \frac{1}{2}\lambda_n$, with λ_n the Gell–Mann matrices shown in Eq. (B.16). Making this replacement in the Coulomb potential, we find the chromostatic potential between two quarks in the proton to be

$$V_{qq}(r) = C_F \frac{\alpha_s(m_q)}{r}, \tag{10.47}$$

where C_F is the color factor on the right-hand side of (10.46).

How different is Georgi's toy world from our own? In fact, we are not so far off. Because the quarks are nonrelativistic, the binding energy can be estimated from a formula analog to the one for the ground state energy of the hydrogen atom

$$\Delta E \sim \alpha_s(m_q)^2 m_q \simeq 16\,\text{keV}. \tag{10.48}$$

Therefore $\Delta E \ll m_q$ and the mass of the proton is essentially the sum of the masses of the quarks. This means that we can fine tune m_q to have $m_{\text{proton}} = 938$ MeV while preserving (10.44). As for the proton size, it is set by the corresponding Bohr radius

$$R_{\text{proton}} \sim \frac{1}{m_q \alpha_s(m_q)} \simeq 90\,\text{fm}. \tag{10.49}$$

Although almost two orders of magnitude above the real proton radius, it is still about 500 times smaller than the radius of the hydrogen atom. Thus we can expect the electronic structure of the atoms not to be radically changed. Notice that now the size of the hadron is dictated by perturbative effects, as opposed to real hadrons where the relevant physics is nonperturbative and their size is determined by the length scale $\Lambda_{\rm QCD}^{-1}$.

The main advantage of this toy model is that in it, unlike in the real world, QCD computations are "easy". In particular, the length scale at which confinement takes place is macroscopic. With $\Lambda_{\rm QCD} \sim 10^{-42} m_q$ and $m_q \sim 300$ MeV we find that

$$\Lambda_{\rm QCD}^{-1} \sim 10^3\text{Mpc} \tag{10.50}$$

and free quarks could be observed! In fact, in this imaginary world the constituent masses are the physical quark masses and the nonrelativistic quark model is the correct QCD description of hadrons.

With this example we wanted to make an important point: confinement itself is not at the bottom of the difficulties with QCD, but the fact that quarks are much lighter than the energy scale at which confinement occurs. This is illustrated also in "real" QCD with heavy quarks, those whose mass is much larger than $\Lambda_{\rm QCD}$. This is the case of the *b*, *c* and *t* quarks, although the short lifetime of the latter prevents it from forming hadrons. Then the strong coupling constant is small at the quark mass scale and the bound state of heavy quarks is amenable to QCD perturbation theory. Moreover, applying (10.42) we have

$$\langle \mathbf{p}^2 \rangle \ll m_q^2 \qquad \text{(heavy quarks)}, \tag{10.51}$$

and therefore heavy quarks inside hadrons are nonrelativistic.

References

1. Djouadi, A.: The anatomy of electroweak symmetry breaking. Tome I: the Higgs boson in the standard model. Phys. Rept. **457**, 1 (2008)
2. Grojean, C.: New approaches to electroweak symmetry breaking. Phys. Usp. **50**, 1 (2007)
3. Stueckelberg, E.C.G.: Die Wechselwirkungs Kräfte in der Elektrodynamik und in der Feldtheorie der Kernkräfte. Helv. Phys. Acta **11**, 225 (1938)
4. Ruegg H., Ruiz-Altaba M.: The Stückelberg field. Int. J. Mod. Phys. **A19**, 3265 (2004) [arXiv:hep-th/0304245]
5. Hooft, G.'t: Renormalizable Lagrangians for massive Yang–Mills fields. Nucl. Phys. **B35**, 167 (1971)
6. Hooft, G.'t, Veltman, M.J.G.: Regularization and renormalization of gauge fields. Nucl. Phys. **B44**, 189 (1972)
7. Hooft, G.'t.: Nobel lecture: a confrontation with infinity. Rev. Mod. Phys. **72**, 333 (2000)
8. Veltman, M.J.G.: Nobel lecture: from weak interactions to gravitation. Rev. Mod. Phys. **72**, 341 (2000)
9. Georgi, H.: Weak Interactions and Modern Particle Theory (revised and updated edition). Dover, New York (2009)

Chapter 11
Symmetries II: Discrete Symmetries

This is probably the most technical chapter of this book. Discrete symmetries play a fundamental role in modern particle physics and cosmology. We have delayed their study until now to be able to develop all the tools required to explore some or their fascinating consequences. Specifically, we present an outline of the derivation of the CPT theorem from first principles and some of the consequences of the proof, in particular, the connection between spin and statistics.

11.1 Discrete Symmetries in Classical Mechanics and Field Theory

In relativistic mechanics, the basic equations describing the dynamics of a system of N charged particles coupled to the electromagnetic field are the Maxwell equations supplemented by the Lorentz force

$$\begin{aligned} \partial_\mu F^{\mu\nu} &= j^\nu, \\ \varepsilon^{\mu\nu\sigma\lambda}\,\partial_\nu F_{\sigma\lambda} &= 0, \\ m_k \frac{d^2 x_k^\mu}{d\tau^2} &= F^{\mu\nu}(x_k) j_\nu(x_k), \end{aligned} \tag{11.1}$$

where $k = 1, \ldots, N$ and the current is given by

$$j^\mu(x) = \sum_{k=1}^{N} q_k \int d\tau \frac{dx_k^\mu}{d\tau} \delta^{(4)}\Big[x^\mu - x_k^\mu(\tau)\Big]. \tag{11.2}$$

Besides their Lorentz covariance, these equations are preserved by three discrete symmetries: parity (P), charge conjugation (C) and time reversal (T). Parity acts by reversing the sign of the spatial coordinates

$$\mathrm{P} : (x^0, \mathbf{x}) \longrightarrow (x^0, -\mathbf{x}). \tag{11.3}$$

L. Álvarez-Gaumé and M. Á. Vázquez-Mozo, *An Invitation to Quantum Field Theory*, Lecture Notes in Physics 839, DOI: 10.1007/978-3-642-23728-7_11,

Physically, this transformation corresponds to mirror reflection combined with a rotation of π around an axis normal to the plane of the mirror. The corresponding transformation of the electromagnetic field is then

$$\mathrm{P}: \begin{cases} A_0(x^0, \mathbf{x}) \longrightarrow A_0(x^0, -\mathbf{x}) \\ \mathbf{A}(x^0, \mathbf{x}) \longrightarrow -\mathbf{A}(x^0, -\mathbf{x}) \end{cases}. \tag{11.4}$$

Charge conjugation acts by reversing the signs of all charges

$$\mathrm{C}: q_a \longrightarrow -q_a, \tag{11.5}$$

and the electromagnetic potential

$$\mathrm{C}: A_\mu(x) \longrightarrow -A_\mu(x). \tag{11.6}$$

Finally, the effect of time reversal is to reverse the sign of the time coordinate

$$\mathrm{T}: (x^0, \mathbf{x}) \longrightarrow (-x^0, \mathbf{x}), \tag{11.7}$$

while acting on the electromagnetic field potential as

$$\mathrm{T}: \begin{cases} A_0(x^0, \mathbf{x}) \longrightarrow A_0(-x^0, \mathbf{x}) \\ \mathbf{A}(x^0, \mathbf{x}) \longrightarrow -\mathbf{A}(-x^0, \mathbf{x}) \end{cases}. \tag{11.8}$$

This transformation implies that the electric field is invariant while the magnetic field changes sign, as it is heuristically expected: the magnetic field is generated by moving charges whose momenta are reversed by T. This changes the sign of the magnetic field and leaves the electric field unmodified.

Discrete symmetries can be implemented in classical field theory as well. Let us take the simplest example of a complex scalar field $\phi(x)$ coupled to a U(1) gauge field with action

$$S = \int d^4x \left[(D_\mu \phi)^* D^\mu \phi - \frac{1}{4} F_{\mu\nu} F^{\mu\nu} \right], \tag{11.9}$$

where $D_\mu = \partial_\mu + ieA_\mu$ is the covariant derivative (cf. Sect. 4.3). From of the gauge field under the discrete symmetries found above, we have to find out how the scalar field transforms under P, C and T in such a way that the action, and hence the field equations, remains invariant. Since the Maxwell action is invariant under these transformations, we only have to take care of the first term in (11.9).

Let us begin with parity. From (11.4) we find the action to be invariant provided

$$\mathrm{P}: \phi(x^0, \mathbf{x}) \longrightarrow \eta_P \phi(x^0, -\mathbf{x}). \tag{11.10}$$

Here we allow for a complex phase η_P. For a real field this phase reduces to a global sign, $\eta_P = \pm 1$. Charge conjugation, on the other hand, reverses the sign of the gauge field and therefore maps D_μ to D_μ^*. To compensate this and leave the action invariant, charge conjugation has to interchange the scalar field with its complex conjugate according to

$$\mathrm{C}:\phi(x)\longrightarrow\eta_C\phi(x)^*. \tag{11.11}$$

Again a phase η_C has been introduced. It is important to notice that in classical field theory charge conjugation only acts on the fields and not on the parameters of the Lagrangian, in particular e. This contrasts with what we saw in mechanics where C reverses the sign of the particle electric charges. Finally, for time reversal the transformation of the covariant derivative

$$\mathrm{T}:\partial_\mu+ieA_\mu(t,\mathbf{x})\longrightarrow-\Big[\partial_\mu-ieA_\mu(-t,\mathbf{x})\Big] \tag{11.12}$$

suggests that T has to interchange the scalar field with its complex conjugate

$$\mathrm{T}:\phi(t,\mathbf{x})\longrightarrow\eta_T\phi(-t,\mathbf{x})^*, \tag{11.13}$$

where $|\eta_T|=1$.

The nonabelian gauge action is invariant under parity with the same transformation of the gauge field as in the abelian case (11.4). For charge conjugation and time reversal, on the other hand, the invariance of the action requires that the transformation also affects the gauge indices of the Lie algebra valued gauge field potential. This can be seen by noticing that transforming $A_\mu(t,\mathbf{x})$ as in (11.6) and (11.7) results in a change of the relative sign between the derivative and commutator terms in the nonabelian field strength (4.51). This is compensated by a transposition of the gauge indices in such a way that, under charge conjugation,

$$\mathrm{C}:A_\mu(x)\longrightarrow-A_\mu(x)^T. \tag{11.14}$$

This results in $F_{\mu\nu}(x)\to-F_{\mu\nu}(x)^T$ that leaves the action invariant. Similarly, for time reversal we have

$$\mathrm{T}:\begin{cases}A_0(x^0,\mathbf{x})\longrightarrow A_0(-x^0,\mathbf{x})^T\\ \mathbf{A}(x^0,\mathbf{x})\longrightarrow-\mathbf{A}(-x^0,\mathbf{x})^T\end{cases}. \tag{11.15}$$

The nonabelian electric and magnetic fields $\mathbf{E}=\mathbf{E}^aT^a$ and $\mathbf{B}=\mathbf{B}^aT^a$ transform with the same signs as their abelian counterparts plus a transposition in the gauge indices.

Discrete symmetries become more interesting when applied to spinor fields. In this case the transformation has to act on the spinor indices as well. Thus, the transformation of a Dirac spinor under P, C and T has the form

$$\psi_\alpha(x)\longrightarrow\Gamma_{\alpha\beta}\psi_\beta(x')\qquad\text{or}\qquad\psi_\alpha(x)\longrightarrow\Gamma_{\alpha\beta}\psi_\beta(x')^* \tag{11.16}$$

where $x'^\mu=(x^0,-\mathbf{x})$ for P, $x'^\mu=x^\mu$ for C and $x'^\mu=(-x^0,\mathbf{x})$ for T. Viewing these as active transformations on the fields, we require that the transformed spinors satisfy the Dirac equation with respect to the coordinates x^μ. Thus, the complex matrix $\Gamma_{\alpha\beta}$ has to be chosen such that

$$\left(i\gamma^\mu\partial_\mu - m\right)\psi(x) = 0 \longrightarrow \begin{cases} \left(i\gamma^\mu\partial_\mu - m\right)\Gamma\psi(x') = 0 \\ \left(i\gamma^\mu\partial_\mu - m\right)\Gamma\psi(x')^* = 0 \end{cases}. \tag{11.17}$$

Let us begin with parity. The invariance of the Dirac equation means that Γ has to commute with γ^0 and anticommute with γ^i. This leads to

$$\mathrm{P}: \psi(x^0, \mathbf{x}) \longrightarrow \eta_P\gamma^0\psi(x^0, -\mathbf{x}), \tag{11.18}$$

including an arbitrary phase. It can be immediately checked that this transformation keeps the Dirac action unchanged.

As in the case of the complex scalar field, the transformation of the Dirac spinor under charge conjugation involves the complex conjugate spinor,[1] $\psi(x) \to \Gamma\psi(x)^*$. The transformed spinor satisfies the Dirac equation provided

$$\Gamma^{-1}\gamma^\mu\Gamma = -\gamma^{\mu*}. \tag{11.19}$$

The form of the matrix Γ depends on the representation of the Dirac algebra. For the one shown in Eq. (A.7) we can take $\Gamma = -i\gamma^2$ and the transformation of the spinor is then

$$\mathrm{C}: \psi(x) \longrightarrow \eta_C(-i\gamma^2)\psi(x)^*, \tag{11.20}$$

with $|\eta_C| = 1$. Applying this transformation to the Dirac Lagrangian we find

$$\mathscr{L}_{\text{Dirac}} = \overline{\psi}\left(i\partial\!\!\!/ - m\right)\psi \xrightarrow{\mathrm{C}} \mp\mathscr{L}_{\text{Dirac}} + \text{total derivative}. \tag{11.21}$$

The signs $\mp$ correspond respectively to the case where the spinors are commutating and anticommuting fields. It comes about because of the matrix identity $(AB)^T = \pm B^T A^T$, where the sign depends on whether the entries of the matrix A commute (+) or anticommute (−) with those of the matrix B. At the classical level it does not matter whether the spinors are taken to commute or anticommute, since a global sign in the action does not change the equations of motion. The relevance of this sign for the quantum theory will be discussed in the next section.

The transformation of the Dirac spinor under T also involves the complex conjugate field, $\psi(x) \to \Gamma\psi(x)^*$, with

$$\Gamma^{-1}\gamma^0\Gamma = \gamma^{0*}, \quad \Gamma^{-1}\gamma^i\Gamma = -\gamma^{i*}. \tag{11.22}$$

In the Dirac algebra representation (A.7) the matrix $\Gamma = -\gamma^1\gamma^3$ satisfies the required property and therefore we have

---

[1] Had we tried $\psi(x) \to \Gamma\psi(x)$ we would find, using Schur's lemma, that the Dirac equation is preserved only if Γ is proportional to the identity, so the transformation is trivial.

$$\mathrm{T}: \psi(t, \mathbf{x}) \longrightarrow \eta_T(-\gamma^1\gamma^3)\psi(-t, \mathbf{x})^*, \tag{11.23}$$

with η_T a phase. The Dirac Lagrangian, on the other hand, transforms as

$$\overline{\psi}(x)\Big(i\partial\!\!\!/ - m\Big)\psi(x) \xrightarrow{\mathrm{T}} \pm\overline{\psi}(x')\Big(i\partial\!\!\!/' - m\Big)\psi(x') + \text{total derivative}, \tag{11.24}$$

where $x'^{\mu} = (-x^0, \mathbf{x})$ and the signs $\pm$ corresponds respectively to commuting and anticommuting spinor fields. Integrating we find that the action remains invariant up to a global sign, $S_{\text{Dirac}} \to \pm S_{\text{Dirac}}$.

In short we have shown that if $\psi(t, \mathbf{x})$ is a solution of the Dirac equation so are the transformed fields

$$\begin{aligned}
\psi^P(x^0, \mathbf{x}) &= \eta_P\gamma^0\psi(x^0, -\mathbf{x}),\\
\psi^C(x^0, \mathbf{x}) &= \eta_C(-i\gamma^2)\psi(x^0, \mathbf{x})^*,\\
\psi^T(x^0, \mathbf{x}) &= \eta_T(-\gamma^1\gamma^3)\psi(-x^0, \mathbf{x})^*.
\end{aligned} \tag{11.25}$$

This remains true also when they are coupled to the electromagnetic field, provided it is also transformed according to (11.4), (11.6) or (11.8). The phases η_P, η_C and η_T cannot be determined a priori.

Finally, the question remains whether the transformed fields (11.25) do transform as spinors under Lorentz transformations. To see that they do, we only have to realize that the following three sets of matrices

$$\begin{aligned}
\gamma^{\mu\dagger} &= \gamma^0\gamma^\mu\gamma^0,\\
-\gamma^{\mu *} &= (-i\gamma^2)\gamma^\mu(-i\gamma^2)^{-1},\\
\gamma^{\mu T} &= (-\gamma^1\gamma^3)\gamma^\mu(-\gamma^1\gamma^3)^{-1}
\end{aligned} \tag{11.26}$$

are three representations of the Dirac algebra. Then, using (3.39), one can construct the corresponding representations of the generators of SO(1,3) in which $\psi^P(x)$, $\psi^C(x)$ and $\psi^T(x)$ transform.

11.2 Parity and Charge Conjugation in Quantum Field Theory

We turn now to the implementation of discrete symmetries in quantum field theory, dealing first with parity and charge conjugation. The case of time reversal involves some important subtleties and will be deferred to the next section.

Wigner's theorem establishes that symmetries in the quantum theory are realized by unitary or antiunitary operators acting on the Hilbert space. Here we explore the case of Dirac fermions. Other types of fields can be treated in a similar fashion. In the case of parity and charge conjugation, the transformed field operators $\psi^P(x)$ and $\psi^C(x)$ are related to $\psi(x)$ by a similarity transformation by unitary operators $\mathscr{P}$ and $\mathscr{C}$

$$\begin{aligned}\mathscr{P}\psi(x^0,\mathbf{x})\mathscr{P}^{-1} &= \eta_P\gamma^0\psi(x^0,-\mathbf{x}),\\ \mathscr{C}\psi(x^0,\mathbf{x})\mathscr{C}^{-1} &= \eta_C(-i\gamma^2)\psi(x^0,\mathbf{x})^*.\end{aligned} \tag{11.27}$$

In the second identity the complex conjugation is understood to act as hermitian conjugation on the creation-annihilation operators.[2]

From these relations we can derive the transformations of the creation-annihilation operators of particles and antiparticles. We use the expansion of the Dirac field in terms of them given in Eq. (3.52). At this point we have to take into account that all c-numbers (i.e., parameters and wave functions) pass through the unitary operators, which only act on the creation-annihilation operators of particles and antiparticles, $b(\mathbf{k},s)$ and $d(\mathbf{k},s)$. The γ-matrices on the right-hand side of (11.27), however, act on the indices of the wave functions $u_\alpha(\mathbf{k},s)$ and $v_\alpha(\mathbf{k},s)$.

We begin with parity. It is not difficult to check that

$$\gamma^0 u(\mathbf{k},s) = u(-\mathbf{k},s), \quad \gamma^0 v(\mathbf{k},s) = -v(-\mathbf{k},s). \tag{11.28}$$

This leads to the following transformation of the annihilation operators

$$\begin{aligned}\mathscr{P}b(\mathbf{k},s)\mathscr{P}^{-1} &= \eta_P b(-\mathbf{k},s)\\ \mathscr{P}d(\mathbf{k},s)\mathscr{P}^{-1} &= -\eta_P^* d(-\mathbf{k},s).\end{aligned} \tag{11.29}$$

The corresponding transformation of the creation operators is derived taking the adjoint and using the unitarity of $\mathscr{P}$. From this we infer that the parity operator act on the single particle and antiparticle states by reversing the sign of the spatial momentum and multiplying the state by η_P (particles) or $-\eta_P^*$ (antiparticles). This phase is called the *intrinsic parity* of the state. It appears in the transformation of the one-particle and -antiparticle states

$$\mathscr{P}|\mathbf{k},s\rangle = \eta_P|-\mathbf{k},s;0\rangle, \quad \mathscr{P}|0;\mathbf{k},s\rangle = -\eta_P^*|0;-\mathbf{k},s\rangle. \tag{11.30}$$

This is an elementary but important result: the intrinsic parity of spin-$\frac{1}{2}$ particles and antiparticles are opposite.

We proceed next to study the action of charge conjugation on the creation-annihilation operators of the Dirac field. The relevant identity to be used here is

$$u(\mathbf{k},s) = -i\gamma^2 v(\mathbf{k},s)^*, \quad v(\mathbf{k},s) = -i\gamma^2 u(\mathbf{k},s)^*. \tag{11.31}$$

We find

$$\begin{aligned}\mathscr{C}b(\mathbf{k},s)\mathscr{C}^{-1} &= \eta_C d(\mathbf{k},s),\\ \mathscr{C}d(\mathbf{k},s)\mathscr{C}^{-1} &= \eta_C^* b(\mathbf{k},s),\end{aligned} \tag{11.32}$$

[2] Alternatively, the right-hand side of this expression can be written as $i\gamma^0\gamma^2\overline{\psi}(x)^T$.

so charge conjugation interchanges particles with antiparticles. At the level of the single particle states this means

$$\mathscr{C}|\mathbf{k}, s; 0\rangle = \eta_C^*|0; \mathbf{k}, s\rangle, \quad \mathscr{C}|0; \mathbf{k}, s\rangle = \eta_C|\mathbf{k}, s; 0\rangle. \tag{11.33}$$

The example of the Dirac fermion serves as a template to derive the transformations of other fields under parity and charge conjugation. We only briefly mention the case of the electromagnetic field. In particular, the transformations of the classical electromagnetic potential fixes the phases η_P and η_C. Since photon physical states have transverse polarizations, the single photon state transform under parity as

$$\mathscr{P}|\mathbf{k}, \lambda\rangle = -|-\mathbf{k}, -\lambda\rangle, \tag{11.34}$$

with $\lambda = \pm 1$ the helicity of the state. Thus the intrinsic parity of the photon is $\eta_P = -1$. Moreover, since $A_\mu(x)$ is a Hermitian field the photon is its own antiparticle

$$\mathscr{C}|\mathbf{k}, \lambda\rangle = -|\mathbf{k}, \lambda\rangle. \tag{11.35}$$

11.3 Majorana Spinors

In the light of the discussion of the previous section we have to entertain the possibility of a Fermi field being self-conjugate under C. These are called Majorana spinors and by definition satisfy $\psi^C(x) = \psi(x)$, i.e.

$$\psi(x) = \eta_C(i\gamma^0\gamma^2)\overline{\psi}(x)^T. \tag{11.36}$$

Using the chiral decomposition of $\psi(x)$ shown in (3.31) this Lorentz covariant constraint can be written as

$$\begin{pmatrix} u_+ \\ u_- \end{pmatrix} = \eta_C \begin{pmatrix} i\sigma^2 u_-^* \\ -i\sigma^2 u_+^* \end{pmatrix}. \tag{11.37}$$

This equation is solved by

$$\psi = \frac{1}{\sqrt{2}} \begin{pmatrix} u_+ \\ -i\eta_C\sigma^2 u_+^* \end{pmatrix}. \tag{11.38}$$

The Majorana spinor depends on a single complex two-component function. It has the same number of degrees of freedom as a Weyl spinor, since both of them are related by the identity

$$\psi(x) = \frac{1}{\sqrt{2}}\left[\psi_+(x) + \psi_+^C(x)\right] \quad \text{with} \quad \psi_+ = \begin{pmatrix} u_+ \\ 0 \end{pmatrix}. \tag{11.39}$$

The constraint (11.36) can be seen as a Lorentz covariant way to impose a reality condition on the spinor. In fact, it is possible to find a representation of the Dirac algebra where all γ-matrices are purely imaginary and where the Majorana condition simply reads $\psi(x) = \psi(x)^*$.

Imposing (11.36) on the expansion (3.52) of the field $\psi(x)$ leads to the identification of the creation and annihilation operators of particles and antiparticles, $b(\mathbf{k}, s) = \eta_C d(\mathbf{k}, s)$. Majorana fermions are therefore their own antiparticles. It is instructive to write the Dirac Lagrangian for a Majorana spinor

$$\begin{aligned}\mathscr{L}_{\text{Dirac}} &\equiv \overline{\psi}(i\partial\!\!\!/ - m)\psi \\ &= \frac{i}{2}u_+^\dagger \sigma_+^\mu \overleftrightarrow{\partial}_\mu u_+ - \frac{m}{2}\Big[\eta_C^* u_+^T (i\sigma^2) u_+ + \text{h.c.}\Big]\end{aligned} \tag{11.40}$$

Taking into account that $i\sigma^2_{ab} = \varepsilon_{ab}$, we see that we have recovered the Majorana mass term constructed in Eq. (3.27) on purely group theoretical grounds. In terms of the Weyl spinor ψ_+ of Eq. (11.39) the Majorana mass term can be written as

$$\Delta\mathscr{L} = -\frac{m}{2}\Big(\overline{\psi}_+ \psi_+^C + \overline{\psi_+^C}\psi_+\Big). \tag{11.41}$$

The current $j^\mu = \overline{\psi}\gamma^\mu\psi$ can also be computed with the result

$$\overline{\psi}\gamma^\mu\psi = \frac{1}{2}\Big(u_+^\dagger \sigma_+^\mu u_+ + u_+^T \sigma_+^{\mu T} u_+^*\Big) = 0, \tag{11.42}$$

where we have to use the anticommuting character of u_+. The vanishing of the current indicates that a Majorana fermion has zero electric charge, as corresponds to a particle that is its own antiparticle. In fact we can see that the condition (11.36) is not preserved by a U(1) phase rotation of $\psi(x)$.

11.4 Time Reversal

The implementation of time reversal in the quantum theory requires some additional considerations. In the previous section we have seen how parity and charge conjugation are implemented by unitary operators acting on the Hilbert space. We will see now that such a choice for the time reversal symmetry leads to inconsistencies.

We come back momentarily to classical mechanics. Time reversal changes the sign of the canonical momenta while preserving that of the coordinates. If this is a symmetry of the Hamiltonian, applying T to the initial conditions $\mathrm{T} : (\mathbf{q}_0, \mathbf{p}_0) \to (\mathbf{q}_0, -\mathbf{p}_0)$ has to be equivalent to evolving the system time t from $(\mathbf{q}_0, \mathbf{p}_0)$, applying time reversal and evolving again a time t. This is coded in the following "commutative diagram"

$$\begin{array}{ccc} \mathbf{q}_0, \mathbf{p}_0 & \xrightarrow{\mathrm{T}} & \mathbf{q}_0, -\mathbf{p}_0 \\ \downarrow t & & \uparrow t \\ \mathbf{q}(t), \mathbf{p}(t) & \xrightarrow{\mathrm{T}} & \mathbf{q}(t), -\mathbf{p}(t) \end{array} \tag{11.43}$$

In the quantum theory this diagram gives the action of the operator $\mathscr{T}$ that implements time reversal on the evolution operator, namely

$$e^{-itH}\mathscr{T}e^{-itH} = \mathscr{T}, \tag{11.44}$$

where H is the Hamiltonian of the theory. Infinitesimally this implies that iH anti-commutes with time reversal,

$$\{iH, \mathscr{T}\} = 0. \tag{11.45}$$

According to Wigner's theorem we are now faced with a double choice: $\mathscr{T}$ being a symmetry it has to be either a unitary or an antiunitary operator. In the first case we find that

$$H\mathscr{T} + \mathscr{T}H = 0 \quad (\mathscr{T} \text{ unitary}). \tag{11.46}$$

If $\mathscr{T}$ is antiunitary, on the other hand, complex numbers are conjugated in passing through the time reversal operator. This implies

$$H\mathscr{T} - \mathscr{T}H = 0 \quad (\mathscr{T} \text{ antiunitary}). \tag{11.47}$$

In fact, taking $\mathscr{T}$ unitary is a source of trouble. From (11.46) we conclude that if $|\psi_E\rangle$ is a Hamiltonian eigenstate with eigenvalue E, then $\mathscr{T}|\psi_E\rangle$ is an eigenstate as well with eigenvalue $-E$. Since the original spectrum is unbounded from above, the transformed theory does not have a ground state (i.e., the spectrum is unbounded below). If the time reversal operator $\mathscr{T}$ is antiunitary we see that both $|\psi_E\rangle$ and $\mathscr{T}|\psi_E\rangle$ have the same eigenvalue and the spectrum is invariant. The antiunitarity of $\mathscr{T}$ is also necessary for the invariance of the canonical commutation relations under time reversal.

Any antiunitary operator can be written, in a basis-dependent way, as the product $\mathscr{U}\mathscr{K}$ of a unitary operator $\mathscr{U}$ and a second operator $\mathscr{K}$ that acts by complex conjugation on the numerical coefficients. With the help of this decomposition we can easily show that the square of an antiunitary operator is unitary. Indeed, using that $\mathscr{K}^2 = \mathbf{1}$ we find that

$$(\mathscr{U}\mathscr{K})^2 = \mathscr{U}\mathscr{U}^* \tag{11.48}$$

which is a unitary operator.

Let us briefly study the consequences of time reversal invariance in nonrelativistic quantum mechanics. From the transformation of the angular momentum operator

$$\mathscr{T}\mathbf{J}\mathscr{T}^{-1} = -\mathbf{J} \tag{11.49}$$

we conclude that in the Hilbert space of spin-$\frac{1}{2}$ particles the time reversal operator can be written as[3]

$$\mathscr{T} = i\sigma_2\mathscr{K} = e^{\frac{i\pi}{2}\sigma_2}\mathscr{K}. \tag{11.50}$$

Hence, $\mathscr{T}$ is equivalent to a rotation of π and a complex conjugation. For its square we have $\mathscr{T}^2 = \sigma_2\sigma_2{}^* = -\mathbf{1}$. This is expected, since from (11.50) we learn that $\mathscr{T}^2$ is a rotation of 2π that multiplies the spin wave function by -1. The result can be easily generalized to a system of N spin-$\frac{1}{2}$ particles to give

$$\mathscr{T}^2 = (-1)^N. \tag{11.51}$$

An important result in this context is Kramers theorem. It states that in a theory with an odd number of spin-$\frac{1}{2}$ particles and invariant under time reversal the spectrum presents a double degeneracy. The proof is very simple. We have seen that if $|\psi_E\rangle$ is an energy eigenstate so is $\mathscr{T}|\psi_E\rangle$ and, moreover, with the same eigenvalue. To show the double degeneracy of the spectrum we only have to prove that these two states are linearly independent. Assuming that they are not, i.e., $\mathscr{T}|\psi_E\rangle = \lambda|\psi_E\rangle$ for some complex $\lambda \neq 0$ we find

$$(-1)^N|\psi_E\rangle = \mathscr{T}^2|\psi_E\rangle = \lambda^*\mathscr{T}|\psi_E\rangle = |\lambda|^2|\psi_E\rangle. \tag{11.52}$$

Now, if N is odd this leads to a contradiction and, as a consequence, in this case $|\psi_E\rangle$ and $\mathscr{T}|\psi_E\rangle$ are linearly independent. Kramers' degeneracy also appears in the presence of an external electric field, but it is lifted by a magnetic field because it breaks time reversal (remember that the transformation T preserves the electric field and changes the sign of the magnetic field).

We move next to the implementation of time reversal in quantum field theory. As an example we work out the case of Dirac fields in some detail, other fields being treated in a similar fashion. The transformation of the field operator is

$$\mathscr{T}\psi(x^0,\mathbf{x})\mathscr{T}^{-1} = \eta_T(-\gamma^1\gamma^3)\psi(-x^0,\mathbf{x}). \tag{11.53}$$

When considering free fields we have to bear in mind that $\mathscr{T}$ not only acts on the creation-annihilation operators but also complex conjugates the one-particle wave functions. As for parity and charge conjugation, we will need some identities for the one-particle wave functions, in this case

$$\begin{aligned} u(-\mathbf{k},-s) &= -(-1)^{\frac{1}{2}-s}\gamma^1\gamma^3 u(\mathbf{k},s)^*, \\ v(-\mathbf{k},-s) &= (-1)^{\frac{1}{2}-s}\gamma^1\gamma^3 v(\mathbf{k},s)^*. \end{aligned} \tag{11.54}$$

[3] For the sake of simplicity we ignore other degrees of freedom. Their inclusion does not change the result.

Using them we arrive at the following transformation for the annihilation operators of particles and antiparticles

$$\begin{aligned}\mathscr{T} b(\mathbf{k}, s)\mathscr{T}^{-1} &= (-1)^{\frac{1}{2}-s}\eta_T b(-\mathbf{k}, -s), \\ \mathscr{T} d(\mathbf{k}, s)\mathscr{T}^{-1} &= -(-1)^{\frac{1}{2}-s}\eta_T^* d(-\mathbf{k}, -s).\end{aligned} \tag{11.55}$$

We see how, up to a global phase factor, time reversal changes the signs of both the momentum and the third components of spin. Using hermitian conjugation the corresponding identities for creation operators are obtained. From them it is straightforward to prove that

$$\mathscr{T}^2 = (-1)^F, \tag{11.56}$$

with F the fermion number operator that counts the number of spin-$\frac{1}{2}$ particles plus antiparticles in the state.

In the discussion carried out in the last two sections we have focussed entirely on the operator formalism. It is worthwhile to comment briefly on discrete symmetries in the path integral formalism. Correlation functions are expressed as functional integrals over classical field configuration, with the proviso that fermionic fields are taken to be anticommutative objects. For parity and charge conjugation we have shown that the Dirac action is invariant under the replacement of all fields by their transformed ones (remember that in the case of charge conjugation the anticommutativity of the Dirac fields is crucial for this result), so the invariance of the quantum theory is guaranteed. For time reversal, on the other hand, we found that when the Dirac fields are anticommutative the action changes sign. This might seem to pose a problem for the invariance of the quantum theory. It is not so, however, since the change of sign in the action only results in a global phase after path integration. This is irrelevant for the computation of the correlation functions, and the theory is invariant under T.

11.5 CP Symmetry and CP Violation

Having introduced parity, charge conjugation and time reversal we pass to study now the discrete transformation obtained by combining the first two, called CP. Its action on the quantum fields is easy to obtain from previous results. In the case of a complex scalar field $\phi(x)$ we have

$$(\mathscr{C}\mathscr{P})\phi(x^0, \mathbf{x})(\mathscr{C}\mathscr{P})^{-1} = \eta_{CP}\phi(x^0, -\mathbf{x})^\dagger, \tag{11.57}$$

where η_{CP} is a phase. At the level of the creation-annihilation operators the transformation interchanges particles with antiparticles, while changing the sign of the momentum at the same time. In the case of the Dirac field the result is

$$
\begin{aligned}
(\mathscr{C}\mathscr{P})b(\mathbf{k},s)(\mathscr{C}\mathscr{P})^{-1} &= \eta_{CP}d(-\mathbf{k},s),\\
(\mathscr{C}\mathscr{P})d(\mathbf{k},s)(\mathscr{C}\mathscr{P})^{-1} &= -\eta_{CP}^{*}b(-\mathbf{k},s).
\end{aligned}
\tag{11.58}
$$

Since CP reverses the sign of the momentum without messing with the spin indices it changes the helicity of the fermions.

In Chap. 5 we have learned that the electro-weak sector of the standard model is chiral, i.e., left- and right-handed fermions couple to the $W^{\pm}$ and Z^0 bosons in different ways. A consequence of this is that both parity and charge conjugation are maximally violated. This feature is most glaring in the case of the neutrinos. The transformation of the left-handed neutrino under $\mathscr{P}$ and $\mathscr{C}$ gives respectively a right-handed neutrino and a left-handed antineutrino, neither particle being observed in Nature. This is, however, not the case under $\mathscr{C}\mathscr{P}$, that transforms a left-handed neutrino into a right-handed antineutrino, a particle that is known to exist.

This led to the expectation that CP might be a good symmetry of *all* fundamental forces. This however is not the case. The evidence for CP violation came from the study of the system formed by the neutral kaon K^0 and its antiparticle $\overline{K}^0$. These particles we have already encountered in Chap. 5 (see page 89) as part of the octet of pseudoscalar mesons. Choosing the phase $\eta_{CP} = 1$, the CP transformation of the kaon states at rest is given by

$$
\mathscr{C}\mathscr{P}|K^0\rangle = |\overline{K}^0\rangle, \quad \mathscr{C}\mathscr{P}|\overline{K}^0\rangle = |K^0\rangle.
\tag{11.59}
$$

The two neutral kaons are only distinguished by their strangeness quantum number, $S = 1$ for K^0 and $S = -1$ for $\overline{K}^0$. Since strangeness is not conserved by weak interactions, the two states can mix with each other. In fact, both decay weakly into two and three pions

$$
K^0, \overline{K}^0 \longrightarrow \begin{cases} \pi^+ + \pi^- \\ \pi^0 + \pi^0 \end{cases}, \quad K^0, \overline{K}^0 \longrightarrow \begin{cases} \pi^+ + \pi^- + \pi^0 \\ \pi^0 + \pi^0 + \pi^0 \end{cases}.
\tag{11.60}
$$

In all channels the final state is a CP eigenstate. In the case of the two pions its eigenvalue is CP $= 1$, whereas for the three pion decay the final state has CP $= -1$.

Instead of $|K^0\rangle$ and $|\overline{K}^0\rangle$ it is convenient to consider the CP (and C) eigenstates

$$
\begin{aligned}
|K_S^0\rangle &= \frac{1}{\sqrt{2}}\Big(|K^0\rangle + |\overline{K}^0\rangle\Big), \quad \mathrm{CP} = 1,\\
|K_L^0\rangle &= \frac{1}{\sqrt{2}}\Big(|K^0\rangle - |\overline{K}^0\rangle\Big), \quad \mathrm{CP} = -1.
\end{aligned}
\tag{11.61}
$$

Were CP a preserved symmetry of the weak interactions, the decays $K_S^0 \to 3\pi$ and $K_L^0 \to 2\pi$ should be forbidden. In 1964, however, experiments showed that the second decay takes place. This was the first evidence that CP is violated in the standard model.

Complex couplings in the action are a potential source of CP violation in a quantum field theory. To see this let us consider a theory where the interaction Hamiltonian has the form

$$H_{\text{int}} = \int d^3x \left[\sum_i g_i \mathscr{O}_i(x) + \sum_i g_i^* \mathscr{O}_i(x)^\dagger \right], \tag{11.62}$$

where g_i are the couplings and $\mathscr{O}_i(x)$ a series of operators. The CP transformation interchanges them with their Hermitian conjugates and inverts the sign of the spatial coordinates. The transformed Hamiltonian is then

$$(\mathscr{C}\mathscr{P}) H_{\text{int}} (\mathscr{C}\mathscr{P})^{-1} = \int d^3x \left[\sum_i g_i \mathscr{O}_i(x)^\dagger + \sum_i g_i^* \mathscr{O}_i(x) \right]. \tag{11.63}$$

Hence, the invariance of the theory requires the couplings to be real $g_i = g_i^*$. This notwithstanding, a theory with complex couplings can still be CP-invariant. If it contains complex fields there is the possibility that the phases of the couplings can be absorbed in a redefinition of the global (irrelevant) phases of the fields.

Let us see how this applies to the standard model. In Chap. 10 (see page 198) we learned that the electroweak Lagrangian contains complex couplings due to the presence of the Cabibbo-Kobayashi-Maskawa (CKM) matrix. This is a unitary three-by-three matrix depending on nine real parameters and can be parametrized in terms of three real "mixing angles" and six complex phases. Given this state of affairs, CP violation could still be avoided provided all six complex phases could be absorbed in the arbitrary phases of the quark fields. Since there are six quark species it might look like all six phases of the CKM matrix could be eliminated this way. This is not the case because the standard model Lagrangian has a U(1) global symmetry that acts as a phase rotation of all the quark fields by the same phase. This means that there are only *five* independent phases we can play with. The consequence is that there is a complex phase in the CKM matrix which cannot be eliminated and therefore CP symmetry is violated.

CP violation here is a direct consequence of the existence of three quark families. Indeed, in a world with only two families the quark mixing matrix would be a two-by-two unitary matrix depending on one mixing angle and three complex phases (four real parameters in total). Since the number of quarks would be four, all three phases can be removed and we are left with a real matrix depending on one mixing angle, the Cabibbo angle. In this case there would be no room for CP violation in the electroweak sector.

The previous analysis has to be repeated in the leptonic sector of the standard model. If the neutrino masses are of the Dirac type then repeating the same arguments as above we conclude that the leptonic CKM matrix has a complex phase that cannot be eliminated by a phase redefinition of the lepton fields. For Majorana neutrinos the situation is very different since the Majorana condition (11.36) is incompatible with a phase redefinition of the field. This means that the complex couplings in the weak lepton-neutrino current can only be absorbed by changing the phase of the three charged leptons. As a result only three of the six complex phases of the leptonic CKM matrix can be disposed of.[4]

Another source of CP violation in the standard model comes from the strong interaction sector. As explained in Chap. 4 (see page 72) there is the possibility of adding to the QCD action the term

$$S_\theta = -\frac{\theta g_{\rm YM}^2}{32\pi^2}\int d^4x\, F^A_{\mu\nu}\widetilde{F}^{\mu\nu A} = -\frac{\theta g_{\rm YM}^2}{16\pi^2}\int d^4x\, \mathbf{E}^A\cdot\mathbf{B}^A. \tag{11.64}$$

Looking at the transformations of the chromoelectric and chromomagnetic fields under P and C derived in Sect. 11.1, we find the action of the CP transformation to be $\mathbf{E}^A(x^0,\mathbf{x}) \to \mathbf{E}^A(x^0,-\mathbf{x})$ and $\mathbf{B}^A(x^0,\mathbf{x}) \to -\mathbf{B}^A(x^0,-\mathbf{x})$. This implies that (11.64) changes sign. This has very important consequences since S_θ adds to the QCD action constructed from (5.27), which indeed *preserves* CP. We have found that

$$S_{\rm QCD} + S_\theta \xrightarrow{\rm CP} S_{\rm QCD} - S_\theta. \tag{11.65}$$

Hence, if $\theta \neq 0$ CP is also violated by the strong interaction.

One of the consequences of the presence of the term (11.64) in the QCD action is that it generates a nonvanishing electric dipole moment for the neutron [1, 2]. The fact that this is not observed experimentally can be used to impose a very strong bound on the value of the θ-parameter,

$$|\theta| < 10^{-9}. \tag{11.66}$$

From a theoretical point of view it is still to be fully understood why θ either vanishes or has a very small value. This is called the *strong CP problem*.

11.6 The CPT Theorem

The CPT transformation, the combination of parity, charge conjugation and time reversal, acts on the quantum fields by replacing them by their Hermitian conjugates while, at the same time, reversing the sign of the space-time coordinates. Since the three discrete symmetries commute with each other, there is no ambiguity in the definition of this transformation. It inherits from $\mathscr{T}$ the property of being implemented by an antiunitary operator that we denote by $\Theta \equiv \mathscr{C}\mathscr{P}\mathscr{T}$. For example, for a complex scalar field we have

$$\Theta\phi(x)\Theta^{-1} = \phi(-x)^\dagger, \tag{11.67}$$

[4] We should bear in mind that the Majorana neutrino break the global U(1) phase symmetry of the Lagrangian and therefore all three arbitrary phases of the charged leptons are independent. Notice also that in this case there is CP violation even with only two families, since having two charged leptons only allows for the elimination of two of the three phases of the mixing matrix.

and for a Dirac spinor

$$\Theta\psi(x)\Theta^{-1} = -i\gamma_5\psi(-x)^*. \tag{11.68}$$

In the latter equation we have to bear in mind that the complex conjugation symbol acts as Hermitian conjugation on the creation-annihilation operators. As in Eq. (11.27) we refrained from using the dagger symbol to avoid giving the wrong impression that the CPT operation transposes the spin indices.

Acting on the single particle states the CPT operator interchanges particles with antiparticles and reverses the sign of the helicity. The precise form of the transformation can be found using that the field (resp. its Hermitian conjugate) interpolates between the vacuum and the one-particle (resp. one-antiparticle) state. In the case of a Dirac spinor, for example,

$$\begin{aligned}\langle\Omega|\psi_\alpha(t,\mathbf{x})|\mathbf{p},s;0\rangle &= u_\alpha(\mathbf{p},s)e^{-iE_\mathbf{p}t+i\mathbf{p}\cdot\mathbf{x}},\\ \langle\Omega|\psi_\alpha(t,\mathbf{x})^\dagger|0;\mathbf{p},s\rangle &= v_\alpha(\mathbf{p},s)^*e^{-iE_\mathbf{p}t+i\mathbf{p}\cdot\mathbf{x}}.\end{aligned} \tag{11.69}$$

Applying Eq. (7.32) to the CPT operator and using (11.68) we find the following expression for the CPT-transformed one-particle states

$$\begin{aligned}\langle\Omega|\psi_\alpha(t,\mathbf{x})^\dagger|\mathbf{p},s;0\rangle_\Theta &= i(-1)^{\frac{1}{2}+s}v_\alpha(\mathbf{p},-s)^*e^{-iE_\mathbf{p}t+i\mathbf{p}\cdot\mathbf{x}},\\ \langle\Omega|\psi_\alpha(t,\mathbf{x})|0;\mathbf{p},s\rangle_\Theta &= -i(-1)^{\frac{1}{2}+s}u_\alpha(\mathbf{p},-s)e^{-iE_\mathbf{p}t+i\mathbf{p}\cdot\mathbf{x}}.\end{aligned} \tag{11.70}$$

In deriving this expression we have combined the wave function identities (11.28), (11.31) and (11.54). Comparing now (11.69) with (11.70) we finally arrive at the CPT transformation of the single particle states

$$\begin{aligned}\Theta|\mathbf{p},s;0\rangle &= i(-1)^{\frac{1}{2}+s}|0;\mathbf{p},-s\rangle,\\ \Theta|0;\mathbf{p},s\rangle &= -i(-1)^{\frac{1}{2}+s}|\mathbf{p},-s;0\rangle.\end{aligned} \tag{11.71}$$

The analysis for other types of fields can be carried out along the same lines.

The CPT operator satisfies a number of other interesting properties. It squares to the fermion number, $\Theta^2 = (-1)^F$. Since it involves time reversal it is rather intuitive that Θ has to transform "in" into "out" states and vice versa, namely

$$\Theta|\alpha;\text{in}\rangle = |\Theta\alpha;\text{out}\rangle, \tag{11.72}$$

where by $\Theta\alpha$ we indicate the CPT-transformed state. Using the antiunitarity of Θ we arrive at the following transformation of the S-matrix elements

$$\langle\alpha;\text{out}|\beta;\text{in}\rangle = \langle\Theta\beta;\text{out}|\Theta\alpha;\text{in}\rangle. \tag{11.73}$$

Now, taking into account that $S|\alpha;\text{in}\rangle = |\alpha;\text{out}\rangle$, we find the following identity for the S-matrix

$$\Theta S = S^{-1}\Theta. \tag{11.74}$$

CPT invariance has a very special status in quantum field theory. Unlike the three individual discrete symmetries, the invariance of a theory under the combination of them is related to very fundamental physical principles. A special role is also played by the spin-statistic connection implying that particles with integer (resp. half-integer) spin follow the Bose-Einstein (resp. Fermi-Dirac) statistics.

Our aim here is to study the so-called *CPT theorem*: any local quantum field theory that is Poincaré invariant is also invariant under the combination of $\mathscr{P}$, $\mathscr{C}$ and $\mathscr{T}$ (in any order). This result was first proved by Lüders and Pauli [3, 4]. Here, however, we will sketch a very elegant and general proof due to Jost [5] (see also [6, 7] for details). This is based on an axiomatic formulation of quantum field theory whose postulates are a kind of sophisticated generalization of the properties we listed in Chap. 2 for a free scalar field (see page 16):

- The states of a quantum field theory form a Hilbert space that carries a unitary representation of the Poincaré group including space-time translations and proper, orthochronous Lorentz transformations $\mathfrak{L}_+^\uparrow$, whose double cover is[5] SL(2, $\mathbb{C}$). In this Hilbert space there is a distinguished state $|\Omega\rangle$, the vacuum, invariant under the Poincaré group

$$\mathscr{U}(A, a)|\Omega\rangle = |\Omega\rangle. \tag{11.75}$$

  Here A denotes an SL(2, $\mathbb{C}$) transformation and a^μ is a four-vector associated with a space-time translation. Finally, the states in the Hilbert space have non-negative mass squared and positive energy. In technical terms this means that the spectrum of the momentum operator P^μ is the forward cone $p_\mu p^\mu = m^2 \geq 0,\ p^0 \geq 0$.
- For the benefit of the mathematically-inclined, let us mention that the naive quantum fields $\Phi(x)$ we have dealt with so far are rather singular objects, and so are their products at the same space-time point. To be rigorous one should smear the field operators by test functions

$$\Phi(f) = \int d^4x f(x)\Phi(x), \tag{11.76}$$

  where *f(x)* and its derivatives are taken to decrease faster than any power when $x^\mu \to \infty$. What this means is that fields are operator-valued distributions in Minkowski space-time.
- The quantum fields must have well-defined transformations under the Poincaré group,

$$\mathscr{U}(A, a)\Phi(x)\mathscr{U}(A, a)^{-1} = M(A^{-1})\Phi\big(\Lambda(A)x + a\big). \tag{11.77}$$

  The field $\Phi(x)$ has a number of undotted and dotted indices that we have omitted to keep the expression simple. It is on these indices that $M(A^{-1})$ acts. From the

---

5 See Appendix B (Sect. B.4).

discussion in Sect. B.4 [see Eq. (B.41)] it follows that $M(A^{-1})$ is a monomial in A^{-1} and $(A^{-1})^*$.

- Fields defined on causally-disconnected regions of space-time cannot interfere with one another. This is expressed by the statement that, depending on their bosonic or fermionic character,

$$\left[\Phi_i(f), \Phi_j(g)\right] = 0 \quad \text{or} \quad \{\Phi_i(f), \Phi_j(g)\} = 0 \tag{11.78}$$

whenever the support of the test functions *f(x)* and *g(x)* lie in regions of space time that are space-like separated (cf. Fig. 1.4 and the associated discussion). This property is called in the literature *microcausality* or *local commutativity*

We now formulate the CPT theorem using these axioms: it states that there is a antiunitary operator Θ uniquely defined by the following transformation of a general quantum field with a number *(n,m)* of undotted and dotted indices [cf. Eqs. (11.67) and (11.68)]

$$\Theta \Phi_{a_1 \ldots a_n \dot{b}_1 \ldots \dot{b}_m}(x) \Theta^{-1} = (-1)^m (-i)^F \Phi^\dagger_{\dot{a}_1 \ldots \dot{a}_n b_1 \ldots b_m}(-x), \tag{11.79}$$

where $F = 0$ for bosons and $F = 1$ for fermions, and the fact that it preserves the vacuum, $\Theta|\Omega\rangle = |\Omega\rangle$. Notice that the indices appearing on the right-hand side of Eq. (11.79) are those of the Hermitian conjugated field $\Phi^\dagger$. We have to remember that Hermitian conjugation transforms dotted into undotted indices and vice versa.

The Proof

To prove[6] the CPT theorem we are going to work with Wightman correlation functions, defined as the vacuum expectation value of the product of a number of quantum fields

$$W_n(x_1, \ldots, x_n) = \langle \Omega | \Phi(x_1) \ldots \Phi(x_n) | \Omega \rangle. \tag{11.80}$$

For simplicity, here and in the following we consider a single type of field and omit the indices. We also ignore the fact that these functions have to be understood as distributions acting on some test functions. It can be shown that any quantum field theory can be reconstructed from the knowledge of its Wightman functions. Notice that the time-ordered correlation functions used in Chap. 6 to compute S-matrix amplitudes can be written as a linear combination of Wightman functions with step-function coefficients.

Let us explore the consequences of CPT invariance on the Wightman functions. From the antilinearity of Θ and the CPT-invariance of the vacuum it follows [see Eq. (7.32)]

---

[6] This section is substantially more mathematical than the rest of the text and it can be safely skipped. Its purpose is to prove Eq. (11.82).

$$\begin{aligned}\langle\Omega|\Phi(x_1)\ldots\Phi(x_n)|\Omega\rangle =&\langle\Omega|\Theta\Phi(x_n)^\dagger\Theta^{-1}\ldots\Theta\Phi(x_1)^\dagger\Theta^{-1}|\Omega\rangle \\ =&(-1)^{M+F}\langle\Omega|\Phi(-x_n)\ldots\Phi(-x_1)|\Omega\rangle.\end{aligned} \tag{11.81}$$

Here M is the *total* number of dotted field indices in the correlation function and $(-1)^F$ is the sign that comes form inverting the order of the fields. Remarkably, the identity we just derived

$$W_n(x_1,\ldots,x_n) = (-1)^{M+F}W_n(-x_n,\ldots,-x_1) \tag{11.82}$$

is not just a necessary condition for CPT symmetry, but it is also sufficient.

Hence, to prove the CPT theorem one shows that Eq. (11.82) holds for any quantum field theory satisfying the list of axioms listed above. The strategy of Jost's proof is to perform an analytic continuation of the Wightman function to complex values of the arguments and then show that there is a corner of the analyticity domain in which this relation is satisfied.

We begin then by studying the analyticity properties of the Wightman functions. The first thing to notice is that, as a consequence of translational invariance, they are function of the $n-1$ coordinate differences

$$W_n(x_1,\ldots,x_n) = \mathscr{W}_n(\xi_1,\ldots,\xi_{n-1}) \quad \text{where } \xi_i = x_i - x_{i+1}. \tag{11.83}$$

To prove this one has to use the Poincaré invariance of the vacuum together with $\Phi(x) = e^{iP\cdot x}\Phi(0)e^{-iP\cdot x}$, where P^μ is the momentum operator of the theory. To perform the analytic continuation on ξ_i we start by writing $\mathscr{W}_n(\xi_1,\ldots,\xi_{n-1})$ in terms of its Fourier transform as

$$\mathscr{W}_n(\xi_1,\ldots,\xi_n) = \int\left(\prod_{i=1}^{n-1}d^4\xi_i\right)\widetilde{\mathscr{W}_n}(q_1,\ldots,q_{n-1})\exp\left(-i\sum_{i=1}^{n-1}\xi_i\cdot q_i\right). \tag{11.84}$$

In fact, $\widetilde{\mathscr{W}_n}(q_1,\ldots,q_{n-1})$ has the interesting property that it is equal to zero whenever one of the arguments lies outside the spectrum of P^μ. This is clear if we write the Fourier transform as

$$\begin{aligned}\widetilde{\mathscr{W}_2}(q_1,\ldots,q_2) =&\langle\Omega|\Phi(0)\delta^{(4)}(P-q_1)\Phi(0)\delta^{(4)}(P-q_2) \\ &\cdots\Phi(0)\delta^{(4)}(P-q_{n-1})\Phi(0)|\Omega\rangle.\end{aligned} \tag{11.85}$$

This last property tells us how to analytically continue the Wightman function to complex arguments. Since q_i is confined to be in the forward light-cone ($q_i^2 \geq 0$, $q_i^0 > 0$), the integral (11.84) will define an analytic function if we replace

$$\xi_i \longrightarrow \zeta_i \equiv \xi_i - i\eta_i, \tag{11.86}$$

provided η_i lies inside the forward light-cone, $V_+ = \{\eta_i \in \mathbb{R}^4 | \eta_i^2 > 0,\ \eta_i^0 > 0\}$. Indeed, when this condition is satisfied the integrand has a damping exponential factor guaranteeing that the resulting integral defines a holomorphic function. With this we have found the primitive domain of holomorphy of the Wightman functions

$$\mathfrak{T}_{n-1} = \{(\zeta_1, \ldots, \zeta_{n-1}) \in \mathbb{C}^{4(n-1)} | -\text{Im}\zeta_i \in V_+\}. \tag{11.87}$$

Notice that, by definition, this domain does not contain any real points. The original Wightman function $\mathscr{W}_n(\xi_1, \ldots, \xi_{n-1})$ is retrieved as the boundary value of the analytic function on $\mathfrak{T}_{n-1}$ when $\text{Im}\zeta_i \to 0$.

This is not the largest analyticity domain of the Wightman function. Proper complex Lorentz transformations (see Sect. B.4) have a well defined, continuous action on the Wightman functions. They can be used to further analytically continue them into the extended domain $\mathfrak{T}'_{n-1}$ obtained by acting with arbitrary proper complex Lorentz transformations on points in $\mathfrak{T}_{n-1}$,

$$\mathfrak{T}'_{n-1} = \{(\Lambda\zeta_1, \ldots, \Lambda\zeta_{n-1}) | (\zeta_1, \ldots, \zeta_{n-1}) \in \mathfrak{T}_{n-1}, \Lambda \in \mathfrak{L}_+(\mathbb{C})\}. \tag{11.88}$$

The extension to $\mathfrak{T}'_{n-1}$ has a double advantage. First of all, space-time inversion $\mathscr{P}\mathscr{T}$ is an element of $\mathfrak{L}_+(\mathbb{C})$. This means that if $(\zeta_1, \ldots, \zeta_{n-1})$ belongs to $\mathfrak{T}'_{n-1}$ so does $(-\zeta_1, \ldots, -\zeta_n)$. Moreover, applying the transformation of $\mathscr{W}_n(\zeta_1, \ldots, \zeta_{n-1})$ under space-time inversion we find [cf. Eq. (B.45)]

$$\mathscr{W}_n(\zeta_1, \ldots, \zeta_{n-1}) = (-1)^M \mathscr{W}_n(-\zeta_1, \ldots, -\zeta_{n-1}), \tag{11.89}$$

with M the number of dotted indices.

Equation (11.89) comes short of the CPT theorem by the fact that the arguments are complex and not in the right order. A second property of the domain $\mathfrak{T}'_{n-1}$ comes now handy: unlike $\mathfrak{T}_{n-1}$, it contains real points. These are the so-called Jost points, the real $(\zeta_1, \ldots, \zeta_{n-1}) \in \mathfrak{T}'_{n-1}$ such that

$$\left(\sum_{i=1}^{n-1} \lambda_i \zeta_i\right)^2 < 0 \quad \text{for all } \lambda_i \geq 0 \text{ and such that } \sum_{i=1}^{n-1} \lambda_i > 0. \tag{11.90}$$

This condition implies that, in particular, the Jost points satisfy $\zeta_i^2 < 0$. Notice that since the Jost points are in $\mathfrak{T}'_{n-1}$ the Wightman function *are* holomorphic there. This point will be crucial in the following.

Let us consider now $(x_1, \ldots, x_n)$ such that $\zeta_i \equiv x_i - x_{i+1}$ is a Jost point. Since $\zeta_i^2 < 0$ we can use the microcausality postulate stating that quantum fields evaluated at x_i either commute or anticommute. This leads to the following relation between the Wightman functions evaluated at the Jost points

$$\begin{aligned}\mathscr{W}_n(\zeta_1, \ldots, \zeta_{n-1}) &= \langle\Omega|\Phi(x_1) \ldots \Phi(x_n)|\Omega\rangle \\ &= (-1)^F \langle\Omega|\Phi(x_n) \ldots \Phi(x_1)|\Omega\rangle \\ &= (-1)^F \mathscr{W}_n(-\zeta_n, \ldots, -\zeta_1).\end{aligned} \tag{11.91}$$

Now, as $\mathscr{W}_n(\zeta_1, \ldots, \zeta_{n-1})$ is holomorphic at the Jost points, the previous relation between Wightman functions has to hold in a neighborhood of those points and, by analytic continuation, in the whole extended domain $\mathfrak{T}'_{n-1}$. Combining this with Eq. (11.89) the following identity is obtained

$$\mathscr{W}_n(\zeta_1, \ldots, \zeta_{n-1}) = (-1)^{M+F}\mathscr{W}_n(\zeta_{n-1}, \ldots, \zeta_1), \tag{11.92}$$

valid throughout the whole extended domain $\mathfrak{T}'_{n-1}$ and therefore on $\mathfrak{T}_{n-1}$. Going finally to the boundary of $\mathfrak{T}_{n-1}$ by taking $\text{Im}\,\zeta_{n-1} \to 0$ the CPT identity (11.82) is obtained. This concludes the proof.

The most appealing feature of the above proof of the CPT theorem is its generality. It avoids entering into the details of the theory, such as the terms in the action, and instead only appeals to very general and fundamental properties of quantum field theory. In particular it makes very transparent how the invariance under CPT depends on three crucial ingredientes: Poincaré invariance, positivity of the energy and causality, the latter encoded in the local commutativity postulate. This is what gives CPT the very special status it enjoys in theoretical physics since, unlike other discrete symmetries, it cannot be broken without destroying what are believed to be very fundamental physical principles.

Implications of the Theorem

To close this section we discuss some consequences of the CPT theorem for the spectrum of the theory. It was first proved by Lüders and Zumino [8] that CPT invariance implies the equality of the masses and decay widths of particles and antiparticles. To see how this comes about look at a theory with a CPT-invariant Hamiltonian $\Theta H \Theta^{-1} = H$ and assume that it can be split into two pieces $H = H_s + H_w$. The first piece H_s contains the free Hamiltonian and the strong interactions. H_w represents the weakly coupled interactions of the theory.

Denote by $|\psi_n\rangle$ the energy eigenstates of H_s and define the resolvent

$$G_n(\lambda) = \langle\psi_n|\frac{1}{\lambda - H}|\psi_n\rangle. \tag{11.93}$$

In the limit where the weakly interacting terms H_w are switched off the state $|\psi_n\rangle$ becames an energy eigenstate of the theory and the function $G_n(\lambda)$ has a pole located at $\lambda = m_n$, where m_n is the mass of the state. When H_w is included, the states $|\psi_n\rangle$ become unstable (they are no longer energy eigenstates) and the pole of the resolvent originally located at $\lambda = m_n$ migrates into the complex plane. The real part of the pole gives its mass, and its imaginary part its width (i.e., the inverse of its lifetime).

We now repeat the analysis for the CPT transformed state $\Theta|\psi_n\rangle \equiv |\Theta\psi_n\rangle$. The corresponding resolvent is

$$G_n^{\Theta}(\lambda) = \langle\Theta\psi_n|\frac{1}{\lambda - H}|\Theta\psi_n\rangle = \langle\Theta\psi_n|\Theta\Theta^{-1}\frac{1}{\lambda - H}\Theta|\psi_n\rangle. \tag{11.94}$$

Using now the antilinearity of Θ and the invariance of the Hamiltonian we can write

$$\Theta^{-1}\frac{1}{\lambda - H}\Theta = \frac{1}{\lambda^* - H}. \tag{11.95}$$

Applying one last time the antiunitarity of Θ we arrive at the conclusion that the resolvent is CPT invariant

$$\begin{aligned} G_n^{\Theta}(\lambda) =& \langle\Theta\psi_n|\Theta\frac{1}{\lambda^* - H}|\psi_n\rangle \\ =& \langle\psi_n|\frac{1}{\lambda^* - H}|\psi_n\rangle^* = G_n(\lambda). \end{aligned} \tag{11.96}$$

We know that CPT acting on one-particle states exchanges particles and antiparticles. Since the pole structure of $G_n(\lambda)$ and $G_n^{\Theta}(\lambda)$ are the same, we are led to conclude that particles and antiparticles have the same mass and decay widths.

11.7 Spin and Statistics

When dealing with the canonical quantization of the Dirac field in Chap. 3 we were forced to replace the canonical Poisson brackets with anticommutators, instead of the commutators used for scalar and gauge fields. The reason for this choice was compelling. Had we used commutators the resulting quantum theory would have had an energy spectrum unbounded from below.

This relation between spin and (anti)commutators is dictated by the spin-statistics theorem. It states that fields with integer spin have to be quantized using canonical commutation relations, whereas fields with half-integer spin require the use of anticommutators. In other words, particles with integer and half-integer spin follow the Bose-Einstein and Fermi-Dirac statistics respectively.[7]

As the CPT theorem, the spin-statistic theorem in a consequence of locality and Poincaré invariance. In fact it can be easily proved using the techniques applied in the previous section to the CPT theorem. Let us thus consider a quantum field theory with a single quantum field $\Phi(x)$ and look at the two-point Wightman function

$$W(x_1, x_2) = \langle\Omega|\Phi(x_1)^\dagger\Phi(x_2)|\Omega\rangle. \tag{11.97}$$

Applying now the CPT identity (11.82) to this function we have

$$\langle\Omega|\Phi(x_1)^\dagger\Phi(x_2)|\Omega\rangle = (-1)^{M+F}\langle\Omega|\Phi(-x_2)\Phi(-x_1)^\dagger|\Omega\rangle \tag{11.98}$$

---

[7] The reader surely has noticed by now that in this book we have been assuming the spin-statistics theorem all the time since we have been consistently using the term "fermion" or "fermionic" to refer to spinor fields.

The sign $(-1)^F$ comes from reversing the order of the two operators when they are spatially separated, and therefore it is equal to $+1$ for bosonic and -1 for fermionic fields. Furthermore, M is equal to the number of dotted indices in the Wightman function. Since Hermitian conjugation transforms dotted into undotted indices and vice versa, we find that M equals to total number of dotted *plus* undotted indices of the field $\Phi(x)$.

At this point we have to recall that, according to the field theory postulates, quantum fields are in fact distributions. Smearing the fields with an arbitrary test function $f(x)$, we can write the identity (11.98) as

$$\|\Phi(f)|\Omega\rangle\|^2 = (-1)^{M+F}\|\Phi(\overline{f})|\Omega\rangle\|^2, \qquad (11.99)$$

where $\overline{f}(x) = f(-x)$. Now, since the Hilbert space of the quantum theory only contains states with positive norm we are led to conclude that

$$(-1)^{M+F} = 1. \qquad (11.100)$$

Fields with integer spin have an even number of dotted plus undotted indices, and from this identity we see that $\Phi(x_1)^\dagger$ and $\Phi(x_2)$ have to commute when $x_1 - x_2$ is space-like. If, on the other hand, $\Phi(x)$ is a half-integer spin field the number of dotted plus undotted indices has to be odd and, as a consequence of (11.100), the fields have to be anticommuting. This proves the spin-statistics theorem.

References

1. Ramond, P.: Journeys Beyond the Standard Model. Perseus Books, Cambridge (1999)
2. Mohapatra, R.N.: Unification and Supersymmetry: The Frontiers of Quark-Lepton Physics. Springer, New York (2003)
3. Lüders, G.: On the equivalence of the invariance under time reversal and under particle-anti-particle conjugation. Dansk. Mat. Fys. Medd. **28**, 5 (1954)
4. Pauli, W.: Exclusion principle, Lorentz group and reflection of space-time and charge. In: Pauli, W. (eds) Niels Bohr and the Development of Physics, Pergamon, London (1955)
5. Jost, R.: Eine Bemerkung zum CTP Theorem. Helv. Phys. Acta **30**, 409 (1957)
6. Streater, R.F., Wightman, A.S.: Spin and Statistics and All That. Princeton University Press, Princeton (1964)
7. Haag, R.: Local Quantum Physics. Fields, Particles, Algebras. Springer, Berlin (1992)
8. Lüders, G., Zumino, B.: Some consequences of TCP-invariance. Phys. Rev. **106**, 385 (1957)

Chapter 12
Effective Field Theories and Naturalness

Effective field theories are among the most powerful instruments in the toolbox of contemporary physics. Although the concept of effective field theory has been already discussed in Chap. 8, here we are going to provide a relatively elementary description of the relevant technology. Although rather unrealistic, the examples of effective field theories studied next serve the purpose of illustrating the relevant physics involved. The chapter will be closed with a discussion of the concept of naturalness, which plays a central role in modern particle physics. The reader is advised not to be scared by the technicalities of the Feynman diagram computations contained in the chapter. Most of the conclusions can be reached without caring too much about the precise value of the numerical prefactors.

12.1 Energy Scales in Quantum Field Theory

When introducing the renormalization group in Chap. 9 we did not go into a detailed evaluation of Feynman diagrams. In the present chapter we will provide some more details, particularly the computation of loop diagrams and how to extract divergences from them. The interpretation of the results coming from particle colliders like the Tevatron at Fermilab or the LHC at CERN requires extensive loop computations in the standard model. Simple cut-off methods or other regularization schemes are inadequate for this task.

The calculation of quantum corrections is necessary also when trying to derive effective field theories valid at low energies from more fundamental descriptions at high energies. This raises the important question of the separation of scales in quantum field theory. In the standard model, for example, we have widely separated energy scales. We have the mass of the $W^{\pm}$ boson $m_W \sim 10^2$ GeV characterizing the scale of the electroweak processes. At the same time there is another energy scale that can be constructed from the three fundamental constants of physics, Planck's constant $\hbar$, the speed of light c and Newton's constant G_N. This is the Planck mass defined by

L. Álvarez-Gaumé and M. Á. Vázquez-Mozo, *An Invitation to Quantum Field Theory*, Lecture Notes in Physics 839, DOI: 10.1007/978-3-642-23728-7_12,

$$M_P = \sqrt{\frac{\hbar c}{G_N}} \sim 10^{19}\,\text{GeV}. \tag{12.1}$$

The theory might have additional "intermediate" high energy scales, such as the grand unification (GUT) energy at about 10^{15}–10^{16} GeV.

The important question to be addressed is how physical processes at energies below the lower scale, $E < m_W$, depend on higher scales such as the Planck or the GUT scale. The analysis of this question often invokes the *naturalness criterion* to be explained later. It should be stressed, however, that naturalness rather than a law of Nature should be seen as a good guiding principle to understand the sizes of the dimensionful parameters, specially masses, in the low-energy theory.

Later on we will argue that the fact that light fermions, such as the electron or the muon, have masses much smaller than the Planck scale should not be considered unnatural. The situation is different for scalar particles, such as the so-far hypothetical Higgs boson. In this case the fact that the Higgs mass is expected to be light with respect to other higher energy scales is quite unnatural. To see this let us take the point of view that the standard model is a valid physical description up to some energy scale Λ, that we can take to be the Planck, GUT or any other relevant scale. Identifying then Λ as the momentum cutoff in the quantum theory, one finds that the correction to the Higgs mass depends quadratically on it

$$\delta m^2 \sim \frac{g^2}{16\pi^2}\Lambda^2. \tag{12.2}$$

The conclusion is that the Higgs mass should be very sensitive to the details of the physics at higher energies since it strongly depends on the energy scale Λ at which the standard model should be replaced by a more complete description. This simple remark is the kernel of the famous *hierarchy problem*. Due to the strong dependence of the Higgs mass corrections on the scale of new physics, keeping m_{H} and Λ widely apart

$$m_{\text{H}}^2 = m_0^2 + \delta m^2 \ll \Lambda^2 \tag{12.3}$$

requires a fantastic fine tuning of m_0^2. This is what is meant by the statement that a light Higgs mass is unnatural.

The questions we just introduced will be formulated in more detail later on. We still do not have answers to many questions about naturalness. Their resolution will bring deep breakthroughs in our understanding of Nature at short and long distances.

12.2 Dimensional Regularization

To get started we present the method of dimensional regularization and renormalization that is universally used specially in theories with local gauge symmetries like the standard model. Introduced in the early seventies [1, 2], it is a remarkably clever way to regulate the integrals appearing in the perturbative calculation of scattering amplitudes.

In Chap. 8 we learned in a particular example how the calculation of Feynman diagrams with loops leads to divergent integrals over the momenta running in them, that should be regularized. In all cases the integrand has the structure of a rational function of the loop momenta. Dimensional regularization (DR) prescribes these integrals to be carried out in an arbitary dimension d instead of $d = 4$. The integration gives a finite result depending on d that can then be analytically continued to complex values of the dimension. The divergence of the original loop integral reappears as poles of different order in $d - 4$. This is a rather economical and efficient way of regularizing divergences.

Apart from its computational simplicity DR presents various advantages. The most important is that it automatically preserves the symmetries of the theory whenever they admit an extension to higher dimensions. This is the case of vector-like gauge theories like QCD where the gauge symmetry can be formulated in any dimension. This means that DR regularizes the theory without breaking gauge invariance.

Chiral (gauge) symmetries, on the other hand, are more problematic. The reason is that the notion of chirality is very "four-dimensional", and chiral gauge theories like the electroweak sector of the standard model require a very special treatment. The fact that chiral symmetries do not admit an extension to higher dimensions is related to the existence of the chiral anomaly studied in Chap. 9. The fact that DR frequently preserves the original symmetry is a big advantage with respect to other regularization procedures. Among its few drawbacks one should mention that so far all attempts to find a nonperturbative formulation of DR have been unsuccessful.

Since our presentation only requires the calculation of a few one-loop diagrams, we illustrate the use of DR with the basic integral

$$I_n(d, m^2) = \int \frac{d^d p}{(2\pi)^d} \frac{1}{(p^2 - m^2 + i\varepsilon)^n}. \tag{12.4}$$

For $d = 4$ this integral is divergent when $n = 0, 1, 2$. To evaluate it we begin with the integration over p^0. Due to the $i\varepsilon$ prescription, the integrand

$$\frac{1}{(p^2 - m^2 + i\varepsilon)^n} = \frac{1}{[(p^0)^2 - E_{\mathbf{p}}^2 + i\varepsilon]^n} \tag{12.5}$$

has poles located just above and below the real p^0 axis with real parts $\pm E_{\mathbf{p}}$. The integral over the real axis can be carried out applying Cauchy's theorem to the contour shown in Fig. 12.1. Since the contour encloses no poles and the integrand vanishes as $|p^0| \to \infty$, the integration over real p^0 can be expressed as the integral over the imaginary axis, namely

$$\int_{-\infty}^{\infty} \frac{dp^0}{2\pi} \frac{1}{[(p^0)^2 - E_{\mathbf{p}}^2 + i\varepsilon]^n} = i(-1)^n \int_{-\infty}^{\infty} \frac{dp_E^0}{2\pi} \frac{1}{[(p_E^0)^2 + E_{\mathbf{p}}^2]^n}. \tag{12.6}$$

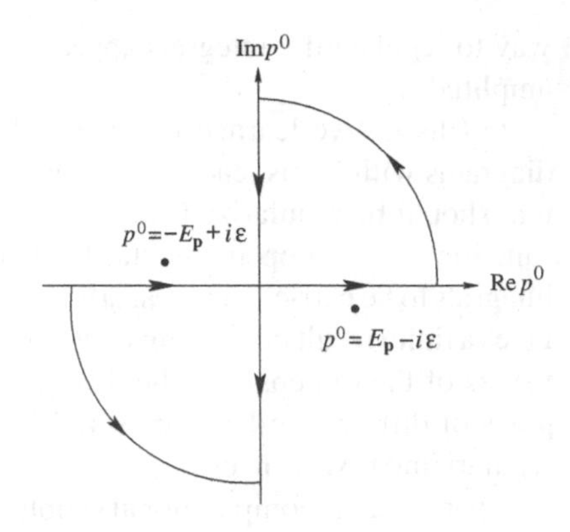

Fig. 12.1 Contour of integration used to evaluate the integral $I_n(d,m^2)$. The absence of poles in the first and third quadrants allows the integration contour along the real axis to be rotated to the imaginary axis (Wick rotation)

For $m \neq 0$ the poles are away from the imaginary axis and the $i\varepsilon$ prescription drops out of the second integral. Hence, Eq. (12.4) becomes

$$I_n(d,m^2) = i(-1)^n \int \frac{d^d p_E}{(2\pi)^d} \frac{1}{(p_E^2+m^2)^n}, \tag{12.7}$$

where p_E is an Euclidean d-dimensional momentum, $p_E^2 = (p_E^0)^2 + \mathbf{p}^2$.

To compute of the Euclidean integral (12.7) one exponentiates the integrand with the help of the identity

$$\frac{1}{a^n} = \frac{1}{\Gamma(n)} \int_0^\infty dt\, t^{n-1} e^{-az} \quad (a>0), \tag{12.8}$$

to write

$$\int \frac{d^d p_E}{(2\pi)^d} \frac{1}{(p_E^2+m^2)^n} = \frac{1}{\Gamma(n)} \int_0^\infty dt\, t^{n-1} e^{-tm^2} \int \frac{d^d p_E}{(2\pi)^d} e^{-tp_E^2}. \tag{12.9}$$

The Euclidean character of p_E ensures the convergence of the momentum integral. Using

$$\int \frac{d^d p_E}{(2\pi)^d} e^{-p_E^2} = \pi^{\frac{d}{2}}, \tag{12.10}$$

we arrive at the final result for the integral (12.7)

$$I_n(d,m^2) = \frac{i(-1)^n}{(4\pi)^{2+\frac{d-4}{2}}} \frac{\Gamma(n-2-\frac{d-4}{2})}{\Gamma(n)(m^2)^{n-2-\frac{d-4}{2}}}. \tag{12.11}$$

The dependence of $I_n(d,m^2)$ on the dimension can be analytically continued to complex values of d. The pole structure is derived from the properties of the Euler gamma function $\Gamma(z)$, which can be found in any book on special functions. Here we just notice that $\Gamma(z)$ has poles at nonpositive integer values of the argument, $z = 0, -1, -2, \ldots$ In the case at hand, for $n > 2$ the integral converges as $d \to 4$. When $n = 1, 2$, on the contrary, the expression contains either a factor of $\Gamma(1 - d/2)$ or $\Gamma(2 - d/4)$ in the numerator, both functions diverging in the limit $d \to 4$.

To find the behavior of $I_n(d, m^2)$ around $d = 4$ we use the Laurent expansion of the gamma function around its poles

$$\Gamma(-k+\varepsilon) = \frac{(-1)^k}{k!}\left[\frac{1}{\varepsilon} + \psi(k+1) + \mathscr{O}(\varepsilon)\right] \quad k \in \mathbb{N}, \tag{12.12}$$

where $\psi(z)$ is the dilogarigthm function, defined as the logarithmic derivative of the gamma function

$$\psi(z) = \frac{d}{dz}\log\Gamma(z), \quad \psi(k+1) = -\gamma + \sum_{n=1}^{k}\frac{1}{n}, \tag{12.13}$$

with $\gamma = -\psi(1) = 0.5772\ldots$ the Euler-Mascheroni constant. Applying this to our integral we find its behavior as $d \to 4$ to be

$$I_n(d, m^2) \xrightarrow{d\to 4} -\frac{i(m^2)^{2-n}}{16\pi^2}\frac{2}{d-4} + \text{finite part}, \quad n = 1, 2. \tag{12.14}$$

Other integrals can be computed along similar lines. One of special interest is

$$I_n(d, m^2, q) = \int \frac{d^d p}{(2\pi)^d}\frac{1}{(p^2 + 2p\cdot q - m^2 + i\varepsilon)^n}. \tag{12.15}$$

Upon completing the square in the denominator and shifting the integration variable (a legitimate step once the integral is regularized), it reduces to an integral of the type (12.7)

$$I_n(d, m^2, q) = I_n(d, m^2 + q^2). \tag{12.16}$$

In the case of the integrals arising in the calculation of diagrams with higher loops we would find a collection of gamma functions that would produce poles at $d = 4$ of various orders

$$\frac{1}{(d-4)^L}, \frac{1}{(d-4)^{L-1}}, \ldots, \frac{1}{d-4}, \tag{12.17}$$

with L the number of loops in the diagram. It is quite remarkable in fact that the second and higher order poles are determined by the first order poles as we will show later on. This provides a very useful check on multi-loop computations.

The analytical continuation of the dimension in the integrals can be formulated in an axiomatic way. All we need is to define the operation of d-dimensional integration (with d complex) preserving the basic properties of multi-dimensional integrals, namely

$$\begin{aligned}\int d^d p\Big[f(p)+g(p)\Big] &= \int d^d p f(p)+\int d^d p g(p),\\ \int d^d p f(\lambda p) &= \lambda^{-d}\int d^d p f(p) \quad (\text{with } \lambda\in\mathbb{C}),\\ \int d^d p f(p+k) &= \int d^d p f(p).\end{aligned} \tag{12.18}$$

Since rational functions can be turned into exponentials by using identities like (12.8), the evaluation of any integral appearing in the computation of a Feynman diagram can be done using the previous properties together with the action of the d-dimensional integration on Gaussian functions, given by Eq. (12.10).

A simple perusal of the DR calculation carried out above shows that both $I_1(d,m^2)$ and $I_2(d,m^2)$ diverge exactly in the same way as $d\to 4$, namely with a simple pole. The conclusion is that DR is only sensitive to logarithmic divergences and all polynomial divergences are regularized to zero. As a matter of fact, in DR we can write

$$\int d^d p(p^2)^n = 0. \tag{12.19}$$

One way to see how this identity comes about is to consider $I_n(d,m^2)$ with $n\geq -1$ and take the limit $m\to 0$ in the region $\mathrm{Re}\, d\geq 2n$ and away from the poles of the gamma function in (12.11). The limit is then equal to zero and the result can be analytically continued to the whole complex d plane. DR is also useful in handling infrared divergences.

The fact that DR eliminates quadratic divergences might seem surprising in the light of the previous discussion of the hierarchy problem. Indeed, as DR regularizes the quadratic divergences to zero it seems that the whole hierarchy problem results from using a clumsy regulator, and that by using DR we could shield the Higgs mass from the scale of new physics. This is not the case, but for interesting reasons. In spite of DR the Higgs mass is still sensitive to high energy scales. If it is ever found with a low mass, we will also get relevant information on what shields its mass from the higher scales. Before explaining the interesting reasons, we need to develop more theory.

12.3 The ϕ^4 Theory: A Case Study

To get a better understanding of how a quantum field theory is regularized using DR we look into a very simple field theory: a massive real scalar field $\phi(x)$ with a quartic self-interaction

$$\mathcal{L} = \frac{1}{2}\partial_\mu\phi\partial^\mu\phi - \frac{1}{2}m^2\phi^2 - \frac{\lambda}{4!}\phi^4. \tag{12.20}$$

As we learned in Chap. 6 the perturbative expansion is constructed using the Feynman rules. In this case we only have to specify one propagator and one vertex

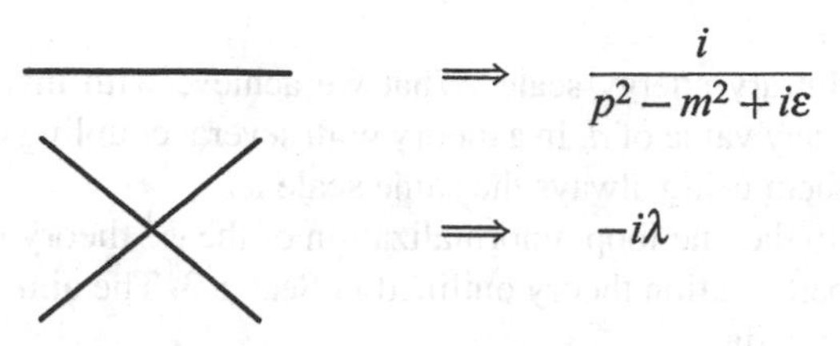

together with the delta function conservation $(2\pi)^4\delta^{(4)}(p_1 + p_2 + p_3 + p_4)$, where we use the convention that all momenta in the vertex are incoming. Since the scalar field is real, it does not carry charge and therefore the lines of the Feynman diagrams do not have orientation.

The quantization using DR requires defining the theory in d dimensions

$$S = \int d^d x \mathcal{L}(\phi, \partial_\mu\phi). \tag{12.21}$$

Since the dimensions of the fields and parameters in the action depend on d, it is useful to stop for a moment and carry out some dimensional analysis. In natural units $\hbar = c = 1$ the action is dimensionless and looking at the kinetic term we can fix the energy dimensions of the scalar field[1]

$$D_\phi = \frac{d-2}{2}. \tag{12.22}$$

The same analysis can be done for fermions and gauge fields with the respective result

$$D_\psi = \frac{d-1}{2}, \quad D_A = \frac{d-2}{2}. \tag{12.23}$$

The energy dimensions of the parameter of the scalar theory (12.20) are

$$D_m = 1, \quad D_\lambda = 4 - d. \tag{12.24}$$

In the case of scalar field theories with cubic self-interaction and/or Yukawa couplings to Dirac fermions, the dimension of the corresponding coupling constants are

$$\begin{aligned} \lambda'\phi^3 \quad &\Longrightarrow \quad D_{\lambda'} = 1 + \frac{4-d}{2} \\ g\phi\overline{\psi}\psi \quad &\Longrightarrow \quad D_g = \frac{4-d}{2} \end{aligned} \tag{12.25}$$

---

[1] Our choice of natural units allows us to specify the dimensions of all quantities in terms of powers of energy. Thus, for the coordinates we have $[x^\mu] = E^{-1}$, which we denote by $D_x = -1$.

In the particular case of the ϕ^4 example the dependence of the energy dimension of λ with d suggests replacing the coupling constant in the action (and therefore in the Feynman rules) by

$$\lambda \longrightarrow \mu^{4-d}\lambda, \tag{12.26}$$

where μ is an arbitrary energy scale. What we achieve with this is that λ is kept dimensionless for any value of d. In a theory with several couplings we would do the same with all of them using always the same scale μ.

We apply DR to the one loop renormalization of the ϕ^4 theory using the method of renormalized perturbation theory outlined in Sect. 8.3. The aim is to compute the renormalized Lagrangian

$$\mathscr{L}_{\text{ren}} = \frac{1}{2}\partial_\mu\phi_0\partial^\mu\phi_0 - \frac{m_0^2}{2}\phi_0^2 - \frac{\lambda_0}{4!}\phi_0^4, \tag{12.27}$$

depending on the bare parameters and fields. This can be written as $\mathscr{L}_{\text{ren}} = \mathscr{L} + \mathscr{L}_{\text{ct}}$ where

$$\mathscr{L} = \frac{1}{2}\partial_\mu\phi\partial^\mu\phi - \frac{1}{2}m^2\phi^2 - \frac{\lambda}{4!}\mu^{4-d}\phi^4 \tag{12.28}$$

depends only on renormalized couplings and fields and the counterterms have the structure

$$\mathscr{L}_{\text{ct}} = \frac{1}{2}A(d-4)\partial_\mu\phi\partial^\mu\phi - \frac{1}{2}m^2B(d-4)\phi^2 - \frac{\lambda}{4!}\mu^{4-d}C(d-4)\phi^4. \tag{12.29}$$

The functions $A(d-4)$, $B(d-4)$ and $C(d-4)$ contain all the dependence of $\mathscr{L}_{\text{ren}}$ on the regulator d and are related to the bare quantities by

$$\begin{aligned}
\phi_0(x) &\equiv \sqrt{Z_\phi(d-4)}\phi(x) = \sqrt{1+A(d-4)}\phi(x), \\
m_0^2(d-4) &= m^2\frac{1+B(d-4)}{1+A(d-4)}, \\
\lambda_0(d-4) &= \lambda\mu^{4-d}\frac{1+C(d-4)}{[1+A(d-4)]^2}.
\end{aligned} \tag{12.30}$$

The time-ordered Green's functions of the renormalized fields computed using the Lagrangian $\mathscr{L}_{\text{ren}}$ are finite in the limit $d \to 4$. The field $\phi(x)$ interpolates between the vacuum and the one particle states and therefore the scattering amplitudes computed in terms of these Green's functions are finite as well. The renormalization conditions are then used to express the renormalized mass m and coupling constant λ in terms of measurable quantities.

We see now how this program is implemented. The first divergent Feynman diagram appears in the one-loop calculation of the two-point function

$$= \frac{1}{2}\lambda\mu^{4-d}\int\frac{d^dp}{(2\pi)^d}\frac{1}{p^2-m^2+i\varepsilon}$$

$$= \frac{1}{2}\lambda\mu^{4-d}I_1(d,m^2). \tag{12.31}$$

The factor of $\frac{1}{2}$ is a symmetry factor. We can take advantage of the calculations made in the previous section to isolate the divergent part of the diagram as $d \to 4$

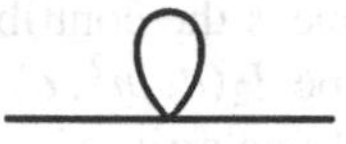

$$= -i\frac{\lambda m^2}{16\pi^2}\frac{1}{d-4} + \text{finite part.} \tag{12.32}$$

To cancel this divergence we add a counterterm $-\frac{1}{2}\delta m^2\phi^2$ to the Lagrangian density where δm^2 is given by

$$\delta m^2 = -\frac{\lambda m^2}{16\pi^2}\frac{1}{d-4}. \tag{12.33}$$

Adding this counterterm means to include in the Feynman rules a new vertex with two external legs

$$= -i\delta m^2(2\pi)^4\delta^{(4)}(p_1+p_2). \tag{12.34}$$

Its contribution to the two point function to order λ exactly cancels the divergent part of the one loop diagram (12.31). There is of course an ambiguity in the definition of the counterterm because in addition to the pole we could also have subtracted a finite part. For the time being, however, we choose not to do so.

The next divergent diagram in the ϕ^4 theory comes from the one-loop calculation of the four-point function. In fact there are three diagrams contributing at order λ^2

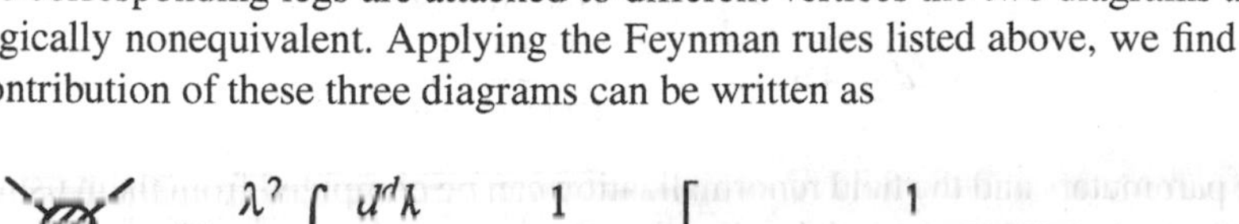

p_1 p_3 p_1 p_3 p_1 p_4

$$\equiv \quad + \quad + \tag{12.35}$$

p_2 p_4 p_2 p_4 p_2 p_3

The last two diagrams differ in a permutation of the momenta p_3 and p_4. Since the corresponding legs are attached to different vertices the two diagrams are topologically nonequivalent. Applying the Feynman rules listed above, we find that the contribution of these three diagrams can be written as

$$= \frac{\lambda^2}{2}\int\frac{d^dk}{(2\pi)^d}\frac{1}{k^2-m^2+i\varepsilon}\left[\frac{1}{(k+p_1+p_2)^2-m^2+i\varepsilon}\right.$$
$$\left.+\frac{1}{(k+p_1+p_3)^2-m^2+i\varepsilon}+\frac{1}{(k+p_1+p_4)^2-m^2+i\varepsilon}\right] \tag{12.36}$$

The $\frac{1}{2}$ in front of the integral is again a combinatorial factor associated with the symmetry of each of the three diagrams.

A look at the integrals (12.36) shows that for $d = 4$ they diverge logarithmically. To extract their divergent part we begin by exploiting the identity

$$\frac{1}{a_1 a_2} = \int_0^1 \frac{dx}{[xa_1 + (1-x)a_2]^2}, \tag{12.37}$$

where x is called a Feynman parameter. This reduces the contribution of the three diagrams to a combination of integrals of the type $I_2(d, m^2, q)$ computed above. Using the expansion of the integrals around $d = 4$ one arrives at the result

$$= \mu^{4-d}\frac{3i\lambda^2}{16\pi^2}\left(-\frac{2}{d-4} + \text{finite part}\right). \tag{12.38}$$

As with the two-point function, we can remove the divergence (12.38) by adding a counterterm $-\frac{\delta\lambda}{4!}\mu^{4-d}\phi^4$ to the Lagrangian density of the theory with

$$\delta\lambda = -\frac{3\lambda^2}{16\pi^2}\frac{1}{d-4}. \tag{12.39}$$

We can also incorporate this couterterm in the Feynman rules by adding the new vertex

$$= -i\delta\lambda(2\pi)^4\delta^{(4)}(p_1 + p_2 + p_3 + p_4). \tag{12.40}$$

We have computed the only two one-loop 1PI diagrams that are divergent. It is not difficult to show that the one-loop contribution to the six-point function is finite. Thus, our calculation determines, at one loop order, the functions appearing in the counterterm Lagrangian (12.29)

$$\begin{aligned} A(d-4)_{1-\text{loop}} &= 0, \\ B(d-4)_{1-\text{loop}} &= -\frac{\lambda}{16\pi^2}\frac{1}{d-4}, \\ C(d-4)_{1-\text{loop}} &= -\frac{3\lambda}{16\pi^2}\frac{1}{d-4}. \end{aligned} \tag{12.41}$$

The bare parameters and the field renormalization can be computed from them using Eq. (12.29). In particular, we find that $Z_\phi(d-4)_{1-\text{loop}} = 1$ and the scalar field does not get renormalized at one loop.

Having reached this point, some remarks are in order. We notice that the construction of counterterms is intrinsically ambiguous because together with the divergent part we can also subtract a finite contribution. In our analysis we just removed the pole parts without imposing any renormalization condition at a particular value of the momenta. This is called minimal subtraction (MS). Most frequently, however, one uses a modified version of MS denoted by $\overline{\text{MS}}$ consisting in subtracting, together with the pole, also the term $\gamma - \log(4\pi)$. This results in a simplification of the expressions.

Both MS and $\overline{\text{MS}}$ are examples of *mass independent subtraction schemes* [3–5]. They owe their name to the fact that by subtracting just the pole at $d = 4$ (plus maybe other numerical constants) we get counterterms that are independent of any mass scale of the theory. As we will see very soon, this feature is very convenient for computational purposes and their use is crucial in the construction of effective field theories to be presented later.

In the study of the renormalization of QED carried out in Chap. 8 we learned that it is enough to consider the contribution of the 1PI irreducible diagrams, since all other diagrams can be written in terms of them. Here the situation is exactly the same, and we can focus our attention on the correlation functions obtained by summing all amputated 1PI Feynman diagrams with n external legs and incoming momenta $p_1, \ldots, p_n$. The correlation functions computed from the Lagrangian $\mathscr{L} + \mathscr{L}_{\text{ct}}$ are finite in the limit $d \to 4$, since the counterterms are constructed to cancel all divergences. These bare 1PI correlation functions depend on the bare parameters m_0 and λ_0 but are independent of the energy scale μ, which does not enter in (12.27).

The bare 1PI correlation functions can be written as [cf. (8.90) and the associated discussion]

$$\Gamma_n(p_i; m_0, \lambda_0, d-4)_0 = Z_\phi(d-4)^{-\frac{n}{2}}\Gamma_n(p_i; m, \lambda, \mu, d-4), \tag{12.42}$$

where in the right-hand side we have the correlation function computed with the renormalized fields. This depends only on the renormalized couplings m and λ, as well as on μ. Since the bare parameters do not depend on μ, the left-hand side is independent of this arbitrary scale, while the μ-dependence of the right-hand side is both explicit and implicit through the renormalized parameters. Taking the derivative with respect to μ we can write the analog of the renormalization equations discussed in Chap. 8

$$\left[\mu\frac{\partial}{\partial\mu} + \beta\left(\lambda, \frac{m}{\mu}, d-4\right)\frac{\partial}{\partial\lambda} + \gamma_m\left(\lambda, \frac{m}{\mu}, d-4\right)m\frac{\partial}{\partial m}\right.$$
$$\left. -n\gamma\left(\lambda, \frac{m}{\mu}, d-4\right)\right]\Gamma_n(p_i; m, \lambda, \mu, d-4) = 0. \tag{12.43}$$

Since they only involve renormalized quantities these equations are regular in the limit $d \to 4$. The functions appearing on the left-hand side are defined by

$$\beta\left(\lambda, \frac{m}{\mu}, d-4\right) = \mu\frac{\partial\lambda}{\partial\mu},$$
$$\gamma_m\left(\lambda, \frac{m}{\mu}, d-4\right) = \frac{\mu}{m}\frac{\partial m}{\partial\mu}, \tag{12.44}$$
$$\gamma\left(\lambda, \frac{m}{\mu}, d-4\right) = \frac{1}{2}\mu\frac{\partial}{\partial\mu}\log Z_\phi.$$

They measure the change of λ, m and Z_ϕ with the scale μ.

The main reason to introduce the energy scale μ was to keep the dimensions of the expressions correct. It would be interesting to trade the evolution of the renormalized couplings with respect to μ by their evolution with respect to some physically meaningful energy scale, such as the one characterizing the process under study. To this end, we rescale all momenta in the correlation functions by a common factor $p_i \to sp_i$ and study how they change with s.

The 1PI function $\Gamma_n(sp_i; \lambda, m, \mu, d-4)$ has canonical dimension

$$D_n = 4 - n - \frac{d-4}{2}(n-2). \tag{12.45}$$

This formula can be heuristically justified by dimensional analysis of the contribution of a one-loop 1PI diagram with $n = 2k$ external legs. The canonical dimension D_n determines how the correlation function changes when we change the energy units in which the dimensionful parameters p_i, m and μ are measured. In mathematical terms this means that the correlation function is a homogeneous function of weight D_n, namely

$$\Gamma_n(\xi sp_i; \lambda, \xi m, \xi\mu, d-4) = \xi^{D_n}\Gamma_n(sp_i; \lambda, m, \mu, d-4), \tag{12.46}$$

for any $\xi > 0$. Euler's well known theorem for homogeneous functions implies that

$$\left(\mu\frac{\partial}{\partial\mu} + s\frac{\partial}{\partial s} + m\frac{\partial}{\partial m} - D_n\right)\Gamma_n(sp_i; \lambda, m, \mu, d-4) = 0. \tag{12.47}$$

This equation is useful because it can be combined with Eq. (12.43) evaluated at sp_i to eliminate the derivative with respect to μ. Indeed, subtracting the two equations and taking the limit $d \to 4$ we arrive at the Callan–Symanzik equation

$$\left\{-s\frac{\partial}{\partial s} + \beta\left(\lambda, \frac{m}{\mu}\right)\frac{\partial}{\partial\lambda} + \left[\gamma_m\left(\lambda, \frac{m}{\mu}\right) - 1\right]m\frac{\partial}{\partial m}\right.$$
$$\left. + 4 - n\left[1 + \gamma\left(\lambda, \frac{m}{\mu}\right)\right]\right\}\Gamma_n(sp_i; m, \lambda, \mu) = 0. \tag{12.48}$$

It is at this point that the advantage of using a mass independent renormalization schemes becomes evident. Since the couterterms are mass independent, the functions appearing in the renormalization group equations only depend on the coupling constant λ

$$\left\{-s\frac{\partial}{\partial s}+\beta(\lambda)\frac{\partial}{\partial\lambda}+\left[\gamma_m(\lambda)-1\right]m\frac{\partial}{\partial m}\right.$$
$$\left.+4-n\left[1+\gamma(\lambda)\right]\right\}\Gamma_n(sp_i;m,\lambda,\mu)=0. \tag{12.49}$$

This fact makes it possible to formally integrate the equations in s to find how the 1PI correlation functions behaves under a simultaneous rescaling of *all* momenta. We introduce the functions $\overline{\lambda}(s)$ and $\overline{m}(s)$ solving the differential equations

$$s\frac{\partial}{\partial s}\overline{\lambda}(s)=\beta\left(\overline{\lambda}(s)\right),\qquad \frac{s}{\overline{m}(s)}\frac{\partial}{\partial s}\overline{m}(s)=\gamma_m\left(\overline{\lambda}(s)\right)-1 \tag{12.50}$$

with the initial conditions $\overline{\lambda}(1)=\lambda$, $\overline{m}(1)=m$. Since both $\beta(\lambda)$ and $\gamma_m(\lambda)$ can be computed order by order in renormalized perturbation theory, these equations completely determine the functions $\overline{\lambda}(s)$ and $\overline{m}(s)$.

Although not explicitly indicated, the function $\overline{\lambda}(s)$ also depends on the initial condition λ, whereas $\overline{m}(s)$ depends on both λ and m. This can be seen by rewriting (12.50) in integral form

$$\log s=\int_{\lambda}^{\overline{\lambda}(s)}\frac{dt}{\beta(t)},\qquad \overline{m}(s)=m\exp\left[\int_{\lambda}^{\overline{\lambda}(s)}dt\frac{\gamma_m(t)-1}{\beta(t)}\right]. \tag{12.51}$$

Differentiating with respect to λ and m we find that

$$\frac{\partial\overline{\lambda}}{\partial\lambda}=\frac{\beta(\overline{\lambda})}{\beta(\lambda)},\qquad \frac{\partial\overline{m}}{\partial m}=\frac{\overline{m}}{m},\qquad \frac{\partial\overline{m}}{\partial\lambda}=\overline{m}\left[\frac{\gamma_m(\overline{\lambda})-\gamma_m(\lambda)}{\beta(\lambda)}\right]. \tag{12.52}$$

Using these identities it is not difficult to show that the renormalization group equations (12.49) are formally solved by

$$\Gamma_n(sp_i;m,\lambda,\mu)=s^{4-n}\Gamma_n\left(p_i,\overline{m}(s),\overline{\lambda}(s),\mu\right)\exp\left[-n\int_1^s\frac{ds'}{s'}\gamma\left(\overline{\lambda}(s')\right)\right] \tag{12.53}$$

This solution gives the dependence of the 1PI Green's functions on the momentum rescaling factor s and can be used to determine their high-energy behavior by taking the limit $s\to\infty$. To get some intuition about the meaning of Eq. (12.53), we consider a massless theory sitting at a fixed point of the renormalization group flow where $\beta(\lambda^*)=0$. In this case (12.53) takes the simple form

$$\Gamma_n(sp_i;\lambda^*,\mu)=s^{4-n(1+\gamma^*)}\Gamma_n(p_i;\lambda^*,\mu). \tag{12.54}$$

This result would be the one expected from dimensional analysis in a theory where the scalar field $\phi(x)$ has energy dimensions $D_\phi=1+\gamma^*$ instead of the canonical value $D_\phi=1$. For this reason γ^* is called the anomalous dimension of the field. This name is also given by extension to the function $\gamma(\lambda, m/\mu)$ introduced in Eq. (12.44), although it should be noticed that, strictly speaking, it only admits such an interpretation at the fixed point.

12.4 The Renormalization Group Equations in Dimensional Regularization

So far we have studied some general properties of the renormalization group equations. To extract more precise information from them one needs to know the functions $\beta(\lambda)$, $\gamma(\lambda)$ and $\gamma_m(\lambda)$. This is going to be our next task, namely we are going to learn how to compute these functions using DR. The techniques we describe next were developed in [3, 4].

Our analysis is going to be as general as possible. For this reason we consider a theory characterized by a number of bare couplings denoted collectively by λ_{0k}, with $k = 1, \ldots, N$. They include coupling constants, masses and the wave function renormalization of the different fields. As prescribed by DR, we define the theory in dimension d and proceed to construct the counterterms required to cancel the divergences order by order in renormalized perturbation theory. A divergent L-loop diagram give rise to poles of the type shown in Eq. (12.17). Collecting the contributions to all orders in perturbation theory the bare couplings can be expressed as a Laurent series in $d-4$, namely

$$\lambda_{0k} = \mu^{D_k}\left[\lambda_k + \sum_{\nu=1}^{\infty}\frac{a_k^{(\nu)}(\lambda_\ell)}{(d-4)^\nu}\right], \tag{12.55}$$

where the coefficients $a_k^{(\nu)}(\lambda_\ell)$ depend on the renormalized couplings λ_ℓ. The energy dimensions D_k of the bare couplings are generically of the form

$$D_k = D_k^{(0)} + (d-4)D_k^{(1)}, \tag{12.56}$$

where $D_k^{(0)}$ is the dimension of λ_{0k} in four dimensions. The renormalized couplings have been rescaled by the appropriate powers of μ in order to make them dimensionless.

Each coefficient $a_k^{(\nu)}(\lambda_\ell)$ in the expansion (12.55) receives contributions from divergent L-loop diagrams with $L \geq \nu$. What makes these expressions still useful is that all coefficients $a_k^{(\nu)}(\lambda_\ell)$ with $\nu > 1$ can be expressed in terms of $a_k^{(1)}(\lambda_\ell)$, as we show next.

Let us differentiate Eq. (12.55) with respect to μ. Since the bare couplings are independent of this energy scale, the left hand side gives zero. The right-hand side, however, depends on μ both explicitly and implicitly through the renormalized couplings λ_k. We can take the derivative of the couplings to have the form

$$\mu \frac{\partial \lambda_k}{\partial \mu} = A_k(\lambda_\ell) + (d-4) B_k(\lambda_\ell) \tag{12.57}$$

and solve order by order in $d-4$. To begin with we get two terms proportional to $d-4$ that have to cancel. This determines the coefficient $B_k(\lambda_\ell)$ to be

$$B_k(\lambda_\ell) = -\lambda_k D_k^{(1)}. \tag{12.58}$$

Repeating this for the terms of order $(d-4)^0$ we find

$$\begin{aligned} A_k(\lambda_\ell) &= -\lambda_k D_k^{(0)} - D_k^{(1)} \left(1 - \lambda_k \frac{\partial}{\partial \lambda_k}\right) a_k^{(1)}(\lambda_\ell) \\ &\quad + \sum_{j \neq k}^{N} D_j^{(1)} \lambda_j \frac{\partial}{\partial \lambda_j} a_k^{(1)}(\lambda_\ell). \end{aligned} \tag{12.59}$$

Finally, from the coefficients of the poles at $d=4$ the following recursion relation is obtained

$$\begin{aligned} 0 &= \left(D_k^{(0)} + A_k \frac{\partial}{\partial \lambda_k}\right) a_k^{(\nu)}(\lambda_\ell) + D_k^{(1)} \left(1 - \lambda_k \frac{\partial}{\partial \lambda_k}\right) a_k^{(\nu+1)}(\lambda_\ell) \\ &\quad + \sum_{j \neq k}^{N} \left[A_j(\lambda_\ell) \frac{\partial}{\partial \lambda_j} a_k^{(\nu)}(\lambda_\ell) - D_j^{(1)} \lambda_j \frac{\partial}{\partial \lambda_j} a_k^{(\nu+1)}(\lambda_\ell) \right], \end{aligned} \tag{12.60}$$

where $A_j(\lambda_\ell)$ is given above in terms of $a_k^{(1)}(\lambda_\ell)$.

There are several interesting conclusions to be extracted from the relations just derived. First, taking the limit $d \to 4$ we find that the running of the coupling λ_k is fully determined by the coefficients of the single poles at $d=4$, namely

$$\begin{aligned} \mu \frac{\partial \lambda_k}{\partial \mu} &= -\lambda_k D_k^{(0)} - D_k^{(1)} \left(1 - \lambda_k \frac{\partial}{\partial \lambda_k}\right) a_k^{(1)}(\lambda_\ell) \\ &\quad + \sum_{j \neq k}^{N} D_j^{(1)} \lambda_j \frac{\partial}{\partial \lambda_j} a_k^{(1)}(\lambda_\ell). \end{aligned} \tag{12.61}$$

At this point we have to recall that all couplings have been made dimensionless by rescaling them with powers of μ. This can be undone by a new rescaling

$$\lambda_k \longrightarrow \mu^{-D_k^0} \lambda_k \tag{12.62}$$

that has the effect of canceling the first term on the right-hand side of Eq. (12.61). This equation becomes even simpler when using a mass independent subtraction scheme. In this case the counterterms do not depend on the mass couplings of the theory and as a consequence they do not appear in the sum on the right-hand side of (12.61).

As we already explained, generically $a_k^{(1)}(\lambda_\ell)$ receives contributions to all loops. However, as promised, the recursion relations (12.60) give a way to compute $a_k^{(\nu)}(\lambda_\ell)$ with $\nu = 2, 3, \ldots$ in terms of $a_k^{(1)}(\lambda_\ell)$. This fact provides a very convenient method to check loop computations since, for instance, the coefficient of the two-loop $1/(d-4)^2$ pole is determined by the one-loop $1/(d-4)$ pole, and so on.

We go back now to the ϕ^4 theory we studied in the previous section. At one loop we only have to worry about two couplings m^2 and λ, since there is no wave function renormalization at this order. From Eqs. (12.30) and (12.41) we find the following expression for the bare couplings at one loop

$$\begin{aligned} m_0^2 &= m^2\left(1-\frac{\lambda}{16\pi^2}\frac{1}{d-4}\right), \\ \lambda_0 &= \mu^{4-d}\left(\lambda-\frac{3\lambda^2}{16\pi^2}\frac{1}{d-4}\right). \end{aligned} \tag{12.63}$$

We have used a mass-independent scheme, and the coefficients of the Laurent expansions of the bare couplings do not depend on the renormalized mass m. This simplifies the calculation of the renormalization group functions $\beta(\lambda)$ and $\gamma_{m^2}(\lambda)$

$$\begin{aligned} \beta(\lambda) &\equiv \mu\frac{\partial\lambda}{\partial\mu} = \frac{3\lambda^2}{16\pi^2} \\ \gamma_{m^2}(\lambda) &\equiv \frac{\mu}{m^2}\frac{\partial m^2}{\partial\mu} = \frac{\lambda}{16\pi^2} \end{aligned} \tag{12.64}$$

With this result we find that the beta function vanishes at zero coupling but it is positive for $\lambda > 0$. Applying what we learned from our analysis of Sect. 8.2 we conclude that $\lambda = 0$ is an infrared fixed point of the renormalization group flow. The one-loop beta function equation in (12.64) can be integrated to give

$$\lambda(\mu) = \frac{\lambda(\mu_0)}{1-\frac{3}{16\pi^3}\lambda(\mu_0)\log\left(\frac{\mu}{\mu_0}\right)}, \tag{12.65}$$

where μ_0 is an arbitrary reference energy scale. This shows that the renormalized coupling grows with the energy μ. The coupling decreases as we go to lower energies, so perturbation theory becomes more and more reliable in this regime. This also indicates that the operator ϕ^4, which was originally marginal, becomes irrelevant once quantum corrections are included. Finally, we notice that the one-loop result (12.65) blows up at the nonperturbative energy scale

$$\mu = \mu_0 \exp\left[\frac{16\pi^3}{3\lambda(\mu_0)}\right]. \tag{12.66}$$

This Landau pole, similar to the one discussed for QED in Sect. 8.2, indicates that the theory becomes strongly coupled at high energies.

12.5 The Issue of Quadratic Divergences

We can return now to the problem of quadratic divergences introduced in Sect. 12.1. Instead of using DR we regularize the Euclidean integral (12.7) for $d = 4$ using a sharp cutoff $|p_E| < \Lambda$ and find the following leading behavior as $\Lambda \to \infty$

$$\int_{|p_E|<\Lambda} \frac{d^4 p_E}{(2\pi)^4} \frac{1}{(p_E^2 + m^2)^n} \sim \begin{cases} \frac{m^2}{8\pi^2}\left[\frac{\Lambda^2}{m^2} - \log\left(\frac{\Lambda^2}{m^2}\right)\right] & n = 1 \\ \frac{1}{8\pi^2}\left[\log\left(\frac{\Lambda^2}{m^2}\right) - \frac{1}{2}\right] & n = 2 \\ \frac{m^{4-2n}}{8\pi^2(n-1)(n-2)} & n > 2 \end{cases} \tag{12.67}$$

where we have dropped all terms that go to zero in this limit. These expressions contrasts with the DR result (12.11), where the integral diverges in the same way for $n = 1$ and $n = 2$, namely with a simple pole at $d = 4$.

The one-loop renormalization of the ϕ^4 field theory described in the previous section can now be implemented using the cutoff regularization. The cancelation of the divergent part of the diagram (12.31) gives the bare mass and coupling constant

$$\begin{aligned} m_0(\Lambda)^2 &= m^2\left\{1 - \frac{\lambda}{16\pi^2}\left[\frac{\Lambda^2}{m^2} - \log\left(\frac{\Lambda^2}{m^2}\right)\right]\right\}, \\ \lambda(\Lambda) &= \lambda\left[1 - \frac{3\lambda}{16\pi^2}\log\left(\frac{\Lambda^2}{m^2}\right)\right]. \end{aligned} \tag{12.68}$$

We can invert the first equation to write the renormalized mass in terms of the bare parameters to first order in the bare coupling constant

$$m^2 = m_0(\Lambda)^2 + \frac{\lambda_0(\Lambda)}{16\pi^2}\left[\Lambda^2 - \log\frac{\Lambda^2}{m_0(\Lambda)^2}\right]. \tag{12.69}$$

Would the scalar theory be valid for arbitrary high energies, this would be the end of the story. The cutoff Λ would be an artifact of the quantization that should disappear at the end of the calculations. Physical results would only depend on the renormalized quantities m and λ. The situation is however different if we have reasons to believe that our theory is only valid up to certain energy scale at which new physics is expected to play a role. Then Eq. (12.69) has to be interpreted in Wilsonian terms (see Sect. 8.5) by regarding Λ as the energy above which our ϕ^4 theory is replaced by some unknown new dynamics. Just below this scale the leading part of the theory (not including irrelevant operators) is defined by the Lagrangian (12.27), the effect of the high energy degrees of freedom is codified in the cutoff dependence of the bare field and parameters $\phi_0(x, \Lambda)$, $m_0(\Lambda)$ and $\lambda_0(\Lambda)$. From this point of view m and λ are the parameters characterizing the theory at energies well below the cutoff scale, $E \ll \Lambda$.

The relation (12.69) between the low energy (renormalized) mass and the high energy bare parameters shows a strong dependence of the former on the cutoff Λ. Indeed, due to the term proportional to Λ^2 the value of the mass m will be determined by the cutoff scale unless the value of $m_0(\Lambda)$ is carefully chosen to cancel the contribution of the quadratic term up to many decimal places. The conclusion is that the preservation of the hierarchy $m \ll \Lambda$ requires an important fine tuning of the mass at the cutoff scale. This is the hierarchy problem.

Any theory with fundamental scalars is afflicted with this problem, including the standard model due to the presence of the Higgs field. The only exception are theories with Nambu-Goldstone bosons. They only include derivatives couplings preserving the invariance $\phi(x) \rightarrow \phi(x) +$ constant, thus forbidding any mass term in the action.[2]

We have seen in Sect. 12.2 that using DR there are no quadratic (or polynomial) divergences. The momentum integral of the one-loop self-energy diagram (12.31) has the same divergent behavior when $d \rightarrow 4$ as the milder logarithmically divergent integral appearing in the calculation of the four-point function. In fact, quadratically divergent integrals are not signaled in DR by higher order poles at $d = 4$, but by additional poles for $d < 4$. We can see this from Eq. (12.11). For the logarithmically divergent integral $I_2(d, m^2)$ the Gamma function in the numerator has a single pole for real positive d, namely at $d = 4$. In the case of $I_1(d, m^2)$, on the other hand, $\Gamma(\frac{2-d}{2})$ has, besides the pole at $d = 4$, another one at $d = 2$. Generically [6], in the integrals arising from L-loop diagrams these additional poles occur for fractional values of the dimension, $d = 4 - \frac{2}{L}$. This is how quadratic divergences are identified using DR.

The previous discussion might lead us to believe that the hierarchy problem is a regularization artifact that can be disposed of by a smart choice of the regulator. The whole thing, however, is more complicated. Integrating the second equation in (12.64) we find

$$m(\mu)^2 = m(\mu_0)^2 \exp\left[\int_{\lambda(\mu_0)}^{\lambda(\mu)} \frac{dx}{\beta(x)} \gamma_{m^2}(x)\right]. \tag{12.70}$$

This expression shows that the mass at the scale μ is proportional to the initial condition $m(\mu_0)$. This however does not clarify the issue, since in order to decide whether the ultraviolet sensitivity of the low energy parameters persists in DR we should find out how the initial condition $m(\mu_0)$ depends on the high energy scales.

We arrive at the conclusion that in order to understand the low energy role of quadratic divergences in DR we should look into the more general problem of understanding this regularization procedure in a Wilsonian setup. Therefore we turn to a brief description of effective field theories to address systematically the question of the separation of scales and how low-energy properties can be derived in a theory with

---

[2] The other known way of canceling quadratic divergences is to have supersymmetry (see Sect. 13.2), where the quadratically divergent corrections to the scalar masses are cancelled by the contribution of diagrams with fermion loops.

natural energy scales that are widely separated. As a bonus we will clarify the role played in Physics by nonrenormalizable field theories and arrive at the formulation of a criterion for naturalness [7] in quantum field theory.

12.6 Effective Field Theories: A Brief Introduction

One of the main reasons behind our progress in the understanding of physical processes is the fact that at a given length scale, and using the correct variables, our description of the physical phenomena is to a large extend independent of the physics at much shorter distances. This is a glaring fact in the history of Physics. For example, thermodynamics was formulated well before a microscopic description of the thermal processes in terms of statistical mechanics and atomic theory was available. Similarly, a moderately accurate calculation of the energy levels of the hydrogen atom is possible without concerning ourselves with the internal structure of the proton. Details such as its spin or charge radius have indeed an effect on the hydrogen spectral lines, but these are subleading corrections.

These examples, that can be multiplied at will, illustrate the basic fact that Physics largely deals with the formulation of effective theories describing physical phenomena within a certain range of energy scales with an acceptable accuracy. This is also the case in quantum field theory. At energies below the Fermi energy $1/\sqrt{G_F}$ weak interactions can be faithfully described using the Fermi theory. Only when higher energies become available experimentally, the effective low energy description has to be replaced by (or embedded in) a more general theory that takes into account the new degrees of freedom relevant for the exploration of new phenomena.

The basic ingredients in the building of effective field theories are the light degrees of freedom and the relevant symmetries of the problem. The latter provide the guiding principle to write a Lagrangian that would be the starting point for the calculation of observables. In Sect. 8.5 we learned that the infrared physics is dominated by relevant and marginal operators. A Lagrangian constructed using only these operators defines a renormalizable theory, such as QED, QCD or ϕ^4. Observables can be computed in terms of a limited number of parameters associated to the renormalized couplings of the relevant and marginal operators in the action.

The description in terms of relevant and marginal operators is very accurate in the deep infrared region. However, if we want to include the corrections due to new physics above an energy scale M we have to include irrelevant operators. These, generically, will appear in the action suppressed by the necessary powers of the scale M at which the new degrees of freedom become excited. This is precisely what happens in the case of Fermi's theory of weak processes, where β-decay is described by the four-fermion interaction of the Fermi theory discussed in Sect. 8.5 and M is set by the Fermi energy. Another example is provided by the description of the low energy properties of leptons and light quarks. At energies $E \ll m_b, m_t$ we do not have to include the b and t quarks as dynamical fields. They make themselves noticed, however, through irrelevant operators in the effective action for the light fields suppressed by powers of $1/m_b$ and $1/m_t$.

These examples should be enough to illustrate how as we go to higher energies (i.e., as we increase the power of our "magnifying glass") effective theories are replaced by a more complete description. In general this process repeats as we increase the energy. For example, the standard model is itself expected to be embedded at higher energies in a more general theory (maybe a GUT). Going to arbitrarily high energies would require a theory of everything that is not yet available.

The presence of irrelevant operators means that effective field theories are nonrenormalizable. This might seem to be a problem. The common lore states that nonrenormalizable theories do not have predictive power. The cancelation of the divergences in loop diagrams requires the introduction of an infinite number of different couterterms. Thus, in order to calculate observables one would need to specify an infinite number of parameters associated with the infinite number of irrelevant operators generated by quantum corrections.

This argument, however, is far too naive. In effective field theories we are interested in physical phenomena taking place in a range of energies much below the scale of new physics, $E \ll M$, and the contributions of the nonrenormalizable counterterms to a physical processes are weighted by powers of E/M. As it turns out, to a given degree of accuracy, there are only a few irrelevant operators that need be taken into account in our computations. Therefore, when looked at in the right way, nonrenormalizable theories are respectable and predictive.

It is important to realize that frequently it is experimental information that forces us to introduce higher-dimensional operators in a renormalizable theory. The classic example is the discovery of neutrino masses and mixing. The simplest way to accommodate this experimental fact in the standard model would be to include for each generation a sterile right-handed neutrino that is a singlet under the standard model gauge group $\mathrm{SU}(3) \times \mathrm{SU}(2) \times \mathrm{U}(1)_Y$. This would generate Dirac mass terms for the neutrinos while preserving lepton number conservation. One can, however, take the widely accepted point of view that global symmetries such as lepton number conservation are mere accidental symmetries of the low energy theory that do not have to be preserved at high energies. In this case Majorana mass terms for the neutrinos are allowed.[3] The simplest way to generate these terms is by adding to the standard model Lagrangian the following dimension-five operator

$$\Delta\mathscr{L}_{\mathrm{SM}} = -\frac{1}{M}\sum_{i,j=1}^{3} g_{ij}\left(\overline{\mathbf{L}_i^C}\sigma^2\mathbf{H}\right)\left(\mathbf{H}^T\sigma^2\mathbf{L}_j\right) + \text{h.c.} \tag{12.71}$$

where g_{ij} are dimensionless coupling constants, $\mathbf{L}_i$ are the three lepton doublets introduced in Table 5.1, and $\mathbf{H}$ is the Higgs doublet (10.3). This term is gauge invariant,

[3] As a matter of fact, once we decide that lepton number conservation is not a fundamental symmetry we can also introduce, in addition to the Dirac masses, Majorana mass terms for the right-handed neutrinos.

as can easily be shown using the definition of the charge conjugated spinors and the identity (10.6). Since this operator has dimension five it comes suppressed by M, the energy scale at which new physics is expected. Upon symmetry breaking, the Higgs doublet develops the vacuum expectation value (10.11) and the new term generates a Majorana mass for the three neutrinos [cf. equation (11.41)]

$$\Delta\mathscr{L}_{\text{SM}} = -\frac{\mu^2}{M}\sum_{i,j=1}^{3} g_{ij}\overline{\nu_i^C}\nu_j + \text{h.c.} \tag{12.72}$$

Using the experimental value of μ and the bounds for the neutrino masses it turns out that M can be as high as $M \sim 10^{15}$ GeV $\gg m_W$. The discovery of neutrino masses may provide a hint to new physics at some high energy scale M. This would indicate that the standard model is an effective field theory, and therefore we should also include irrelevant operators to describe the effect of its ultraviolet completion.

After this long digression we study some basic features of effective field theories. Detailed introductions to the subject can be found in Ref. [8–11]; here we follow mainly the presentations of [12–14]. To illustrate our discussion we consider two unphysical toy models that however contain all the main features of more realistic effective field theories. The first is a non-renormalizable theory of a single Dirac spinor with a four-fermion interaction

$$\mathscr{L} = \overline{\psi}(i\partial\!\!\!/ - m)\psi - \frac{a}{\Lambda^2}(\overline{\psi}\psi)^2 + \cdots \tag{12.73}$$

where a is a dimensionless coupling and the dots stand for higher-dimensional operators that we ignore. Using this Lagrangian we can study the effect of loop corrections induced by non-renormalizable interaction. These are the ones that, allegedly, would render the theory non-predictive. From our previous discussion we know that Λ sets the energy scale at which our nonrenormalizable Lagrangian should be completed with new degrees of freedom. We quantize the theory using this scale as a cutoff. The Feynman rules contain a single four-fermion vertex

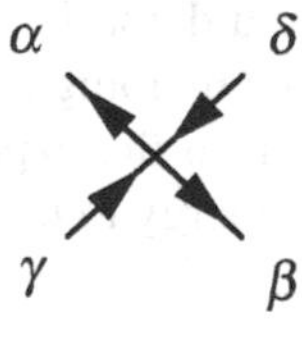

$$\alpha,\ \delta,\ \gamma,\ \beta \quad = \frac{2ia}{\Lambda^2}\Big(\delta_{\alpha\delta}\delta_{\beta\gamma} - \delta_{\alpha\gamma}\delta_{\beta\delta}\Big), \tag{12.74}$$

and the only one-loop diagram contributing to the fermion self-energy is given by

$$
\begin{aligned}
-i\Sigma_{\alpha\beta}(\not{p},\Lambda) &= \alpha \quad\quad \beta \\
&= -\frac{6am}{\Lambda^2}\delta_{\alpha\beta}\int^{\Lambda}\frac{d^4q}{(2\pi)^4}\frac{1}{q^2-m^2+i\varepsilon}.
\end{aligned}
\tag{12.75}
$$

As explained in Chap. 8 the physical mass of the fermion is defined by the zero of $\not{p} - m - \Sigma(\not{p}, \Lambda)$. Since the one-loop fermion self-energy is independent of the momentum, the mass correction is simply given by

$$
\begin{aligned}
\delta m &= -\frac{6iam}{\Lambda^2}\int^{\Lambda}\frac{d^4q}{(2\pi)^4}\frac{1}{q^2-m^2+i\varepsilon} = -\frac{3am}{4\pi^2}\left[1+\frac{m^2}{\Lambda^2}\log\left(\frac{\Lambda^2}{m^2}+1\right)\right] \\
&\sim -\frac{3am}{4\pi^2} \quad (\text{when}\, m \ll \Lambda).
\end{aligned}
\tag{12.76}
$$

To compute the integral we have performed a Wick rotation, as explained in Sect. 12.2, and integrated the Euclidean momentum in the range $|q_E| < \Lambda$. We have found that the leading correction to the mass is of order $(m/\Lambda)^0$, i.e., it is not suppressed by powers of Λ. This is not a peculiarity of the dimension-six operator chosen here. It also occurs for other higher-dimension operators.

This is a disastrous result. It leads to the conclusion that the fermion mass gets corrections from higher-dimensional operators that are not suppressed by the scale Λ and therefore are large even when we are considering energies much below the scale of new physics. What we are saying is that the value of low energy parameters is strongly influenced by what is going on at arbitrarily high energies. Thus, in order to compute the corrections to the mass of the fermion in our theory we would need to know the details of the dynamics at energies above the scale Λ.

The reason behind our failure in separating low energy physics from the details of the theory at high energies lies in the fact that we did not renormalize the theory. Instead of cutting off the integrals at the scale Λ we are going to regularize it using DR and a mass independent subtraction scheme. From the expressions derived in previous sections we can compute the fermion self-energy to be

$$
\begin{aligned}
\Sigma(\not{p}, \Lambda) &= -\frac{6iam}{\Lambda^2}\int\frac{d^dq}{(2\pi)^d}\frac{1}{q^2-m^2+i\varepsilon} \\
&= -\frac{3am}{8\pi^2}\left(\frac{m}{\Lambda}\right)^2\left[\frac{2}{d-4}+\gamma+\log\left(\frac{m^2}{4\pi\mu^2}\right)+\cdots\right].
\end{aligned}
\tag{12.77}
$$

In the $\overline{\text{MS}}$ subtraction scheme we add a counterterm that cancels the pole in $d = 4$ together with the constants $\gamma - \log(4\pi)$. Then we find the following correction to the fermion mass

$$\delta m = -\frac{3am}{8\pi^2}\left(\frac{m}{\Lambda}\right)^2 \log\left(\frac{m^2}{\mu^2}\right). \tag{12.78}$$

This is a much nicer result. The mass correction is suppressed by powers of m/Λ, small in the regime where the effective field theory is applicable, $m \ll \Lambda$. In addition, the expression only depends logarithmically on μ. This energy scale is an artifact of the regularization and therefore should be absent of all physical quantities.

It is a general result that in a mass independent subtraction scheme, effective field theories produce a well defined expansion in powers of m/Λ or E/Λ, where E is the characteristic energy of the process under consideration. This means that to a given numerical accuracy only a few terms in the expansion should be considered. It is in this sense that effective field theories can be considered as predictive as renormalizable quantum field theories.

The reader might be puzzled at the comparison of the different results we have obtained for the mass renormalization using a cutoff, and DR plus a mass independent subtraction scheme. In fact, there is no contradiction between them. Physical predictions cannot depend on the way we choose to regularize and renormalize the theory. Cutting off the integrals at the scale Λ results in an infinite number of contributions to each order in $1/\Lambda$. Were we able to resum these terms we would obtain a result agreeing with the expression found using a mass independent scheme. The latter method provides a systematic way of organizing the $1/\Lambda$ contributions. As a consequence there is only a finite number of operators contributing to a given degree of accuracy.

Before closing our discussion of the Lagrangian (12.73) we mention the fact that the mass correction (12.78) is proportional to the mass m and therefore vanishes for $m = 0$. This apparently innocuous fact has a deep underlying explanation based on a symmetry enhancement of the theory at $m = 0$. Indeed, in the massless case both the kinetic term and the four-fermion interaction are invariant under the discrete chiral transformation

$$\psi \longrightarrow \gamma_5\psi, \quad \overline{\psi} \longrightarrow -\overline{\psi}\gamma_5. \tag{12.79}$$

This is however not a symmetry of the mass term, changing sign under it. Thus, the theory at $m = 0$ has an addition symmetry protecting the fermion from acquiring a mass through quantum corrections. This is why, in general, one can say that having a light fermion is *natural* in spite of the presence of a large energy scale Λ in the theory.

The second example we want to study is a renormalizable theory of two interacting real scalar fields with masses m and M, with $m \ll M$ and Lagrangian

$$\mathscr{L} = \frac{1}{2}\partial_\mu\phi\partial^\mu\phi - \frac{m^2}{2}\phi^2 + \frac{1}{2}\partial_\mu\Phi\partial^\mu\Phi - \frac{M^2}{2}\Phi^2 - \frac{g}{2}\phi^2\Phi. \tag{12.80}$$

From the inspection of the interaction term we find that the Feynman rules contain the vertex

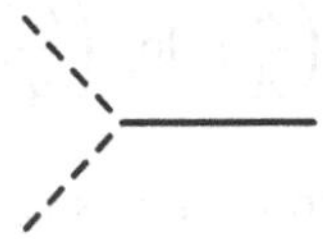

$$= -ig, \tag{12.81}$$

where the dashed and continuous lines represent respectively the light and heavy fields. The propagators are the usual ones for scalar fields with the appropriate values of the mass. To find the leading correction to the mass of the light field due to the heavy field, we consider the following diagram

$$\begin{aligned} &= -g^2\mu^{4-d}\int \frac{d^dq}{(2\pi)^d}\frac{i}{q^2-m^2+i\varepsilon}\frac{i}{(q+p)^2-M^2+i\varepsilon}. \\ &= g^2\mu^{4-d}\int_0^1 dx\, I_2\Big(d, xm^2+(1-x)M^2-x(1-x)p^2\Big). \end{aligned} \tag{12.82}$$

To write the second equality we have colleted together the two propagators using the trick (12.37), thus reducing the expression to a single integral of the type $I_2(d,m^2)$.

Let us take $m = 0$. To compute the leading correction to the mass it is enough to evaluate the previous diagram at zero momentum. An explicit calculation of the integral around $d = 4$ gives

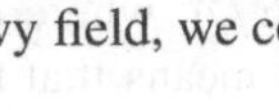

$$\Bigg|_{p^2=m^2=0} = -\frac{ig^2}{16\pi^2}\left[\frac{2}{d-4}+\gamma+\log\left(\frac{M^2}{4\pi\mu^2}\right)+\cdots\right], \tag{12.83}$$

and in the $\overline{\text{MS}}$ subtraction scheme the mass corrections is found to be

$$\delta m^2 = \frac{g^2}{16\pi^2}\log\left(\frac{M^2}{\mu^2}\right), \tag{12.84}$$

which is nonzero even if we set $m = 0$. This simple calculation illustrates the important point that the mass of the scalar field is not protected against quantum corrections. The interaction with the heavy scalar produces a correction to the mass whose scale is set by the dimensionful coupling constant g, while depending only logarithmically on the mass of the heavy scalar. This means that having a scalar with mass well below $g/(4\pi)$ is *unnatural*.

A similar result would be obtained in the case of a light scalar field $\phi(x)$ coupled to a heavy fermion $\psi(x)$ of mass M through the Yukawa interaction

$$\mathscr{L}_{\text{int}} = g'\phi\overline{\psi}\psi. \tag{12.85}$$

Now, since g' is dimensionless, the scalar field mass m gets quantum corrections whose scale is set by the mass of the heavy fermion. This can be seen by computing the one-loop fermion correction to the scalar two-point function. Taking again $m = 0$ and using DR and the $\overline{\text{MS}}$ subtraction scheme the scalar acquires a mass of order

$$\delta m^2 \sim g'^2 M^2 \log\left(\frac{M^2}{\mu^2}\right). \tag{12.86}$$

Therefore there is no natural way to keep the mass of the scalar away from the scale of the heavy fermion.

The theories defined by the Lagrangians (12.73) and (12.80) show that while chiral symmetry makes light fermions *natural*, it is very difficult is to prevent scalar fields from acquiring masses of the order of the higher energy scales in the theory. This illustrates, in a toy model, the hierarchy problem: unlike fermions, low energy scalars are extremely sensitive to high energy scales.

These examples help also to clarify the issue of quadratic divergences in DR. We have seen that, in spite of the apparent absence of quadratic divergences in the Feynman integrals, the scalar masses in general are still quadratically sensitive to high energy scales. The hierarchy problem is therefore not an artifact of the regularization procedure.

12.7 Remarks on Naturalness

There are a number of hints, such as neutrinos masses and oscillations, indicating that the standard model is an effective theory. Since light scalars are unnatural in the presence of higher energy scales, it is necessary to explain what keeps the value of the Higgs mass below the scale 10^{15}–10^{19} GeV where new physics is expected, unless we are willing to accept an unnatural fine tuning of the Higgs mass. This can be rephrased as the fact that the ratio between the GUT or Planck mass scale (i.e., the scale of "new physics" M_{NP}) and the standard model energy scale $M_{\text{SM}} \sim 100$ GeV is a large number

$$\frac{M_{\text{NP}}}{M_{\text{SM}}} \sim 10^{13} - 10^{16}. \tag{12.87}$$

The problem of accounting for large numbers has been around for a long time. Dirac [15] was worried about the emergence of large numbers in Physics, like the ratio between the strengths of the electromagnetic and gravitational interactions of protons and electrons. In a more modern context we can construct a dimensionless ration between the Fermi and Newton constants

$$\frac{G_F c^2}{G_N \hbar^2} \simeq 1.73 \times 10^{33}. \tag{12.88}$$

It is to Dirac's credit that he did not invoke any anthropic explanation. In his large number hypothesis he assumed that all these large dimensionless ratios should be related in a simple way to a single large number which he chose to be the age of the Universe. This led him to conclude that the fundamental constants of Nature vary with time.

We do not want to dwell any further on this subject. Excelent expositions of the notion of naturalness in high energy physics are available in the literature (see, for example, [16, 17]). In view, however, of the examples discussed in the previous section we find it necessary to state a *naturalness criterion* that probably most physicists would find acceptable:

> At any energy scale μ, a physical parameter or a set of physical parameters $\alpha_i(\mu)$ is allowed to be very small only if the replacement $\alpha_i(\mu) = 0$ would increase the symmetry of the system.

This criterion, originally formulated by Wilson [18] and further elaborated among others by 't Hooft [7] and Susskind [19], has been a guide for nearly four decades in the construction of theories beyond the standard model. The fact that the Higgs particle, if thought as an elementary scalar, has not yet been found adds a good deal of drama associated with naturalness.

This naturalness criterion may apply to particle physics, but in the broader context where gravity is included it is severely violated. In our discussion of effective field theories we have systematically forgotten the identity operators which, having zero dimension, should be dominant in the infrared. The reason why we could afford to ignore this operator so far is that we were not considering gravitational effects. The coupling of the identity operator receives contributions from the zero-point energy of all the quantum fields and, as long as gravity is left out of the game, can be simply ignored.

General relativity teaches us that all forms of energy gravitate, and this applies also to the zero-point energy of the quantum fields. Therefore once gravitational effects are considered there is no way to ignore the coupling of the identity operator to the gravitational field. This term is Einstein's famous cosmological constant Λ_c. Its contribution to the energy density of the Universe

$$\rho_\Lambda = \frac{\Lambda_c}{8\pi G_N} \tag{12.89}$$

can be measured from cosmological observations with the result

$$\rho_\Lambda \simeq (10^{-3}\,\mathrm{eV})^4 = 10^{-48}\,\mathrm{GeV}^4. \tag{12.90}$$

On the other hand, since ρ_Λ has dimensions of $(\text{energy})^4$ and the only natural energy scale in gravity is the Planck mass $M_P \sim 10^{19}$ GeV, naturalness would require the scale of ρ_Λ to be set by M_P, that is

$$\rho_\Lambda \sim M_P^4 \sim 10^{76}\,\mathrm{GeV}^4. \tag{12.91}$$

This means that there is a mismatch of more than 120 orders of magnitude between the natural and the measured value of the cosmological constant. To add to the puzzle, if we compare ρ_Λ with the cosmological critical density at the present time

$$\rho_c = \frac{3H_0^2}{8\pi G_N}, \tag{12.92}$$

we find the two values to be very close, $\rho_\Lambda \simeq 0.74\rho_c$.

Is it just serendipity that the measured value of the cosmological constant is so close to the Universe's critical density today? In view of this it is difficult to avoid the question of whether naturalness should only apply to particle physics in the absence of gravity. In fact, the problem of the apparent fine tuning of the cosmological constant is 60 or 70 orders of magnitude worse than the one for the Higgs mass. This brings in the question: why should gravity be excluded in naturalness arguments?

It is safe to say that currently nobody knows the answer to these questions. For the Higgs naturalness problems some scenarios have been suggested: supersymmetry and technicolor among others, that are likely to be tested soon. In the case of the cosmological constant, apart from anthropic arguments [20] or the string landscape [21, 22], there is very little to say. Large numbers are likely to continue haunting particle physicists and cosmologists for some time to come.

12.8 Coda: Heavy Particles and Decoupling

We have appraised mass independent subtraction schemes as the appropriate way to deal with the renormalization of effective field theories. They have, however, an important disadvantage: heavy particles do not decouple at energies below their masses, as expected from the Appelquist-Carazzone decoupling theorem [23].

A simple example showing this is provided by the calculation in QED of the contribution of a heavy fermion to the photon vacuum polarization (more details can be found in [13, 14]). The basic ingredient is the DR computation of the one loop polarization tensor

$$\mu \sim\!\bigcirc\!\sim \nu = \left(p_\mu p_\nu - p^2\eta_{\mu\nu}\right)\Pi(p^2;d-4) \tag{12.93}$$

where the polarization function $\Pi(p^2; d-4)$ is given by

$$\Pi(p^2;d-4) = -\frac{e^2}{12\pi^2}\left\{\frac{2}{d-4} + \gamma_E - \log(4\pi) + 6\int_0^1 dx\, x(1-x)\log\left[\frac{m_f^2 - x(1-x)p^2}{\mu^2}\right]\right\}. \tag{12.94}$$

Following the analysis of Sect. 8.3, the divergence in the previous expression is subtracted by adding to the QED Lagrangian a counterterm of the form $-\frac{1}{4}C(d-4)F_{\mu\nu}F^{\mu\nu}$. In the $\overline{\text{MS}}$ scheme we obtain

$$\Pi(p^2)_{\overline{\rm MS}} = -\frac{e^2}{2\pi^2}\int_0^1 dx\, x(1-x)\log\left[\frac{m_f^2 - x(1-x)p^2}{\mu^2}\right]. \tag{12.95}$$

At leading order in the renormalized charge e the beta-function can be computed as

$$\beta(e) = \frac{e}{2}\mu\frac{\partial}{\partial\mu}\Pi(p^2)_{\overline{\rm MS}} = \frac{e^3}{12\pi^2} \quad (\overline{\rm MS}\ \ \text{scheme}). \tag{12.96}$$

Alternatively, the same result can be obtained from the expression of the bare charge coupling contant

$$e_0 = e\mu^{\frac{4-d}{2}}\left(1 - \frac{e^2}{12\pi^2}\frac{1}{d-4}\right) + \text{finite part} \tag{12.97}$$

by applying the techniques introduced in Sect. 12.4. Since in a mass independent subtraction scheme the beta function is determined solely by the pole at $d = 4$, it does not depend on the value of the fermion mass. This is surprising because, on physical grounds, one would expect the fermion to decouple in the limit of large mass $m_f \to \infty$. In this limit the theory should have a vanishing beta function.

We repeat now the calculation of the beta function but using a mass dependent scheme. In particular we work in the so-called μ-scheme where the counterterm coefficient $C(d-4)$ is chosen in such a way that its contribution cancels the diagram (12.93) evaluated at the Euclidean momentum $p^2 = -\mu^2$. The renormalized polarization tensor is obtained by adding the contribution of the counterterm to the one-loop diagram, with the result

$$\Pi(p^2)_\mu = -\frac{e^2}{2\pi^2}\int_0^1 dx\, x(1-x)\log\left[\frac{m_f^2 - x(1-x)p^2}{m_f^2 + x(1-x)\mu^2}\right], \tag{12.98}$$

where the subscript μ indicates that we are dealing with the renormalized polarization function in the μ-scheme. The calculation of the beta function then gives

$$\begin{aligned}\beta(e) &= \frac{e}{2}\mu\frac{\partial}{\partial\mu}\Pi(p^2)_\mu \\ &= \frac{e^3}{2\pi^2}\int_0^1 dx\left[\frac{x^2(1-x)^2\mu^2}{m_f^2 + x(1-x)\mu^2}\right] \quad (\mu\text{-scheme}).\end{aligned} \tag{12.99}$$

Unlike the result (12.96) this beta function depends on the quotient between the fermion mass m_f and the subtraction point μ. In particular, when $m_f \ll \mu$ we find that $\beta(e)$ approaches the value (12.96), whereas in the opposite limit $m_f \gg \mu$ it tends to zero quadratically

$$\beta(e) \simeq \frac{e^3}{60\pi^2}\left(\frac{\mu}{m_f}\right)^2. \tag{12.100}$$

This is what we expect physically: a heavy fermion decouples from the low energy theory which asymptotically becomes a theory of free photons with vanishing beta function.

To understand why the $\overline{\text{MS}}$ subtraction scheme (or any mass-independent scheme for that matter) renders an incorrect result for the beta function of QED in the limit of large fermion mass we have to look at the renormalized polarization function (12.95). When the momentum goes below the fermion mass, $p^2 \ll m_f^2$, it approaches the value

$$\Pi(p^2)_{\overline{\text{MS}}} \simeq \frac{e^2}{72\pi^2} \log\left(\frac{m_f^2}{\mu^2}\right). \tag{12.101}$$

When $m_f \gg \mu$ this logarithm is large and, as a result, the perturbative expansion breaks down at low energies. This is the reason why the result obtained for the beta function is not reliable in this regime. The problem is absent in the mass dependent scheme used above, where the renormalized polarization tensor (12.98) vanishes in the limit of large fermion mass

$$\Pi(p^2)_\mu \simeq -\frac{e^2}{60\pi^2}\left(\frac{p^2+\mu^2}{m_f^2}\right) \longrightarrow 0. \tag{12.102}$$

The limit of heavy fermion mass is amenable to perturbation theory and the beta function can be reliably computed in this limit.

The way to deal with heavy particles in mass independent schemes is by integrating them out as we move down the energy ladder. At energies below the mass of a particle we have to use an effective field theory including only the light degrees of freedom at the corresponding scale, while the effects of the heavy fields are felt through higher-dimension operators. It is important to bear in mind that both the high and the low energy theories have the same light particle content, so they share the same infrared properties. They are however different in the ultraviolet, where the dynamics of the heavy particle distorts the high energy behavior of the effective field theory.

It is crucial that the description provided by the two theories be consistent at the threshold energy set by the mass of the particle that is being integrated out. This means, for example, that the scattering amplitudes of light particles cannot depend on whether we compute them using one theory or the other. The way to proceed is to match the Feynman graphs computed from the low energy effective field theory with the corresponding one-light-particle irreducible diagrams in the high energy theory. These are those Feynman graphs having only light particles on the external legs and that cannot be disconnected by cutting an internal light-particle line. These matching conditions implement the effects of heavy particles and high energy modes in the low energy effective field theory.

To summarize, the discussion carried out in this section shows that in the $\overline{\text{MS}}$ subtraction scheme, or any other mass-independent scheme, the decoupling of particles as we run from high to low energies has to be implemented by hand, integrating out the field that become heavy as we lower the energy. Thus, every time a particle threshold is found, the corresponding field has to be integrated out and the appropriate matching conditions on the low energy field theory imposed. Proceeding systematically in this way, we guarantee the correct decoupling of the heavy species while retaining the computational advantages of a mass independent scheme.

References

1. 't Hooft, G., Veltman, M.J.G.: Regularization and renormalization of gauge fields. Nucl. Phys. B **44**, 189 (1972)
2. Bollini, C.G., Giambiagi, J.J.: Dimensional renormalization: the number of dimensions as a regularizing parameter. Nuovo Cim. B **12**, 20 (1972)
3. 't Hooft, G.: Dimensional regularization and the renormalization group. Nucl. Phys. B **61**, 455 (1973)
4. 't Hooft, G.: The renormalization group in quantum field theory. In: 't Hooft, G. (ed.) Under the Spell of the Gauge Principle. World Scientific, Singapore (1994)
5. Weinberg, S.: New approach to the renormalization group. Phys. Rev. D **8**, 3497 (1973)
6. Veltman, M.J.G.: The infrared-ultraviolet connection. Acta Phys. Polon. B **12**, 437 (1981)
7. 't Hooft, G.: Naturalness, chiral symmetry and spontaneous chiral symmetry breaking. In: 't Hooft, G. (ed.) Under the Spell of the Gauge Principle. World Scientific, Singapore (1994)
8. Georgi, H.: Effective field theory. Ann. Rev. Nucl. Part. Sci. **43**, 209 (1993)
9. Polchinski, J.: Effective field theory and the fermi surface. In: Harvey, J., Polchinski, J. (eds.) Recent Directions in Particle Theory: from superstrings and Black Holes to the Standard Model. World Scientific, Singapore (1993) [arXiv:hep-th/9210046]
10. Kaplan, D.: Five Lectures on Effective Field Theory, Lectures Delivered at the 17th National Nuclear Physics Summer School, Berkeley (2005) [arXiv:nucl-th/0510023]
11. Burgess, C.P.: Introduction to effective field theory. Ann. Rev. Nucl. Part. Sci. **57**, 329 (2007) [arXiv:hep-th/0701053]
12. Kaplan, D.B.: Effective field theories, lectures at the 7th Summer School in Nuclear Physics, Seattle (1995) [arXiv:nucl-th/9506035]
13. Manohar, A.V.: Effective field theories. In: Latal, H., Schweiger, W. (eds.) Perturbative and Nonperturbative Aspects of Quantum Field Theory. Springer, Berlin (1997) [arXiv:hep-ph/9606222]
14. Pich, A.: Effective field theory. In: Gupta, R., Morel, A., de Rafael, E., David, F. (eds.) Probing The Standard Model Of Particle Interactions. Elsevier, Amsterdam (1999) [arXiv:hep-ph/9806303]
15. Dirac, P.A.M.: A new basis for cosmology. Proc. Roy. Soc. A **165**, 199 (1938)
16. Nelson, P.: Naturalness in theoretical physics. Am. Sci. **73**, 60 (1985)
17. Giudice, G.F.: Naturally speaking: the naturalness criterion and physics at the LHC. In: Kane, G., Pierce, A. (eds.) Perspectives on LHC Physics. World Scientific, Singapore (2008) [arXiv:0801.2562 [hep-ph]]
18. Wilson, K.G.: The renormalization group and strong interactions. Phys. Rev. D **3**, 1818 (1971)
19. Susskind, L.: Dynamics of spontaneous symmetry breaking in the Weinberg–Salam theory. Phys. Rev. D **20**, 2619 (1979)
20. Weinberg, S.: Anthropic bound on the cosmological constant. Phys. Rev. Lett. **59**, 2607 (1987)
21. Susskind, L.: The anthropic landscape of string theory. In: Carr, B. (ed.) Universe or Multiverse? Cambridge University Press, Cambridge 2007 [arXiv:hep-th/0302219]
22. Polchinski, J.: The cosmological constant and the string landscape. In: Gross, D., Henneaux, M., Sevrin, A. (eds.) The Quantum Structure of Space and Time. World Scientific, Singapore (2007) [arXiv:hep-th/0603249]
23. Appelquist, T., Carazzone, J.: Infrared singularities and massive fields. Phys. Rev. D **11**, 2856 (1975)

Chapter 13
Special Topics

In this closing chapter we have decided to present a few special topics in quantum field theory that have applications in cosmology and particle phenomenology. Given that we have not covered so many things in this book, the number of subjects to choose from is vast. Our choice is to study the creation of particles by classical external sources, including the Schwinger effect: the creation of electron–positron pair in a strong electric field; and then to explore the general properties of supersymmetric theories. Currently a large fraction of theories beyond the standard model are based on various supersymmetric completions. The reader will find the most rudimentary properties of such theories and the basic representation of this new symmetry.

13.1 Creation of Particles by Classical Fields

Particle Creation by a Classical Source

In a free quantum field theory the total number of particles is a conserved quantity. For example, in the case of the quantum scalar field studied in Chap. 2 we have that the number operator commutes with the Hamiltonian

$$\hat{n} \equiv \int \frac{d^3k}{(2\pi)^3} \frac{1}{2E_{\mathbf{k}}} \alpha^{\dagger}(\mathbf{k})\alpha(\mathbf{k}), \quad [\hat{H}, \hat{n}] = 0. \tag{13.1}$$

This means that any states with a well-defined number of particle excitations will preserve this number at all times. The situation, however, changes as soon as interactions are introduced, since in this case particles can be created and/or destroyed as a result of the dynamics.

Another case in which the number of particles might change is if the quantum theory is coupled to a classical source. The archetypical example of such a situation is the Schwinger effect, in which a classical strong electric field produces the creation of electron–positron pairs out of the vacuum. However, before plunging into this more

L. Álvarez-Gaumé and M. Á. Vázquez-Mozo, *An Invitation to Quantum Field Theory*, Lecture Notes in Physics 839, DOI: 10.1007/978-3-642-23728-7_13,

involved situation we can illustrate the relevant physics involved in the creation of particles by classical sources with the help of the simplest example: a free scalar field theory coupled to a classical external source $J(x)$. The action for such a theory can be written as

$$S = \int d^4x \left[\frac{1}{2}\partial_\mu\phi(x)\partial^\mu\phi(x) - \frac{m^2}{2}\phi(x)^2 + J(x)\phi(x)\right], \tag{13.2}$$

where $J(x)$ is a real function of the coordinates. Its identification with a classical source is obvious once we calculate the equations of motion

$$\left(\partial_\mu\partial^\mu + m^2\right)\phi(x) = J(x). \tag{13.3}$$

Our plan is to quantize this theory but, unlike the case analyzed in Chap. 2, now the presence of the source $J(x)$ makes the situation a bit more involved. The general solution to the equation of motion can be written in terms of the retarded Green function for the Klein–Gordon equation as

$$\phi(x) = \phi_0(x) + i\int d^4x' G_R(x - x')J(x'), \tag{13.4}$$

where $\phi_0(x)$ is a general solution to the homogeneous equation and

$$\begin{aligned} G_R(t, \mathbf{x}) &= \int \frac{d^4k}{(2\pi)^4}\frac{i}{k^2 - m^2 + i\varepsilon\text{sign}(k^0)}e^{-ik\cdot x} \\ &= i\theta(t)\int\frac{d^3k}{(2\pi)^3}\frac{1}{2E_{\mathbf{k}}}\left(e^{-iE_{\mathbf{k}}t+\mathbf{k}\cdot\mathbf{x}} - e^{iE_{\mathbf{k}}t-i\mathbf{k}\cdot\mathbf{x}}\right), \end{aligned} \tag{13.5}$$

with $\theta(x)$ the Heaviside step function. The denominator in the first integral is reminding us that the integration contour over k^0 surrounds the poles at $k^0 = \pm E_{\mathbf{k}}$ from above. Since $G_R(t, \mathbf{x}) = 0$ for $t < 0$, the function $\phi_0(x)$ corresponds to the solution of the field equation at $t \to -\infty$, before the interaction with the external source.[1]

To make the argument simpler we assume that $J(x)$ is switched on at $t = 0$, and only lasts for a time τ, that is

$$J(t, \mathbf{x}) = 0 \quad \text{if } t < 0 \text{ or } t > \tau. \tag{13.6}$$

We are interested in a solution of (13.3) for times after the external source has been switched off, $t > \tau$. In this case the expression (13.5) can be written in terms of the Fourier modes $\tilde{J}(E, \vec{k})$ of the source as

$$\begin{aligned} \phi(t, \mathbf{x}) = \phi_0(t, \mathbf{x}) + i\int\frac{d^3k}{(2\pi)^3}\frac{1}{2E_{\mathbf{k}}}\Big[&\tilde{J}(E_{\mathbf{k}}, \mathbf{k})e^{-iE_{\mathbf{k}}t+i\mathbf{k}\cdot\mathbf{x}} \\ &- \tilde{J}(E_{\mathbf{k}}, \mathbf{k})^* e^{iE_{\mathbf{k}}t-i\mathbf{k}\cdot\mathbf{x}}\Big]. \end{aligned} \tag{13.7}$$

---

[1] We could have taken instead the advanced propagator $G_A(x)$ in which case $\phi_0(x)$ would correspond to the solution to the equation at large times, after the interaction with $J(x)$.

The general solution $\phi_0(t, \mathbf{x})$ has been already computed in Eq. (2.55). Combining this result with Eq. (13.7) we find the following expression for the late time general solution to the Klein–Gordon equation in the presence of the source

$$\phi(t,x) = \int \frac{d^3k}{(2\pi)^3}\frac{1}{2E_{\mathbf{k}}}\left\{\left[\alpha(\mathbf{k}) + i\tilde{J}(E_{\mathbf{k}},\mathbf{k})\right]e^{-iE_{\mathbf{k}}t+i\mathbf{k}\cdot\mathbf{x}} + \left[\alpha^*(\mathbf{k}) - i\tilde{J}(E_{\mathbf{k}},\mathbf{k})^*\right]e^{iE_{\mathbf{k}}t-i\mathbf{k}\cdot\mathbf{x}}\right\}. \tag{13.8}$$

On the other hand, for $t < 0$ we find from Eqs. (13.4) and (13.5) that the general solution is given by Eq. (2.55).

Now we can proceed to quantize the theory. The conjugate momentum $\pi(x) = \partial_0\phi(x)$ can be computed from Eqs. (2.55) and (13.8). Imposing the canonical equal time commutation relations (2.52) we find that $\alpha(\mathbf{k})$, $\alpha^\dagger(\mathbf{k})$ satisfy the creation-annihilation algebra (2.29). From our previous calculation we find that for $t > \tau$ the expansion of the operator $\phi(x)$ in terms of the creation-annihilation operators $\alpha(\mathbf{k})$, $\alpha^\dagger(\mathbf{k})$ can be obtained from the one for $t < 0$ by the replacement

$$\begin{aligned}\alpha(\mathbf{k}) &\longrightarrow \beta(\mathbf{k}) \equiv \alpha(\mathbf{k}) + i\tilde{J}(E_{\mathbf{k}},\mathbf{k}),\\ \alpha^\dagger(\mathbf{k}) &\longrightarrow \beta^\dagger(\mathbf{k}) \equiv \alpha^\dagger(\mathbf{k}) - i\tilde{J}(E_{\mathbf{k}},\mathbf{k})^*.\end{aligned} \tag{13.9}$$

Since $\tilde{J}(E_{\mathbf{k}},\mathbf{k})$ is a c-number, the operators $\beta(\mathbf{k})$, $\beta^\dagger(\mathbf{k})$ satisfy the same algebra as $\alpha(\mathbf{k})$, $\alpha^\dagger(\mathbf{k})$ and therefore can be interprcted as well as a set of creation-annihilation operators. This means that we can define two vacuum states, $|0_-\rangle$, $|0_+\rangle$ associated with both sets of operators

$$\left.\begin{aligned}\alpha(\mathbf{k})|0_-\rangle &= 0\\ \beta(\mathbf{k})|0_+\rangle &= 0\end{aligned}\right\} \quad \text{for all } \mathbf{k}. \tag{13.10}$$

For an observer at $t < 0$, $\alpha(\mathbf{k})$ and $\alpha^\dagger(\mathbf{k})$ are the natural set of creation-annihilation operators in terms of which to expand the field operator $\phi(x)$. After the usual zero-point energy subtraction the Hamiltonian is given by [cf. 2.59]

$$\hat{H}^{(-)} = \frac{1}{2}\int \frac{d^3k}{(2\pi)^3}\alpha^\dagger(\mathbf{k})\alpha(\mathbf{k}) \tag{13.11}$$

and the ground state of the spectrum for this observer is the vacuum $|0_-\rangle$. At the same time, a second observer at $t > \tau$ will also see a free scalar quantum field (the source has been switched off at $t = \tau$), and consequently will expand ϕ in terms of the second set of creation-annihilation operators $\beta(\mathbf{k})$, $\beta^\dagger(\mathbf{k})$. In terms of these operators the Hamiltonian is written as

$$\hat{H}^{(+)} = \frac{1}{2}\int \frac{d^3k}{(2\pi)^3}\beta^\dagger(\mathbf{k})\beta(\mathbf{k}). \tag{13.12}$$

Then for this late-time observer the ground state of the Hamiltonian is the second vacuum state $|0_+\rangle$.

In our analysis we have been working in the Heisenberg picture, where states are time-independent and the time dependence is in the operators. This means that the states of the theory and defined globally in time. Suppose now that the system is in the "in" ground state $|0_-\rangle$. An observer at $t < 0$ will find that there are no particles

$$\hat{n}^{(-)}|0_-\rangle = 0. \tag{13.13}$$

However the late-time observer will find that the state $|0_-\rangle$ contains an average number of particles given by

$$\langle 0_-|\hat{n}^{(+)}|0_-\rangle = \int \frac{d^3k}{(2\pi)^3}\frac{1}{2E_{\mathbf{k}}}\left|\tilde{J}(E_{\mathbf{k}},\mathbf{k})\right|^2. \tag{13.14}$$

Moreover, $|0_-\rangle$ is no longer the ground state for the "out" observer. On the contrary, this state has a vacuum expectation value for $\hat{H}^{(+)}$

$$\langle 0_-|\hat{H}^{(+)}|0_-\rangle = \frac{1}{2}\int \frac{d^3k}{(2\pi)^3}\left|\tilde{J}(E_{\mathbf{k}},\mathbf{k})\right|^2. \tag{13.15}$$

The key to understand what is going on here lies in the fact that the external source breaks the invariance of the theory under space-time translations. In the particular case we have studied here where $J(x)$ has support on a finite time interval $0 < t < \tau$, this implies that the vacuum is not invariant under time translations, so observers at different times will make different choices of vacua that will not necessarily agree with each other. This is clear in our example. An observer in $t < 0$ will choose the vacuum to be the lowest energy state of her Hamiltonian, $|0_-\rangle$. On the other hand, the second observer at late times $t > \tau$ will naturally choose $|0_+\rangle$ as the vacuum. For this second observer, the state $|0_-\rangle$ is not the vacuum of his Hamiltonian, but an excited state that is a superposition of states with well-defined number of particles. In this sense it can be said that the external source has the effect of creating particles out of the "in" vacuum. Besides, this breaking of time translation invariance produces a violation in the energy conservation as we see from Eq. (13.15). Particles are created from the energy pumped into the system by the external source.

The Schwinger Effect

A typical example of creation of particles by external fields is the Schwinger effect [1] consisting in the creation of electron–positron pairs by a strong electric field. To illustrate the main physical features of this effect we use a heuristic argument based on the Dirac sea picture and the WKB approximation.

In the absence of an electric field the vacuum state of a spin-$\frac{1}{2}$ field is constructed by filling all the negative energy states as depicted in Fig. 1.2. We switch on a constant electric field $\mathscr{E}\mathbf{u}_x$ in the range $0 < x < L$. The associated electrostatic potential is taken to be

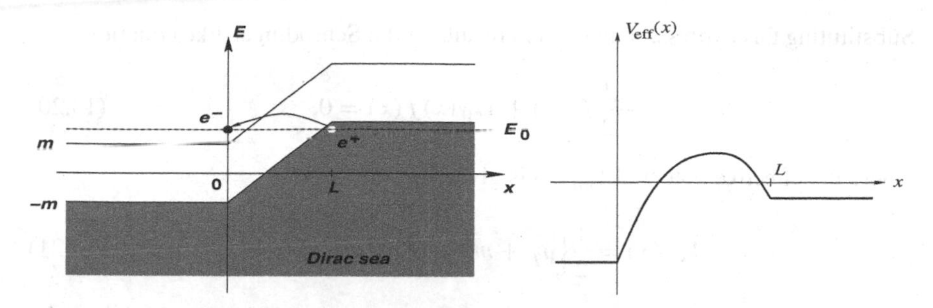

Fig. 13.1 Left: pair creation by a electric field in the Dirac sea picture. Right: effective potential felt by the electron in the x direction. The pair creation corresponds to the tunneling of the particle from the right to the left of the potential bump

$$V(\mathbf{r}) = \begin{cases} 0 & x < 0 \\ -\mathscr{E}x & 0 < x < L \\ -\mathscr{E}L & x > L \end{cases}. \tag{13.16}$$

The Dirac sea is deformed into the shape shown in the left panel of Fig. 13.1 (in drawing this figure we have to bear in mind that electrons have negative electric charge $q = -e$). When the electric field satisfies $e\mathscr{E}L > 2m$ there are states in the Dirac sea with $x > L$ having the same energy as some positive energy states in the region $x < 0$. It is therefore possible for a Dirac sea electron with energy $m \lesssim E_0 \lesssim e\mathscr{E}L - m$ to tunnel through the classically forbidden region leaving a hole behind. The physical interpretation of such process is the production of an electron–positron pair out of the vacuum by the effect of the electric field.

We can make this heuristic picture more precise with a simplified model where electrons are described by a single component wave function $\Psi(x)$ satisfying the equation[2]

$$\left\{\left[i\frac{\partial}{\partial t} + eV(x)\right]^2 + \frac{\partial^2}{\partial x^2} + \nabla_T^2 - m^2\right\}\Psi(t, x, \mathbf{x}_T) = 0. \tag{13.17}$$

This is obtained from the dispersion relation

$$(E + eV)^2 - \mathbf{p}^2 - m^2 = 0 \tag{13.18}$$

using the correspondence principle (1.2). Since the potential only depends on the coordinate x, we have separated it from the transverse coordinates denoted by $\mathbf{x}_T$. This also suggests the following ansatz for the single-particle wave function

$$\Psi(t, x, \mathbf{x}_T) = f(x)e^{-iE_0t + i\mathbf{p}_T\cdot\mathbf{x}_T}. \tag{13.19}$$

[2] Our analysis essentially ignores the effect of the spin of the electron, the two helicities being treated as scalar fields. A more careful treatment of the problem using the Dirac equation can be found in [2].

Substituting this expression in (13.17) results in the Schrödinger-like equation

$$-\frac{1}{2}f''(x) + V_{\text{eff}}(x)f(x) = 0, \tag{13.20}$$

where the effective potential $V_{\text{eff}}(x)$ is given by

$$V_{\text{eff}}(x) = \frac{1}{2}\Big\{\mathbf{p}_T^2 + m^2 - [E_0 + eV(x)]^2\Big\}. \tag{13.21}$$

This effective potential has two flat regions ($x < 0$ and $x > L$) joined by an inverted parabola, as shown in the right panel of Fig. 13.1. In the language of the Schrödinger equation (13.20) the production of particle pairs corresponds to the tunneling of an "analogue particle" of unit mass and zero energy through this potential bump. To solve the problem semiclassically we compute the classical turning points

$$x_{\pm} = \frac{1}{e\mathscr{E}}\Big(E_0 \pm \sqrt{\mathbf{p}_T^2 + m^2}\Big), \tag{13.22}$$

in terms of which the WKB transmission coefficient is given by

$$\begin{aligned} T_{\text{WKB}} &= \exp\left(-2\int_{x_-}^{x_+} dx\sqrt{2|V_{\text{eff}}(x)|}\right) \\ &= \exp\left[-2\int_{x_-}^{x_+} dx\sqrt{m^2 + \mathbf{p}_T^2 - (E_0 - e\mathscr{E}x)^2}\right]. \end{aligned} \tag{13.23}$$

The calculation of the integral yields the result

$$T_{\text{WKB}} = e^{-\frac{\pi}{e\mathscr{E}}(\mathbf{p}_T^2 + m^2)}. \tag{13.24}$$

Integrating the transmission coefficient over transverse momenta gives the number of pairs produced per unit time and unit transverse volume with energies between E_0 and $E_0 + dE$

$$\begin{aligned} \frac{dN}{dt\, d^2x_T} &= 2e^{-\frac{\pi m^2}{e\mathscr{E}}}\left(\frac{dE}{2\pi}\right)\int \frac{d^2p_T}{(2\pi)^2}e^{-\frac{\pi}{e\mathscr{E}}\mathbf{p}_T^2} \\ &= \frac{e\mathscr{E}}{2\pi^2}e^{-\frac{\pi m^2}{e\mathscr{E}}}\left(\frac{dE}{2\pi}\right), \end{aligned} \tag{13.25}$$

where the factor of 2 takes into account the two polarizations of the electron. To find the production rate per unit volume we notice that in the tunneling picture the turning points $x_{\pm}$ are the coordinates at which the two particles of the pair are produced. Shifting the energy by dE results in a change in the positions of the particles by

$dx = \frac{dE}{e\mathscr{E}}$. Using this relation in Eq. (13.25) we find the pair production rate per unit volume to be

$$W = \frac{e^2\mathscr{E}^2}{4\pi^3 c\hbar^2} e^{-\frac{\pi m^2 c^3}{\hbar e \mathscr{E}}}, \tag{13.26}$$

where we have restored the powers of $\hbar$ and c.

The production of electron–positron pairs is exponentially suppressed for "weak" electric fields. This suppression ceases when the exponent becomes of order one, i.e., when the electric field reaches the critical value

$$\mathscr{E}_{\text{crit}} = \frac{m^2 c^3}{\hbar e} \simeq 1.3 \times 10^{16}\,\text{V}\,\text{cm}^{-1}. \tag{13.27}$$

This is indeed a very strong electric field which is extremely difficult to generate in a laboratory. The Schwinger effect can also be produced by time-varying electric fields [3]. It is expected that pair production could be observed in the strong alternating fields produced by lasers.

In QED the decay of the vacuum into electron–positron pairs induced by an external field can be computed from the imaginary part of the effective action $\Gamma[A_\mu]$ in the presence of a classical gauge potential A_μ. In terms of path integrals this quantity is defined by

$$e^{i\Gamma[A_\mu]} \equiv \int \mathscr{D}\overline{\Psi}\mathscr{D}\Psi\, e^{i\int d^4x \overline{\Psi}(i\partial\!\!\!/ - m - eA\!\!\!/)\Psi}. \tag{13.28}$$

Expanding the integrand in powers of the electric charge e gives the diagrammatic expansion

$$i\Gamma[A_\mu] \equiv \quad + \quad + \quad + \ldots \tag{13.29}$$

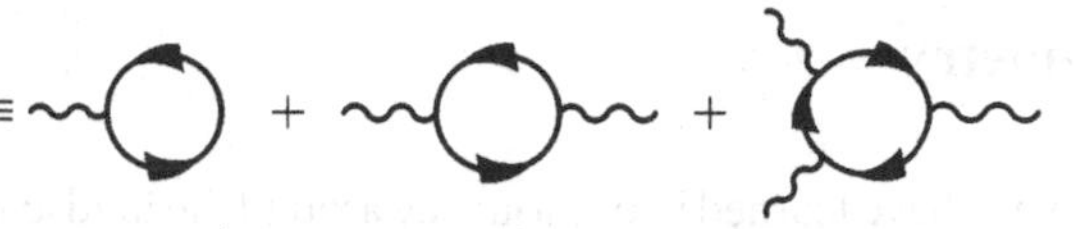

$$= \log\det\left(1 - eA\!\!\!/ \frac{1}{i\partial\!\!\!/ - m}\right).$$

The determinant can be computed using standard heat kernel techniques [1, 3]. The probability of pair production is proportional to the imaginary part of $i\Gamma[A_\mu]$ and gives Schwinger's result

$$W = \frac{e^2\mathscr{E}^2}{4\pi^3} \sum_{n=1}^{\infty} \frac{1}{n^2} e^{-n\frac{\pi m^2}{e\mathscr{E}}}. \tag{13.30}$$

Comparing this with (13.26) we see that our semiclassical analysis only captured the leading term in (13.30). The subleading contributions can also be obtained semiclassically by taking into account the probability of production of several particle pairs, i.e. the tunneling of more than one electron through the barrier.

Here we have illustrated the creation of particles by semiclassical sources in quantum field theory using simple examples. Our results can be summarized as follows: in Minkowski spacetime quantum fields have a vacuum state invariant under the Poincaré group. This, together with the covariance of the theory under Lorentz transformations, implies that all inertial observers agree on the number of particles contained in a quantum state. Coupling the theory to a space- or time-varying external source results in the vacuum not being invariant under space(time) translations. The consequence is that it is no longer possible to define a state which would be recognized as the vacuum by all observers.

This is also the case when fields are quantized on curved backgrounds. If the background is time-dependent (as it happens in a cosmological setup or for a collapsing star) different observers will identify different vacuum states: what one observer calls the vacuum will contain particles for a different one. This is what is behind the phenomenon of Hawking radiation [4]. The emission of particles by a physical black hole formed by gravitational collapse follows from the fact that what an observer in the asymptotic past would identify as the vacuum is full of particles for an observer in the asymptotic future. Thus, a particle detector located far away from the black hole detects a stream of thermal radiation with temperature

$$T_{\text{Hawking}} = \frac{\hbar c^3}{8\pi G_N k M}, \tag{13.31}$$

where M is the mass of the black hole, G_N is Newton's constant and k is Boltzmann's constant. As in the case of the Schwinger effect, particle creation by black holes can be heuristically understood as resulting from quantum tunneling of particles through the barrier created by the black hole gravitational potential [5].

13.2 Supersymmetry

One of the things that we have learned in our journey around the landscape of quantum field theory is that our knowledge of the fundamental interactions in Nature is based on the idea of symmetry, and in particular gauge symmetry. The Lagrangian of the standard model can be written just including all possible renormalizable terms (i.e. with canonical dimension smaller o equal to 4) compatible with the gauge symmetry $\mathrm{SU}(3) \times \mathrm{SU}(2) \times \mathrm{U}(1)_Y$ and Poincaré invariance. All attempts to go beyond start with the question of how to extend the symmetries of the standard model.

As explained in Sect. 6.1, in a quantum field theoretical description of the interaction of elementary particles the basic observable quantity to compute is the scattering or S-matrix giving the probability amplitude for the scattering of a number of incoming particles with a certain momentum into some final products

$$S(\text{in} \to \text{out}) = \langle p_1', \ldots; \text{out}|p_1, \ldots; \text{in}\rangle. \tag{13.32}$$

An explicit symmetry of the theory has to be necessarily a symmetry of the S-matrix. Hence it is fair to ask what is the largest symmetry of the S-matrix.

Let us ask this question in the simple case of the scattering of two particles with incoming four-momenta p_1 and p_2 described by the graph

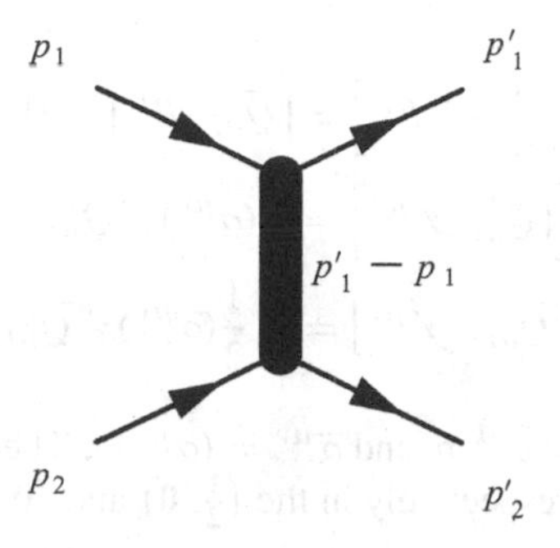

We will make the usual assumptions regarding positivity of the energy and analyticity of the S-matrix. Invariance of the theory under the Poincaré group implies that the amplitude can only depend on the scattering angle ϑ through the square of the transferred momentum $p'_1 - p_1$

$$\begin{aligned} t = (p'_1 - p_1)^2 &= 2\left(m_1^2 - p_1 \cdot p'_1\right) \\ &= 2\left(m_1^2 - E_1 E'_1 + |\mathbf{p}_1||\mathbf{p}'_1| \cos\vartheta\right). \end{aligned} \tag{13.33}$$

We assume now the existence of an extra symmetry with a bosonic conserved charge transforming as a tensor under the Poincaré group. Due to its tensor properties, the charge of the asymptotic states would depend nontrivially on their momentum eigenvalues. Therefore, charge conservation would restrict the scattering angle to a set of discrete values. In this case the S-matrix cannot be analytic, since it would vanish everywhere except for the discrete values selected by the extra symmetry.[3] Thus, the condition of having nontrivial scattering implies that the conserved charges associated with internal symmetries cannot transform as tensors under the Poincaré group (this result is the Coleman–Mandula theorem).

One possible way to extend the symmetry of the theory without renouncing to the analyticity of the scattering amplitudes is to introduce "fermionic" symmetries, i.e. symmetries whose generators are anticommuting objects [6–8]. This means that in addition to the generators of the Poincaré group P^μ, $\mathcal{J}^{\mu\nu}$ and the ones for the internal gauge symmetries G, we can introduce a number of fermionic generators $Q^I_a, \overline{Q}_{\dot{a}I} (I = 1, \ldots, \mathcal{N})$, where $\overline{Q}_{\dot{a}I} = (Q^I_a)^\dagger$. The most general algebra that these generators satisfy is the $\mathcal{N}$-extended supersymmetry algebra [9]

$$\begin{aligned} \{Q^I_a, \overline{Q}_{\dot{b}J}\} &= 2\sigma^\mu_{a\dot{b}} P_\mu \delta^I_J, \\ \{Q^I_a, Q^J_b\} &= 2\varepsilon_{ab} \mathcal{Z}^{IJ}, \\ \{\overline{Q}^I_{\dot{a}}, \overline{Q}^J_{\dot{b}}\} &= 2\varepsilon_{\dot{a}\dot{b}} \overline{\mathcal{Z}}^{IJ}, \end{aligned} \tag{13.34}$$

---

[3] Technically this is correct only if the additional symmetry is additive in the incoming and outgoing Hilbert spaces. If this additivity is violated, then the conclusion does not hold.

where $\mathscr{Z}^{IJ} \in \mathbb{C}$ commute with any other generator and satisfies $\mathscr{Z}^{IJ} = -\mathscr{Z}^{JI}$. We also have the commutators determining the Poincaré transformations of the fermionic generators Q^I_a, $Q_{\dot{a}J}$

$$\begin{aligned}
\left[Q^I_a, P^\mu\right] &= \left[\overline{Q}_{\dot{a}I}, P^\mu\right] = 0, \\
\left[[Q^I_a, \mathscr{J}^{\mu\nu}\right] &= \frac{1}{2}(\sigma^{\mu\nu})_a{}^b Q^I_b, \\
\left[\overline{Q}_{\dot{a}I}, \mathscr{J}^{\mu\nu}\right] &= -\frac{1}{2}(\overline{\sigma}^{\mu\nu})_{\dot{a}}{}^{\dot{b}}\overline{Q}_{\dot{b}I},
\end{aligned} \tag{13.35}$$

where $\sigma^{0i} = -i\sigma_i$, $\sigma^{ij} = \varepsilon^{ijk}\sigma_k$ and $\overline{\sigma}^{\mu\nu} = (\sigma^{\mu\nu})^\dagger$. These identities simply mean that Q^I_a, $\overline{Q}_{\dot{a}J}$ transform respectively in the $\left(\frac{1}{2}, \mathbf{0}\right)$ and $\left(\mathbf{0}, \frac{1}{2}\right)$ representations of the Lorentz group.

We know that the presence of a global symmetry in a theory implies that the spectrum can be classified in multiplets with respect to that symmetry. In the case of supersymmetry we start with $\mathscr{N} = 1$ where there is a single pair of supercharges Q_a, $\overline{Q}_{\dot{a}}$ satisfying the algebra

$$\{Q_a, \overline{Q}_{\dot{b}}\} = 2\sigma^\mu_{a\dot{b}} P_\mu, \quad \{Q_a, Q_b\} = \{\overline{Q}_{\dot{a}}, \overline{Q}_{\dot{b}}\} = 0. \tag{13.36}$$

Notice that in the $\mathscr{N} = 1$ case there is no possibility of having central charges.

We study the representations of the supersymmetry algebra (13.36), starting with the massless case. Given a state $|k\rangle$ satisfying $k^2 = 0$, we can always find a reference frame where the four-vector k^μ takes the form $k^\mu = (E, 0, 0, E)$. Since the theory is Lorentz covariant we can obtain the representation of the supersymmetry algebra in this frame where the expressions are simpler. In particular, the right-hand side of the first anticommutator in Eq. (13.36) is given by

$$2\sigma^\mu_{a\dot{b}} P_\mu = 2(P^0 - \sigma^3 P^3) = \begin{pmatrix} 0 & 0 \\ 0 & 4E \end{pmatrix}. \tag{13.37}$$

Therefore the algebra of supercharges in the massless case reduces to

$$\begin{aligned}
\{Q_1, Q_1^\dagger\} &= \{Q_1, Q_2^\dagger\} = 0, \\
\{Q_2, Q_2^\dagger\} &= 4E.
\end{aligned} \tag{13.38}$$

The commutator $\{Q_1, Q_1^\dagger\} = 0$ implies that the action of Q_1 on any state gives a zero-norm state of the Hilbert space $\|Q_1|\Psi\rangle\| = 0$. If we want the theory to preserve unitarity we must eliminate these null states from the spectrum. This is equivalent to setting $Q_1 \equiv 0$. On the other hand, in terms of the second generator Q_2 we can define the operators

$$a = \frac{1}{2\sqrt{E}} Q_2, \quad a^\dagger = \frac{1}{2\sqrt{E}} Q_2^\dagger, \tag{13.39}$$

which satisfy the algebra of a pair of fermionic creation-annihilation operators, $\{a, a^\dagger\} = 1$, $a^2 = (a^\dagger)^2 = 0$. Starting with a vacuum state $a|\lambda\rangle = 0$ with helicity λ we can build the massless multiplet

$$|\lambda\rangle, \quad \left|\lambda + \frac{1}{2}\right\rangle \equiv a^\dagger |\lambda\rangle. \tag{13.40}$$

Here we consider two important cases:

- Scalar multiplet: we take the vacuum state to have zero helicity and positive parity $|0^+\rangle$ so the multiplet consists of a scalar and a helicity-$\frac{1}{2}$ state

$$|0^+\rangle, \quad \left|\frac{1}{2}\right\rangle \equiv a^\dagger |0^+\rangle. \tag{13.41}$$

  This multiplet is not invariant under the CPT transformation which reverses the sign of the helicity of the states. In order to have a CPT-invariant theory we have to add to this multiplet its CPT-conjugate which can be obtained from a vacuum state with helicity $\lambda = -\frac{1}{2}$

$$|0^-\rangle, \quad \left|-\frac{1}{2}\right\rangle. \tag{13.42}$$

  Putting them together we can combine the two zero helicity states with the two fermionic ones into the degrees of freedom of a complex scalar field and a Weyl (or Majorana) spinor.
- Vector multiplet: now we take the vacuum state to have helicity $\lambda = \frac{1}{2}$, so the multiplet contains also a massless state with helicity $\lambda = 1$

$$\left|\frac{1}{2}\right\rangle, \quad |1\rangle \equiv a^\dagger \left|\frac{1}{2}\right\rangle. \tag{13.43}$$

  As with the scalar multiplet, we add the CPT conjugated obtained from a vacuum state with helicity $\lambda = -1$

$$\left|-\frac{1}{2}\right\rangle, \quad |-1\rangle, \tag{13.44}$$

  which together with (13.43) give the propagating states of a gauge field and a spin-$\frac{1}{2}$ gaugino.

In both cases we see the trademark of supersymmetric theories: the number of bosonic and fermionic states within a multiplet is the same.

In the case of extended supersymmetry we have to repeat the previous analysis for each supersymmetry charge. At the end, we have $\mathcal{N}$ sets of fermionic creation-annihilation operators $\{a^I, a_J^\dagger\} = \delta^I_J$, $(a_I)^2 = (a_I^\dagger)^2 = 0$. Let us work out the case of $\mathcal{N} = 8$ supersymmetry. Since for several reasons we do not want to have states with helicity larger than 2, we start with a vacuum state $|-2\rangle$ of helicity $\lambda = -2$. The rest of the states of the supermultiplet are obtained by applying the eight different creation operators $a_I^\dagger$ to the vacuum:

$$\begin{aligned}
&\lambda = 2 : a_1^\dagger \ldots a_8^\dagger|-2\rangle && \tbinom{8}{8} = 1 \text{ state},\\
&\lambda = \tfrac{3}{2} : a_{I_1}^\dagger \ldots a_{I_7}^\dagger|-2\rangle && \tbinom{8}{7} = 8 \text{ states},\\
&\lambda = 1 : a_{I_1}^\dagger \ldots a_{I_6}^\dagger|-2\rangle && \tbinom{8}{6} = 28 \text{ states},\\
&\lambda = \tfrac{1}{2} : a_{I_1}^\dagger \ldots a_{I_5}^\dagger|-2\rangle && \tbinom{8}{5} = 56 \text{ states},\\
&\lambda = 0 : a_{I_1}^\dagger \ldots a_{I_4}^\dagger|-2\rangle && \tbinom{8}{4} = 70 \text{ states},\\
&\lambda = -\tfrac{1}{2} : a_{I_1}^\dagger a_{I_2}^\dagger a_{I_3}^\dagger|-2\rangle && \tbinom{8}{3} = 56 \text{ states},\\
&\lambda = -1 : a_{I_1}^\dagger a_{I_2}^\dagger|-2\rangle && \tbinom{8}{2} = 28 \text{ states},\\
&\lambda = -\tfrac{3}{2} : a_{I_1}^\dagger|-2\rangle && \tbinom{8}{1} = 8 \text{ states},\\
&\lambda = -2 : |-2\rangle && \tbinom{8}{0} = 1 \text{ state}.
\end{aligned} \tag{13.45}$$

Putting together the states with opposite helicity we find that the theory contains:

- 1 spin-2 field $g_{\mu\nu}$ (a graviton),
- 8 spin-$\frac{3}{2}$ gravitino fields ψ_μ^I,
- 28 gauge fields $A_\mu^{[IJ]}$,
- 56 spin-$\frac{1}{2}$ fermions $\psi^{[IJK]}$,
- 70 scalars $\phi^{[IJKL]}$,

where by $[IJ\ldots]$ we indicated that the indices are antisymmetrized. We see that, unlike the massless multiplets of $\mathcal{N} = 1$ supersymmetry studied above, this multiplet is CPT invariant by itself. As in the case of the massless $\mathcal{N} = 1$ multiplet, here we also find as many bosonic as fermionic states:

$$\begin{aligned}
\text{bosons: } 1 + 28 + 70 + 28 + 1 &= 128 \text{ states},\\
\text{fermions: } 8 + 56 + 56 + 8 &= 128 \text{ states}.
\end{aligned}$$

Now we study briefly the case of massive representations $|k\rangle$, $k^2 = M^2$. Things become simpler if we work in the rest frame where $P^0 = M$ and the spatial components of the momentum vanish. Then, the supersymmetry algebra becomes:

$$\{Q_a^I, \overline{Q}_{\dot{b}J}\} = 2M\delta_{a\dot{b}}\delta_J^I. \tag{13.46}$$

We proceed in a similar way to the massless case by defining the operators

$$a_a^I \equiv \frac{1}{\sqrt{2M}} Q_a^I, \quad a_{\dot{a}I}^\dagger \equiv \frac{1}{\sqrt{2M}} \overline{Q}_{\dot{a}I}. \tag{13.47}$$

The multiplets are found by choosing a vacuum state with a definite spin. For example, for $\mathscr{N} = 1$ and taking a spin-0 vacuum $|0\rangle$ we find three states in the multiplet transforming irreducibly with respect to the Lorentz group:

$$|0\rangle, \quad a_{\dot{a}}^{\dagger}|0\rangle, \quad \varepsilon^{\dot{a}\dot{b}} a_{\dot{a}}^{\dagger} a_{\dot{b}}^{\dagger}|0\rangle, \tag{13.48}$$

which, once transformed back from the rest frame, correspond to the physical states of two spin-0 bosons and one spin-$\frac{1}{2}$ fermion. For $\mathscr{N}$-extended supersymmetry the corresponding multiplets can be worked out in a similar way.

The equality between bosonic and fermionic degrees of freedom is at the root of many of the interesting properties of supersymmetric theories. For example, in Sect. 3 we computed the divergent vacuum energy contributions for each real bosonic or fermionic propagating degree of freedom[4]

$$E_{\text{vac}} = \pm\frac{1}{2}\delta(\mathbf{0}) \int d^3 p \omega_{\mathbf{p}}, \tag{13.49}$$

where the sign $\pm$ corresponds respectively to bosons and fermions. Hence, for a supersymmetric theory the vacuum energy contribution exactly cancels between bosons and fermions. This boson-fermion degeneracy is also responsible for supersymmetric quantum field theories being less divergent than nonsupersymmetric ones.

References

1. Schwinger, J.: On gauge invariance and vacuum polarization. Phys. Rev. **82**, 664 (1951)
2. Wang, R.-C., Wong, C.Y.: Finite size effect in the schwinger particle production mechanism. Phys. Rev. D **38**, 348–359 (1988)
3. Brezin, E., Itzykson, C.: Pair production in vacuum by an alternating field. Phys. Rev. D **2**, 1191 (1970)
4. Hawking, S.W.: Particle creation by black holes. Commun. Math. Phys. **43**, 199 (1975)
5. Parikh, M.K., Wilczek, F.: Hawking radiation as tunneling. Phys. Rev. Lett. **85**, 5042 (2000) (hep-th/9907001)
6. Golfand, Yu.A., Likhtman, E.P.: Extension of the algebra of Poincaré group generators and violations of P-invariance. JETP Lett. **13**, 323 (1971)
7. Volkov, D.V., Akulov, V.P.: Is the neutrino a goldstone particle. Phys. Lett. B **46**, 109 (1973)
8. Wess, J., Zumino, B.: A Lagrangian model invariant under supergauge transformations. Phys. Lett. B **49**, 52 (1974)
9. Haag, R., Ł opuszański, J., Sohnius, M.: All possible generators of supersymmetries of the S-matrix. Nucl. Phys. B **88**, 257 (1975)

[4] For a boson, this can be read off Eq. (2.59). In the case of fermions, the result of Eq. (3.59) gives the vacuum energy contribution of the four real propagating degrees of freedom of a Dirac spinor.

Appendix A
Notation, Conventions and Units

For the benefit of the reader we summarize in this Appendix the main conventions used throughout the book.

A.1 Covariant Notation

We have used the "mostly minus" metric

$$\eta_{\mu\nu} = \begin{pmatrix} 1 & 0 & 0 & 0 \\ 0 & -1 & 0 & 0 \\ 0 & 0 & -1 & 0 \\ 0 & 0 & 0 & -1 \end{pmatrix}. \tag{A.1}$$

Derivatives with respect to the four-vector $x^\mu = (ct, \mathbf{x})$ are denoted by the shorthand

$$\partial_\mu \equiv \frac{\partial}{\partial x^\mu} = \left(\frac{1}{c}\frac{\partial}{\partial t}, \nabla\right).$$

Sporadically we have used the notation

$$f(x)\overleftrightarrow{\partial}_\mu g(x) = f(x)\partial_\mu g(x) - \partial_\mu f(x) g(x). \tag{A.2}$$

As usual space-time indices will be labelled by Greek letters ($\mu, \nu, \ldots = 0, 1, 2, 3$) while Latin indices will be used for spatial directions ($i, j, \ldots = 1, 2, 3$). We reserved α, β for Dirac and $a, b, c, \ldots$ for Weyl spinor indices.

The electromagnetic four-vector potential A^μ is defined in terms of the scalar φ and vector potential $\mathbf{A}$ by

$$A^\mu = (\varphi, \ \mathbf{A}). \tag{A.3}$$

L. Ávarez-Gaumé and M.Á. Vázquez-Mozo, *An Invitation to Quantum Field Theory*, Lecture Notes in Physics 839, DOI: 10.1007/978-3-642-23728-7,

The components of the field strength tensor $F_{\mu\nu} = \partial_\mu A_\nu - \partial_\nu A_\mu$ and its dual $\widetilde{F}_{\mu\nu} = \frac{1}{2}\varepsilon_{\mu\nu\sigma\lambda}F^{\sigma\lambda}$ are given respectively by

$$F_{\mu\nu} = \begin{pmatrix} 0 & E_x & E_y & E_z \\ -E_x & 0 & -B_z & B_y \\ -E_y & B_z & 0 & -B_x \\ -E_z & -B_y & B_x & 0 \end{pmatrix}, \quad \widetilde{F}_{\mu\nu} = \begin{pmatrix} 0 & B_x & B_y & B_z \\ -B_x & 0 & E_z & -E_y \\ -B_y & -E_z & 0 & E_x \\ -B_z & E_y & -E_x & 0 \end{pmatrix}, \tag{A.4}$$

with $\mathbf{E} = (E_x, E_y, E_z)$ and $\mathbf{B} = (B_x, B_y, B_z)$ the electric and magnetic fields. Similar expressions are valid in the nonabelian case.

A.2 Pauli and Dirac Matrices

We have used the notation $\sigma^\mu_\pm = (\mathbf{1}, \pm\sigma_i)$ where σ_i are the Pauli matrices

$$\sigma_1 = \begin{pmatrix} 0 & 1 \\ 1 & 0 \end{pmatrix}, \quad \sigma_2 = \begin{pmatrix} 0 & -i \\ i & 0 \end{pmatrix}, \quad \sigma_3 = \begin{pmatrix} 1 & 0 \\ 0 & -1 \end{pmatrix}. \tag{A.5}$$

They satisfy the identity

$$\sigma_i \sigma_j = \delta_{ij}\mathbf{1} + \varepsilon_{ijk}\sigma_k, \tag{A.6}$$

from where their commutator and anticommutator can be easily obtained.

Dirac matrices have always been used in the chiral representation

$$\gamma^\mu = \begin{pmatrix} 0 & \sigma^\mu_- \\ \sigma^\mu_+ & 0 \end{pmatrix}. \tag{A.7}$$

The chirality matrix is normalized as $\gamma_5^2 = \mathbf{1}$ and defined by $\gamma_5 = -i\gamma^0\gamma^1\gamma^2\gamma^3$. In many places we have used the Feynman's slash notation $\not{a} = \gamma^\mu a_\mu$.

A.3 Units

Unless stated otherwise, we work in natural units $\hbar = c = 1$. Electromagnetic Heaviside-Lorentz units have been used, where the Coulomb and Ampère laws take the form

$$\mathbf{F} = \frac{1}{4\pi}\frac{qq'}{r^3}\mathbf{r}, \quad \frac{dF}{d\ell} = \frac{1}{2\pi c^2}\frac{II'}{d}. \tag{A.8}$$

In these units the fine structure constant is

$$\alpha = \frac{e^2}{4\pi\hbar c}. \tag{A.9}$$

The electron charge in natural units is dimensionless and equal to $e \approx 0.303$.

Appendix B
A Crash Course in Group Theory

Group theory is one of the most useful mathematical tools in Physics in general and in quantum field theory in particular. To make the presentation self-contained we summarize in this Appendix some basic facts about group theory. Here we limit ourselves to the statement of basic results. Proofs and more detailed discussions can be found in the many books on the subject, such as the ones listed in Ref. [1, 2, 3, 4].

B.1 Generalities

Physical transformations have a number of interesting properties. To have an intuitive example in mind let us think of rotations in three-dimensional space. These transformations have interesting properties: if two rotations are performed in sequence the result is another one, and any rotation can be "undone".

Group theory is a way to translate these elementary properties of rotations or any other physical transformations into mathematical terms. A group G is a set of elements among which an operation $G \times G \to G$ is defined that associates to every ordered pair of elements (g_1, g_2) of the group another element, their product $g_1 g_2$. In order to be a group, the set G and the product operation have to satisfy a number of properties:

- The group product should be associative. This means that given three elements $g_1, g_2, g_3 \in G$ they satisfy $g_1(g_2 g_3) = (g_1 g_2)g_3$.
- G has a unit element $\mathbf{1}$ such that $g\mathbf{1} = \mathbf{1}g = g$ for every element g of the group.
- The group G contains together with every element $g \in G$ of the group its inverse, $g^{-1} \in G$, that satisfies the property $g^{-1}g = gg^{-1} = \mathbf{1}$.

In Physics one usually deals with group representations. These are realizations of abstract groups in terms of finite or infinite dimensional matrices. In more

L. Ávarez-Gaumé and M.Á. Vázquez-Mozo, *An Invitation to Quantum Field Theory*, Lecture Notes in Physics 839, DOI: 10.1007/978-3-642-23728-7,

technical terms, it can be said that a representation of a group G is a correspondence between its elements and the set of linear operators acting on a vector space V. This correspondence

$$D(g) : V \longrightarrow V \tag{B.1}$$

has to "mimic" the group product: given $g_1, g_2 \in G$

$$D(g_1)D(g_2) = D(g_1 g_2), \quad D(g_1^{-1}) = D(g_1)^{-1}. \tag{B.2}$$

A representation of a group is a set of operators acting of a certain vector space V. It might well happen that all these operators leave a proper subspace $U \subset V$ (i.e. $U \neq V$ and $U \neq \emptyset$) invariant, $D(g)U \subset U$ for any element $D(g)$ of the representation. When this happens it is said that the representation is reducible. A reducible representation can be decomposed into irreducible ones. These latter are the ones that satisfy that if $D(g)U \subset U$ for any element of the representations then either $U = \emptyset$ or $U = V$.

A very important result concerning irreducible representations is Schur's lemma: if *D(g)* is a irreducible representation of a group G acting on a complex vector space V, and if there is an operator $A : V \rightarrow V$ that commutes with all the elements of this representation, then A must be proportional to the identity, $A = \lambda \mathbf{1}$. Here λ is some complex number.

Schur's lemma can be a useful tool in deciding whether a representation is reducible. If given a group representation we manage to find an operator that, commuting with all elements of such representations, is not proportional to the identity this automatically implies that the representation is reducible. This criterium was used in Chap. 3 to show that Dirac spinors transform in a reducible representation of the Lorentz group.

B.2 Lie Groups and Lie Algebras

Specially interesting for their applications in quantum field theory are the Lie groups whose elements are labelled by a number of continuous parameters. In mathematical terms this means that a Lie group G can be seen as a manifold where the parameters provide a set of (local) coordinates. The simplest example of a Lie group is SO(2), the group of rotations in the plane. Each element $R(\theta)$ is labelled by the rotation angle θ, with the multiplication acting as $R(\theta_1)R(\theta_2) = R(\theta_1 + \theta_2)$. The angle θ is defined modulo 2π, therefore the manifold of SO(2) is a circumference S^1.

One of the interesting properties of Lie groups is that in a neighborhood of the identity any element can be expressed in terms of a set of generators T^A ($A = 1, \ldots, \dim G$) as

$$D(g) = \exp(-i\alpha^A T^A) \equiv \sum_{n=0}^{\infty} \frac{(-i)^n}{n!} \alpha^{A_1} \ldots \alpha^{A_n} T^{A_1} \ldots T^{A_n}, \tag{B.3}$$

where $\alpha^A \in \mathbb{C}$ are a set of coordinates of G in a neighborhood of $\mathbf{1}$. Using the general Baker-Campbell-Haussdorf formula (see for example [5], p. 81–82), the multiplication of two group elements is encoded in the value of the commutator of two generators, that in general has the form

$$[T^A, T^B] = if^{ABC}T^C, \tag{B.4}$$

where $f^{ABC} \in \mathbb{C}$ are called the structure constants. The generators can be normalized in such a way that f^{ABC} is completely antisymmetric in all its indices.

The set of generators T^A with the commutator operation (B.4) define the Lie algebra $\mathfrak{g}$ associated with the Lie group G. Hence, given a representation of the Lie algebra of generators we can construct a representation of the group by exponentiation (at least locally near the identity).

Besides their dimension, i.e. the number of generators, Lie algebras are characterized by their rank. This is defined as the maximal number of generators that commute among themselves. It is easy to see that those commuting generators form a subalgebra, called the Cartan subalgebra of the Lie algebra. The rank of a Lie algebra is therefore equal to the dimension of its Cartan subalgebra.

We illustrate these concepts with three particular examples of physical relevance.

U(1)

This is about the simplest Lie group one can imagine. Its Lie algebra consists of a single generator, T. Group elements can then be written as

$$U(\alpha) = e^{-i\alpha T}. \tag{B.5}$$

with α a real number. This group is abelian and all its irreducible representations are one-dimensional. This last result can be easily proved using Schur's lemma. This means that irreducible representations are of the form

$$D_q(\alpha) = e^{-iq\alpha}, \tag{B.6}$$

where q is a real number labeling the representation. This number is the analog of the electric charge for the U(1) gauge group of QED.

It is useful to make a distinction between noncompact and compact U(1) groups. The difference lies in the fact that in the first case α takes its values over the whole real line. For a compact U(1), on the other hand, the parameter α varies in a compact range. This latter case is realized when all irreducible representations of the group U(1) are characterized by values of q that are integer multiples of some real number q_0, i.e. $q = nq_0$ with $n \in \mathbb{Z}$. If this is the case one has

$$D_q\left(\alpha + \frac{2\pi}{q_0}\right) = D_q(\alpha) \tag{B.7}$$

for every q. This periodicity is not satisfied when the U(1) is noncompact, in which case q can take any real value.

SU(2)

The group SU(2) is well-known from the theory of angular momentum in quantum mechanics. Its Lie algebra has three generators $\{T^1, T^2, T^3\}$ that satisfy

$$[T^k, T^\ell] = i\varepsilon^{k\ell m} T^m. \tag{B.8}$$

The generators

$$T^\pm = \frac{1}{\sqrt{2}}(T^1 \pm iT^2), \quad T^3 \tag{B.9}$$

can alternatively be used to write the SU(2) Lie algebra as

$$[T^3, T^\pm] = \pm T^\pm, \quad [T^+, T^-] = T^3. \tag{B.10}$$

Either form of the algebra shows that no subset of generators is mutually commuting. Therefore the Cartan subalgebra of SU(2) can be taken to be made of a single generator that, by convention, we can take to be T^3.

Using (B.10), the irreducible representations of the Lie algebra of SU(2) can be constructed following the standard techniques familiar from quantum mechanics. They are characterized by their spin s, a nonnegative integer or half-integer, and have dimension $2s + 1$. Here we focus on two basic representations. One is the fundamental two-dimensional representation with spin $s = \frac{1}{2}$. The generators can be written in terms of the Pauli matrices as

$$T^k = \frac{1}{2}\sigma_k, \quad k = 1, 2, 3, \tag{B.11}$$

whereas finite transformations in the connected component of the identity are

$$D_{\frac{1}{2}}(\alpha^k) = e^{-\frac{i}{2}\alpha^k \sigma_k}. \tag{B.12}$$

The second representation of SU(2) that we mention here is the three-dimensional adjoint (or spin 1) representation which can be written as

$$D_1(\alpha^k) = e^{-i\alpha^k J^k}, \tag{B.13}$$

with the generators given by

$$J^1 = \begin{pmatrix} 0 & 0 & 0 \\ 0 & 0 & 1 \\ 0 & -1 & 0 \end{pmatrix}, \quad J^2 = \begin{pmatrix} 0 & 0 & -1 \\ 0 & 0 & 0 \\ 1 & 0 & 0 \end{pmatrix}, \quad J^3 = \begin{pmatrix} 0 & 1 & 0 \\ -1 & 0 & 0 \\ 0 & 0 & 0 \end{pmatrix}. \tag{B.14}$$

The J^k $(k = 1, 2, 3)$ generate rotations around the x, y and z axis respectively.

SU(3)

This group has eight generators and two basic three-dimensional irreducible representations, the fundamental and antifundamental denoted respectively by **3** and $\overline{\mathbf{3}}$. In QCD these representations are associated with the transformation of quarks and antiquarks under the color gauge symmetry SU(3). The elements of these representations can be written as

$$D_{\mathbf{3}}(\alpha^k) = e^{\frac{i}{2}\alpha^k\lambda_k}, \quad D_{\overline{\mathbf{3}}}(\alpha^a) = e^{-\frac{i}{2}\alpha^k\lambda_k^T} \quad (k = 1, \ldots, 8), \tag{B.15}$$

where λ_n are the eight hermitian Gell-Mann matrices

$$\lambda_1 = \begin{pmatrix} 0 & 1 & 0 \\ 1 & 0 & 0 \\ 0 & 0 & 0 \end{pmatrix}, \quad \lambda_2 = \begin{pmatrix} 0 & -i & 0 \\ i & 0 & 0 \\ 0 & 0 & 0 \end{pmatrix}, \quad \lambda_3 = \begin{pmatrix} 1 & 0 & 0 \\ 0 & -1 & 0 \\ 0 & 0 & 0 \end{pmatrix},$$
$$\lambda_4 = \begin{pmatrix} 0 & 0 & 1 \\ 0 & 0 & 0 \\ 1 & 0 & 0 \end{pmatrix}, \quad \lambda_5 = \begin{pmatrix} 0 & 0 & -i \\ 0 & 0 & 0 \\ i & 0 & 0 \end{pmatrix}, \quad \lambda_6 = \begin{pmatrix} 0 & 0 & 0 \\ 0 & 0 & 1 \\ 0 & 1 & 0 \end{pmatrix}, \tag{B.16}$$
$$\lambda_7 = \begin{pmatrix} 0 & 0 & 0 \\ 0 & 0 & -i \\ 0 & i & 0 \end{pmatrix}, \quad \lambda_8 = \begin{pmatrix} \frac{1}{\sqrt{3}} & 0 & 0 \\ 0 & \frac{1}{\sqrt{3}} & 0 \\ 0 & 0 & -\frac{2}{\sqrt{3}} \end{pmatrix}.$$

Hence the generators of the representations $\mathbf{3}$ and $\overline{\mathbf{3}}$ are given by

$$T^k(\mathbf{3}) = \frac{1}{2}\lambda_k, \quad T^k(\overline{\mathbf{3}}) = -\frac{1}{2}\lambda_k^T. \tag{B.17}$$

The rank of SU(3) is 2, its Cartan subalgebra being generated by T^3 and T^8.

Given a representation $D(g)$ of a group G, it is easy to see that the set of operators obtained by complex conjugation $D(g)^*$ are also a representation of the same group. In the case of a Lie group this is reflected in the fact that the generators $-(T^A)^*$ satisfy the Lie algebra relations (B.4) with the same structure constants. In fact, irreducible representations of a Lie algebra can be classified in three types, real, complex and pseudoreal, depending on whether $-(T^A)^*$ is or is not related to the original generators T^A by a similarity transformation:

- Real representations: a representation is said to be real if there is a *symmetric matrix* S which acts as intertwiner between the generators and their complex conjugates, namely

$$(T^A)^* = -ST^AS^{-1}, \quad S^T = S. \tag{B.18}$$

  This is the case of the adjoint representation of SU(2) generated by the matrices (B.14). In this example all the generators are real matrices and the intertwiner is just the identity.

- Pseudoreal representations: are the ones for which an *antisymmetric matrix* S exists with the property

$$(T^A)^* = -ST^AS^{-1}, \qquad S^T = -S. \tag{B.19}$$

As an example we can mention the spin-$\frac{1}{2}$ representation of SU(2) generated by $\frac{1}{2}\sigma_i$. The intertwiner is $S = -i\sigma_2$.

- Complex representations: finally, a representation is complex if the generators and their complex conjugates are not related by a similarity transformation. This is for instance the case of the two three-dimensional representations **3** and $\overline{\mathbf{3}}$ of SU(3).

B.3 Invariants

There are a number of invariants that can be constructed associated with an irreducible representation **R** of a Lie group G and that can be used to label such a representation. Let $T^A_{\mathbf{R}}$ be the generators in a certain representation **R** of the Lie algebra $\mathfrak{g}$. Using the antisymmetry of f^{ABC} it can be proved that the matrix $\sum_{A=1}^{\dim G} T^A_{\mathbf{R}} T^A_{\mathbf{R}}$ commutes with every generator $T^A_{\mathbf{R}}$. Therefore, according to Schur's lemma, it has to be proportional to the identity.[1] This defines the Casimir invariant $C_2(\mathbf{R})$ as

$$\sum_{A=1}^{\dim G} T^A_{\mathbf{R}} T^A_{\mathbf{R}} = C_2(\mathbf{R})\mathbf{1}. \tag{B.20}$$

A second invariant $T_2(\mathbf{R})$ associated with a representation **R** can also be defined by the identity

$$\mathrm{Tr}\, T^A_{\mathbf{R}} T^B_{\mathbf{R}} = T_2(\mathbf{R})\delta^{AB}. \tag{B.21}$$

Taking the trace in Eq. (B.20) and combining the result with (B.21) we find that both invariants are related by

$$C_2(\mathbf{R}) \dim \mathbf{R} = T_2(\mathbf{R}) \dim G, \tag{B.22}$$

with dim **R** the dimension of the representation **R**.

These two invariants appear frequently in quantum field theory calculations with nonabelian gauge fields. For example $T_2(\mathbf{R})$ comes about as the coefficient of the one-loop calculation of the beta-function for a Yang-Mills theory with gauge group G. In the case of SU(N), for the fundamental representation, we find the values

$$C_2(\mathbf{fund}) = \frac{N^2-1}{2N}, \quad T_2(\mathbf{fund}) = \frac{1}{2}, \tag{B.23}$$

[1] Schur's lemma also applies to the representations of a Lie algebra: if a representation is irreducible and there is a matrix of the same dimension as the representation that commutes with all the generators then this element has to be proportional to the identity.

whereas for the adjoint representation the results are

$$C_2(\mathbf{adj}) = N, \qquad T_2(\mathbf{adj}) = N. \tag{B.24}$$

A third invariant $A(\mathbf{R})$ is specially important in the calculation of anomalies. As discussed in Chap. 9, the chiral anomaly in gauge theories is proportional to the group-theoretical factor $\mathrm{Tr}\left[T^A_{\mathbf{R}}\{T^B_{\mathbf{R}}, T^C_{\mathbf{R}}\}\right]$. This leads us to define $A(\mathbf{R})$ as

$$\mathrm{Tr}\left[T^A_{\mathbf{R}}\{T^B_{\mathbf{R}}, T^C_{\mathbf{R}}\}\right] = A(\mathbf{R})d^{ABC}, \tag{B.25}$$

where d^{ABC} is symmetric in its three indices and does not depend on the representation. The cancellation of anomalies in a gauge theory with fermions transformed in the representation $\mathbf{R}$ of the gauge group is guaranteed if the corresponding invariant $A(\mathbf{R})$ vanishes.

It is not difficult to prove that $A(\mathbf{R}) = 0$ if the representation $\mathbf{R}$ is either real or pseudoreal. Indeed, if this is the case, then there is a matrix S (symmetric or antisymmetric) that intertwins the generators $T^A_{\mathbf{R}}$ and their complex conjugates $(T^A_{\mathbf{R}})^* = -ST^A_{\mathbf{R}}S^{-1}$. Then, using the hermiticity of the generators we can write

$$\mathrm{Tr}\left[T^A_{\mathbf{R}}\left\{T^B_{\mathbf{R}}, T^C_{\mathbf{R}}\right\}\right] = \mathrm{Tr}\left[T^A_{\mathbf{R}}\left\{T^B_{\mathbf{R}}, T^C_{\mathbf{R}}\right\}\right]^T = \mathrm{Tr}\left[(T^A_{\mathbf{R}})^*\left\{(T^B_{\mathbf{R}})^*, (T^C_{\mathbf{R}})^*\right\}\right]. \tag{B.26}$$

Now, using (B.18) or (B.19) we have

$$\begin{aligned}\mathrm{Tr}\left[(T^A_{\mathbf{R}})^*\left\{(T^B_{\mathbf{R}})^*, (T^C_{\mathbf{R}})^*\right\}\right] &= -\,\mathrm{Tr}\left[ST^A_{\mathbf{R}}S^{-1}\left\{ST^B_{\mathbf{R}}S^{-1}, ST^C_{\mathbf{R}}S^{-1}\right\}\right] \\ &= -\,\mathrm{Tr}\left[T^A_{\mathbf{R}}\left\{T^B_{\mathbf{R}}, T^C_{\mathbf{R}}\right\}\right],\end{aligned} \tag{B.27}$$

which proves that $\mathrm{Tr}\left[T^A_{\mathbf{R}}\{T^B_{\mathbf{R}}, T^C_{\mathbf{R}}\}\right] = 0$ and therefore $A(\mathbf{R}) = 0$ whenever the representation is real or pseudoreal. Since the gauge anomaly in four dimensions is proportional to $A(\mathbf{R})$, anomalies appear only when the fermions transform in a complex representation of the gauge group.

B.4 A Look at the Lorentz and Poincaré Groups

Finally, we close this Appendix with the review of some features of the Lorentz group used at several places in this book. We avoid getting into detailed proofs. They can be found in a number of textbooks (for example [6, 7]), as well as in reference [6] of Chap. 11.

The Lorentz Group

The Lorentz group SO(1,3) is defined as the group of space-time transformations that preserve the Minkowski metric, that is

$$x'^\mu = \Lambda^\mu{}_\nu x^\nu \quad \text{such that} \quad \eta_{\mu\nu}\Lambda^\mu{}_\sigma\Lambda^\nu{}_\lambda = \eta_{\sigma\lambda}. \tag{B.28}$$

From its very definition we find that $\Lambda^{\mu}{}_{\nu}$ satisfies $\det\Lambda = \pm 1$ and

$$(\Lambda^{0}{}_{0})^2 - \sum_{i=1}^{3}(\Lambda^{i}{}_{0})^2 = 1, \tag{B.29}$$

which follows from the 00 component of the second equation in (B.28). From Eq. (B.29) we find $(\Lambda^{0}{}_{0})^2 \geq 1$ and the Lorentz group can be split into the following four disconnected components

- $\mathfrak{L}^{\uparrow}_{+}$: proper, orthochronous transformations with $\det\Lambda = 1$, $\Lambda^{0}{}_{0} \geq 1$.
- $\mathfrak{L}^{\uparrow}_{-}$: improper, orthochronous transformations with $\det\Lambda = -1$, $\Lambda^{0}{}_{0} \geq 1$.
- $\mathfrak{L}^{\downarrow}_{-}$: improper, non-orthochronous transformations with $\det\Lambda = -1$, $\Lambda^{0}{}_{0} \leq -1$.
- $\mathfrak{L}^{\downarrow}_{+}$: proper, non-orthochronous transformations with $\det\Lambda = 1$, $\Lambda^{0}{}_{0} \leq -1$.

The term (non)-orthochronous refers to whether the Lorentz transformation preserves or not the direction of time. Notice that the identity is included in $\mathfrak{L}^{\uparrow}_{+}$ and therefore this is the only branch of the Lorentz group that forms a subgroup. The other three branches are connected to the orthochronous, proper Lorentz subgroup by parity and time reversal in the following way (see Chap. 11)

$$\mathfrak{L}^{\uparrow}_{+} \xrightarrow{\mathscr{P}} \mathfrak{L}^{\uparrow}_{-}, \quad \mathfrak{L}^{\uparrow}_{+} \xrightarrow{\mathscr{T}} \mathfrak{L}^{\downarrow}_{-}, \quad \mathfrak{L}^{\uparrow}_{+} \xrightarrow{\mathscr{P}\mathscr{T}} \mathfrak{L}^{\downarrow}_{+}. \tag{B.30}$$

We focus then on $\mathfrak{L}^{\uparrow}_{+}$. We are going to see that transformations in this subgroup can be written in terms of complex 2×2 matrices of unit determinant. We consider a four-vector V^{μ} and construct the Hermitian matrix

$$V = V^0\mathbf{1} + \sum_{i=1}^{3} V^i\sigma_i = \begin{pmatrix} V^0 + V^3 & V^1 - iV^2 \\ V^1 + iV^2 & V^0 - V^3 \end{pmatrix}. \tag{B.31}$$

This defines a one-to-one correspondence between four-vectors and Hermitian matrices, whose determinant gives the norm of the vector

$$\det V = \eta_{\mu\nu}V^{\mu}V^{\nu}. \tag{B.32}$$

Now, the determinant is preserved by any SL(2, $\mathbb{C}$) transformation acting as

$$V \longrightarrow AVA^{\dagger}, \quad \det A = 1. \tag{B.33}$$

Since the transformed matrix is also Hermitian, it defines a transformed four-vector V'^{μ} with the same norm. This means that the linear map (B.33) has to act on the components V^{μ} as a Lorentz transformation

$$V^{\mu} \longrightarrow V'_{\mu} = \Lambda^{\mu}{}_{\nu}(A)V^{\nu}. \tag{B.34}$$

That this Lorentz transformation belongs to $\mathfrak{L}^{\uparrow}_{+}$ can be seen as follows: the group SL(2, $\mathbb{C}$) is simply connected and the relation between SL(2,$\mathbb{C}$) and Lorentz transformations continuous. Since it includes the identity, the Lorentz transformation $\Lambda^{\mu}{}_{\nu}(A)$ has to lie in the connected component of the identity, i.e. $\mathfrak{L}_{+}^{\uparrow}$.

The correspondence between $\mathfrak{L}_{+}^{\uparrow}$ and SL(2, $\mathbb{C}$) is in fact two-to-one. This is obvious if we take into account that A and $-A$ define the same Lorentz transformation. This is why SL(2, $\mathbb{C}$) is said to be the double covering of the proper, orthochronous Lorentz group.

The relation between the Lorentz group and SL(2, $\mathbb{C}$) is very important for the definition of spinors. An undotted spinor is a two-component complex object ξ_a (with $a = 1, 2$) that under the Lorentz group transforms as

$$x^{\mu} \longrightarrow \Lambda^{\mu}{}_{\nu}(A)x^{\nu}, \quad \xi_a \longrightarrow A_a{}^b \xi_b. \tag{B.35}$$

Since the spinor ξ_a is a complex objects, its conjugate does not transform with the matrix A but with its complex conjugate A^*. Such objects are called dotted spinor. More precisely, they are two-component complex quantities $\eta_{\dot{a}}$ (with $\dot{a} = \dot{1}, \dot{2}$) that under $\mathfrak{L}_{+}^{\uparrow}$ transforms with the complex conjugate representation, namely

$$x^{\mu} \longrightarrow \Lambda^{\mu}{}_{\nu}(A)x^{\nu}, \quad \eta_{\dot{a}} \longrightarrow (A^*)_{\dot{a}}{}^{\dot{b}} \eta_{\dot{b}}. \tag{B.36}$$

Spinors with upper undotted and dotted indices are defined as objects transforming in the representations $(A^T)^{-1}$ and $(A^{\dagger})^{-1}$ respectively. In fact, these representations are equivalent to A and A^*, as can be seen from the identity

$$(A^T)^{-1} = \varepsilon A \varepsilon^{-1} \quad \text{where} \quad \varepsilon = \begin{pmatrix} 0 & 1 \\ -1 & 0 \end{pmatrix}, \tag{B.37}$$

valid for any $A \in SL(2, \mathbb{C})$. This means that indices can be raised and lowered by contraction with ε^{ab}, $\varepsilon^{\dot{a}\dot{b}}$, ε_{ab} and $\varepsilon_{\dot{a}\dot{b}}$.

Bearing in mind the previous discussion and comparing with (B.33), we see that the matrix V associated with a Lorentz four-vector has an undotted and a dotted index, $V_{a\dot{b}}$. To connect with the SU(2)×SU(2) label of the representations of the Lorentz group introduced in Chap. 3, we notice that undotted spinors correspond to Weyl spinors in the representation $(\frac{1}{2}, \mathbf{0})$. The element of SL(2, $\mathbb{C}$) associated with a Lorentz transformation characterized by a rotation $\theta\mathbf{n}$ and a boost $\boldsymbol{\beta} = (\beta^1, \beta^2, \beta^3)$ can be read from (3.14) to be

$$A = e^{-\frac{i}{2}(\theta\mathbf{n} - i\boldsymbol{\beta})\cdot\sigma}. \tag{B.38}$$

From the same equation we see that a spinor u_- in the representation $(\mathbf{0}, \frac{1}{2})$ transforms with $(A^{\dagger})^{-1}$ and therefore has an upper dotted spinor index. Thus, in the chiral representation of the γ-matrices, a Dirac spinor can be decomposed in dotted and undotted components as

$$\psi = \begin{pmatrix} \xi_a \\ \eta^{\dot{a}} \end{pmatrix}. \tag{B.39}$$

Since all other representations of the Lorentz group can be obtained by decomposing products of the two fundamental representations $(\frac{1}{2},\mathbf{0})$ and $(\mathbf{0},\frac{1}{2})$, any quantity transforming in a irreducible representation of the Lorentz group can be written as a mixed tensor

$$\Phi_{a_1\ldots a_n\dot{b}_1\ldots\dot{b}_m}, \tag{B.40}$$

where all undotted and dotted indices have to be symmetric among themselves.[2] They transform as

$$\Phi'_{a_1\ldots a_n\dot{b}_1\ldots\dot{b}_m} = A_{a_1}{}^{c_1}\ldots A_{a_n}{}^{c_n}(A^*)_{\dot{b}_1}{}^{\dot{d}_1}\ldots(A^*)_{\dot{b}_1}{}^{\dot{d}_1}\Phi_{c_1\ldots c_n\dot{d}_1\ldots\dot{d}_m}, \tag{B.41}$$

that in the language of SU(2) representations corresponds to $(\mathbf{s}_1,\mathbf{s}_2) = (\frac{n}{2},\frac{m}{2})$.

For some technical issues, such as the proof of the CPT theorem outlined in Sect. 11.6, it is necessary to study the complexification of the Lorentz group. This is defined again as in (B.28) but with $\Lambda^{\mu}{}_{\nu}$ complex. The only condition that follows from this equation now is that $\det\Lambda = \pm 1$. Therefore, unlike its real analog, the complexified Lorentz group has two connected components $\mathfrak{L}_{\pm}(\mathbb{C})$ labelled by the sign of the determinant.

Since now coordinates and four-vectors are complex as well, the matrix (B.31) associated to V^{μ} is not Hermitian. This means that

$$V \longrightarrow AVB^T, \qquad A,B \in SL(2,\mathbb{C}), \tag{B.42}$$

defines a complex Lorentz transformation

$$V^{\mu} \longrightarrow V'^{\mu} = \Lambda^{\mu}{}_{\nu}(A,B)V^{\nu}. \tag{B.43}$$

Undotted and dotted spinors transform under $\mathfrak{L}_{+}(\mathbb{C})$ with the matrices A and B belonging to the two factors of SL(2, $\mathbb{C}$)×SL(2, $\mathbb{C}$). For a general tensor (B.40) the transformation is

$$\Phi'_{a_1\ldots a_n\dot{b}_1\ldots\dot{b}_m} = A_{a_1}{}^{c_1}\ldots A_{a_n}{}^{c_n}B_{\dot{b}_1}{}^{\dot{d}_1}\ldots B_{\dot{b}_1}{}^{\dot{d}_1}\Phi_{c_1\ldots c_n\dot{d}_1\ldots\dot{d}_m}. \tag{B.44}$$

Using the same continuity arguments as for the real Lorentz group, we conclude that the correspondence $(A,B)\to\Lambda(A,B)$ defines a two-to-one isomorphism between SL(2, $\mathbb{C}$)×SL(2, $\mathbb{C}$) and the proper complex Lorentz group $\mathfrak{L}_{+}(\mathbb{C})$. The elements (A,B) and $(-A,-B)$ correspond to the same complex Lorentz transformation. An important thing achieved by the complexification of the Lorentz group is that now the space-time inversion $\mathscr{PT} : x^{\mu}\to -x^{\mu}$ lies in the connected component of the identity $\mathfrak{L}_{+}(\mathbb{C})$. This transformation acts on tensors by multiplying it by -1 for each dotted spinor,

[2] Notice that if the quantity were antisymmetric in a pair of dotted or undotted antisymmetric indices these can be eliminated using either ε_{ab} or $\varepsilon_{\dot{a}\dot{b}}$. For example, if $\Phi_{ab\ldots} = -\Phi_{ba\ldots}$ we can write $\Phi_{ab\ldots} = \varepsilon_{ab}\Phi_{\ldots}$ where $\Phi_{\ldots} = \frac{1}{2}\varepsilon^{cd}\Phi_{cd\ldots}$.

$$\mathscr{P}\mathscr{T} : \Phi_{a_1 \ldots a_n \dot{b}_1 \ldots \dot{b}_m} \longrightarrow (-1)^m \Phi_{a_1 \ldots a_n \dot{b}_1 \ldots \dot{b}_m}. \tag{B.45}$$

The Poincaré Group

The Poincaré group $\mathfrak{P}$ is the Lorentz group supplemented by space-time translations

$$\mathfrak{P} : x^\mu \longrightarrow \Lambda^\mu{}_\nu x^\nu + a^\mu. \tag{B.46}$$

The group has ten generators: six of the Lorentz group, $\mathscr{J}_{\mu\nu}$, plus the four of space-time translation, P^μ. In addition to (3.5) its Lie algebra contains the commutators

$$[P_\mu, P_\nu] = 0, \quad [\mathscr{J}_{\mu\nu}, P_\sigma] = i\eta_{\mu\sigma} P_\nu - i\eta_{\nu\sigma} P_\mu. \tag{B.47}$$

Each element of the Poincaré group is labelled by a Lorentz transformation and a four-vector. The restriction of the Lorentz transformations to the proper subgroup $\mathfrak{L}_+^\uparrow \approx SL(2, \mathbb{C})$ defines the proper Poincaré subgroup $\mathfrak{P}_+$.

The unitary irreducible representations of the Poincaré group are labelled by two Casimir operators. The first one is constructed from the generator of translations as

$$M^2 = P_\mu P^\mu. \tag{B.48}$$

The second one is defined by

$$W^2 = W_\mu W^\mu, \tag{B.49}$$

where W^μ is the Pauli-Lubański vector

$$W^\mu = \frac{1}{2} \varepsilon^{\mu\nu\sigma\lambda} J_{\nu\sigma} P_\lambda. \tag{B.50}$$

The representations are classified according to the sign of M in the following three classes:

- *Timelike or massive representations* ($M^2 > 0$). The representation acts on a linear space whose basis we take to be eigenstates of the translation operator P^μ with eigenvalue p^μ. Since $p_\mu p^\mu = M^2 > 0$, we can choose a reference frame where the eigenvalue takes the form $p^\mu = (M, \mathbf{0})$. Then, the Pauli-Lubański vector acting on these states has the form $W^\mu = (0, M\mathbf{J})$, with $\mathbf{J}$ the generator of spatial rotations. The rotation group generated by $\mathbf{J}$ defines the *little group*, i.e. the group preserving the form of the eigenvalue p^μ. The Casimir operator is easily computed to be

$$W^2 = -M^2 s(s+1), \tag{B.51}$$

where s is the spin that takes positive integer or half-integer values. Notice that the second Casimir operator W^2 is a Lorentz scalar and therefore its value is independent of the particular system of coordinates used.

- *Light-like or massless representations* ($M^2 = 0$). We work again in a basis of eigenstates of P^μ. Since the eigenvalues satisfy $p^\mu p_\mu = 0$ the wise choice of reference frame is one where $p^\mu = (M, 0, 0, M)$. It takes a little bit of algebra to check that the transformations preserving this vector are generated by J_3, $K_1 + J_2$ and $K_2 - J_1$. Working out their commutation relations we find that they generate the two-dimensional euclidean group ISO(2) of rotations and translations in a plane. Its unitary finite dimensional representations are one-dimensional and labelled by the eigenvalue of J_3, the helicity, that takes values $\lambda = 0, \pm 1/2, \ldots$ If we want the representation to preserve CPT, we need to include together the positive and negative eigenvalues of J_3. Therefore, the representation associated with a massless particle contains the helicities λ and $-\lambda$. This is the reason why photons or other massless particles come only in two helicity states.
- *Space-like or tachyonic representations* ($M^2 < 0$). There are no known particles transforming under this class of representations. Therefore we will not elaborate on them.

Unitary irreducible representations of the Poincaré group are determined by the eigenvalue of P^2 and the irreducible representation of the corresponding little group [i.e., SO(3) $\approx$ SU(2) for massive and ISO(2) for massless representations]. What we usually call a particle is a state that transforms in one of these irreducible representations, which comes labelled by its mass and spin/helicity.

References

1. Sattinger, D.H., Weaver, O.L.: Lie Groups and Algebras with Applications to Physics, Geometry, and Mechanics. Springer (1986)
2. Georgi, H.: Lie Algebras in Particle Physics, 2nd edn. Perseus Books (1999)
3. Hall, B.C.: Lie Groups, Lie Algebras, and their representations. Springer (2003)
4. Ramond, P.: Group Theory. A Physicist's Survey. Cambridge (2010)
5. Galindo, A., Pascual, P.: Quantum Mechanics I. Springer (1990)
6. Cornwell, J.F.: Group Theory in Physics, vol. II. Academic (1984)
7. Tung, W.-K.: Group Theory in Physics. World Scientific (1985)

Index

L. Ávarez-Gaumé and M.Á. Vázquez-Mozo, *An Invitation to Quantum Field Theory*,
Lecture Notes in Physics 839, DOI: 10.1007/978-3-642-23728-7,